Agricultural Extension

A Reference Manual

(Second Edition)

Edited By

Burton E. Swanson

Food and Agriculture Organization of the United Nations
Rome, 1984

First Edition, 1972

Abridged version of First Edition, 1973

Second Edition, 1984

The Second Edition of **Agricultural Extension: A Reference Manual** was prepared under contract between the Food and Agriculture Organization of the United Nations and the International Program for Agricultural Knowledge Systems (INTERPAKS), University of Illinois at Urbana-Champaign.

> **The designations employed and the presentation of material in this publication do not imply the expression of any opinion whatsoever on the part of the Food and Agriculture Organization of the United Nations concerning the legal status of any country, territory, city or area or of its authorities, or concerning the delimitation of its frontiers or boundaries.**

M-67

ISBN 92-5-101504-X

Table of Contents

Preface

This second edition of Agricultural Extension: A Reference Manual is a substantially revised and up-dated version of the original manual prepared by Addison H. Maunder in 1972 and, subsequently, reissued in abridged form in 1973. In this new edition, most of the original topical areas have been covered; however, the manual has been completely restructured, the chapters rewritten, and new material added. Since the first edition, there have been changes in the conceptualization and practice of extension. These new ideas and approaches to extension have been included in this second edition.

The purposes of this reference manual are several. Firstly, it is designed to provide an historical, developmental and conceptual context for understanding agricultural extension and its role in the agricultural development process. Secondly, it considers the potential clientele to be served by extension in developing nations and, based on other studies, suggests that previous approaches have not adequately reached different client groups and that alternative approaches are needed to achieve broad-based agricultural development. In particular, the needs of small farmers, including women and youth, must be carefully considered, if these important groups of agricultural producers are to be effectively served.

Thirdly, this edition of the manual describes the methods, procedures, and techniques for carrying out extension programmes. In addition, organizational considerations, such as administering, financing and evaluating extension programmes are discussed, as is the need to prepare and up-grade extension workers and support staff as a professional team of educators, communicators and specialists.

The reference manual is primarily directed toward extension practitioners in developing countries, those individuals who have the responsibility of organizing and carrying out extension programmes. In addition, it should be useful to students of agricultural extension, to provide them with an in-depth understanding of the extension process. The manual also directs them to a wide spectrum of new literature that has been written about agricultural extension since the original edition of this manual was published in 1972.

List of Contributors

John H. Behrens is Professor of Agricultural Communications, University of Illinois at Urbana-Champaign.

Robert P. Bentz is Associate Director, Cooperative Extension Service, University of Illinois at Urbana-Champaign.

John B. Claar is Director of the International Programs for Agricultural Knowledge Systems (INTERPAKS), University of Illinois at Urbana-Champaign.

J. Lin Compton is Associate Professor of Extension and Adult Education, Cornell University.

Delbert T. Dahl is Head, Office of Agricultural Communications, University of Illinois at Urbana-Champaign.

James F. Evans is Professor of Agricultural Communications, University of Illinois at Urbana-Champaign.

Frederick C. Fliegel is Professor of Rural Sociology, University of Illinois at Urbana-Champaign.

Thomas H. Henderson is Professor of Agricultural Extension and Director, Department of Agricultural Extension, University of the West Indies, St. Augustine, Trinidad and Tobago.

Janice Jiggins is an Author and Consultant on Women and Agricultural Extension.

Jane S. Johnson is Bibliographer and Librarian for INTERPAKS, University of Illinois at Urbana-Champaign.

Sam H. Johnson III is Assistant Professor of Agricultural Economics, University of Illinois at Urbana-Champaign.

Jae Tae Kang is a Lecturer, Department of Education, Gyung Sang National University, Jinju, Korea.

Earl D. Kellogg is Associate Director of International Agriculture, University of Illinois at Urbana-Champaign.

F. Wilfred Lancaster is Professor of Library and Information Science, University of Illinois at Urbana-Champaign.

Violet M. Malone is Professor of Extension Education, University of Illinois at Urbana-Champaign.

Niels Röling is Professor of Extension Education, Agricultural University, Wageningen.

Abdus Sattar is a Research Assistant, Library and Information Science, University of Illinois at Urbana-Champaign.

Joseph Seepersad is Lecturer, Department of Agricultural Extension, University of the West Indies, St. Augustine, Trinidad and Tobago.

Vickie A. Sigman is Graduate Research Assistant, Agricultural Education, University of Illinois at Urbana-Champaign.

Andrew J. Sofranko is Professor of Rural Sociology, University of Illinois at Urbana-Champaign.

Hae Kyun Song is Professor of Agricultural and Extension Education at Seoul National University.

Burton E. Swanson is Associate Professor of International Agricultural Education and Associate Director of INTERPAKS, University of Illinois at Urbana-Champaign.

Lowell H. Watts is Director, International Extension Center, Colorado State University.

Introduction

This second edition of <u>Agricultural Extension: A Reference Manual</u> has been prepared to reflect the state of affairs in agricultural extension in the first half of the 1980s, as well as current issues and concerns that affect how extension might be organized in the future. First, there is a conscious concern with the role of agricultural extension in technology transfer. New and improved technology is of little value in increasing agricultural productivity and farm incomes unless it is used by farmers. Agricultural extension has an essential role to play in the overall process of technology transfer.

The World Conference on Agrarian Reform and Rural Development (WCARRD), which was held in 1979, recognized that technology was essential to increased productivity. However, it also recognized that technological change could have negative consequences in rural communities, unless the technology was appropriate for all types of agricultural producers. Additionally, agricultural extension must reach all groups of farmers and not just serve the larger, so-called "progressive farmers." These concerns resulted in the theme growth with equity, which is reflected through this manual.

In particular, the technological and educational needs of small farmers, many of whom operate at the subsistence level, must be addressed. Many of these small farmers, and especially those that grow subsistence food crops, are women. Therefore, their specific skills and technological needs and extension programmes appropriate to reach this important group of agricultural producers must also be addressed. Finally, rural youth, who will make up the next generation of farmers, have a need for somewhat different types of agricultural extension programmes. These young men and women will need more agricultural knowledge and skills, as well as positive attitudes towards agriculture and rural life, if they are to increase farm productivity and incomes in the future.

Since the early fifties, there have been different and changing views about the value of agricultural extension. Different approaches have been recommended and tried. In this manual, a functional view of extension is assumed, because no one particular model or system is considered appropriate for a variety of situations. The value of these different models or approaches is that they address different problems, and they are an important source of different perspectives, methods and techniques to improve the effectiveness of national extension systems. Each approach tends to reflect a particular set of problems, objectives, resources, and client groups to be served. Therefore, no particular model is recommended as being better than another. Rather, these different approaches have unique advantages and disadvantages, and they add breadth to our knowledge about agricultural extension and how to more effectively disseminate appropriate technology to agricultural producers.

The structure of this manual is organized into four main sections. The first, entitled the "Evolution of Agricultural Extension," is concerned with the history and development of agricultural extension programmes over the last 100 years or so. After this historical introduction is a brief description of six different models of extension, which reflects the functional perspective being used in this manual. Common dimensions of agricultural extension are then discussed, with a review of the major categories of farmers to be served by extension in developing countries.

The second part of the manual is concerned with the conceptual framework for agricultural extension. Attention is first placed on the organizational setting for agriculture extension, where extension is viewed as an essential component of an agricultural system. In addition, the other institutional components of an agricultural system are described, and the inter-relationships between these different components are discussed.

If agricultural extension is to be effective in the technology transfer process, it is important that there be appropriate technology available for dissemination to each category of farmers in each agro-ecological area served by extension. The chapter, "Extension's Role in Adapting and Evaluating New Technology for Farmers," is concerned primarily with on-farm research, and the role of Subject Matter Specialists (SMSs) and Village Extension Workers (VEWs) in facilitating the modification and testing of new technology to fit the needs of the clientele in each agro-ecological zone. This chapter focuses on the role of extension personnel in maintaining a close working relationship with agricultural research.

The next two chapters deal with the social dimensions of technology transfer. Chapter 4, "Introducing Technological Change: The Social Setting", is concerned with describing the social factors that influence the dissemination of improved agricultural technology. It outlines the basic sociological concepts extension workers should be familiar with as they work in their respective service areas. The next chapter, "Extension Communication and the Adoption Process", looks at the factors that influence farmers to consider, try, and then adopt or reject improved technology. It reviews the different stages farmers go through in the adoption process, and the general proposition that farmers can be differentiated into particular categories, based on how rapidly they adopt new technology.

The third main part of the manual is concerned with improving agricultural extension programmes. These chapters deal with the process of programme development (with an emphasis on participatory methods), and how to plan extension campaigns, as well as descriptions of the various extension methods that can be used to help farmers learn about and use new technology. This part of the manual begins with a chapter, "Extension Strategies for Technology Utilization", which considers the need for a more broad-based approach to technology transfer, and the inclusion of specific groups of clientele, (such as farm women, young farmers, and other small farmers), who are frequently not included in extension programmes. The last chapter in this section deals with organizing an extension information unit to produce extension teaching materials.

The final part of the manual is concerned with "Strengthening the Structure and Function of Agriculture Extension." In this section attention is given to factors such as extension administration and organizational design to improve the effectiveness of national extension systems. In addition, attention is given to evaluation procedures and criteria that can be used to improve the effectiveness of extension programmes and activities. Procedures for both formative and summative evaluations are included in Chapter 13, and one approach to extension evaluation has been highlighted as an example.

The next two chapters (14 and 15) are concerned with the training of extension personnel. The chapter on pre-service training programmes stresses the need for a problem-solving approach to train field-level extension personnel. Then the role and procedures that extension can play in improving the skills and knowledge of extension personnel through in-service training and staff development are outlined. These programmes are concerned with the further improvement of the professional and technical skills of extension workers and increasing the professionalism of these staff members.

The final chapter is concerned with library and information sources to strengthen agriculture extension and training. This chapter discusses the information needs of extension, particularly on the input side, and the types and sources of information that would be useful to extension training centres, subject matter specialists, and other extension personnel. The Epilogue summarizes the main factors that are limiting the effectiveness of national extension systems as were recently identified by 59 national extension directors. These limiting factors suggest some important areas of concern that should be addressed by national programmes and donor agencies.

List of Figures

ix

List of Tables

Chapter 1
The History and Development of Agricultural Extension

B. E. Swanson and J. B. Claar

The purpose of this chapter is to explain briefly what agricultural extension is and what it is not, and to place it in perspective as an essential institutional component in the process of agricultural development. Then the focus will shift to the historical development of agricultural extension in Europe, North America and more recently in the Third World.

Agricultural extension is not a monolithic structure--rather it is an educational process, which has as its goal the communication of useful information to people, then helping them to learn how to use it to build a better life for themselves, their families and their communities. Because extension is a process, it can be organized in different ways: therefore, alternative models have emerged. But these different models or approaches have some common elements, and these will be explored briefly at the end of this chapter.

AGRICULTURAL EXTENSION DEFINED

"Agricultural extension" is a difficult term to define, precisely because it is organized in different ways to accomplish a wide variety of objectives. The term therefore has a variety of meanings to different people, but from this spectrum of different interpretations, there appear to be several common features.

Extension is an on-going process of getting useful information to people (the communication dimension) and then in assisting those people to acquire the necessary knowledge, skills and attitudes to utilze effectively this information or technology (the educational dimension). Generally the goal of the extension process is to enable people to use these skills, knowledge, and information to improve their quality of life. The extension function can be used equally well by both the private and public sectors, although most general agricultural extension organizations are public sector institutions. It can be combined or integrated with other technology transfer functions, as is the case in most commodity development programmes; however, most agricultural extension organizations are expected to concentrate on the educational function. Extension, or non-formal education, as it is sometimes called, can be used effectively in nonagricultural programme areas, such as rural health, family planning or community development, although in this manual the focus is on agricultural programmes. In other words, extension can be successfully used by different types of organizations to reach different groups of people with different messages.

The term, agricultural extension, narrows the focus and defines the areas to which the extension process is applied. Maunder (1973, p. 3), writing in the first edition of this manual defines agricultural extension as "A service or system which assists farm people, through educational procedures, in improving farming methods and techniques, increasing production efficiency and income, bettering their levels of living, and lifting the social and educational standards of rural life."

Some people tend to equate agricultural extension with the term technology transfer. This is incorrect, because technology transfer includes the additional functions of input supply and agri-services. In addition, extension needs to teach farmers management and decision-making skills, as new technology inevitably places more demand on these abilities. Also, agricultural extension should help rural people develop leadership and organizational skills, so they can better organize, operate and/or participate in co-operatives, credit societies and other support organizations, as well as to participate more fully in the development of their local communities. While many of these activities contribute to tech-

1

nology transfer, not all of them are included under this particular function. Therefore, while agricultural extension is an essential and major part of technology transfer (teaching farmers about improved agricultural technology and how to use it), the terms are not synonymous.

IMPORTANCE OF AGRICULTURAL EXTENSION IN THE AGRICULTURAL DEVELOPMENT PROCESS

Agricultural development implies the shift from traditional methods of production to new, science-based methods of production that include new technological components (such as new varieties, cultural practices, commercial fertilizers and/or pesticides), new crops and/or even new farming systems (such as double-cropping in Bangladesh through the inclusion of wheat as a totally new crop grown during the off season). For farmers to adopt these new production technologies successfully, they must first learn about them and then learn how to use them correctly in their farming systems.

Simple changes, such as the adoption of a new variety, may involve a minimal extension input. However, if such a change involves a new time of planting, a higher plant population, more fertilizer or the use of pesticides farmers may have much to learn to adopt the new technology successfully. Furthermore, once the shift to new science-based technology begins, the expectation would be that this is the first step towards more intensive and productive cropping and/or farming systems. This process is the essence of agricultural development, and each step in this process will require an educational and/or communications input. Therefore, the function of agricultural extension, regardless of how it is provided, must be viewed as an essential component in the agricultural development process.

If it is agreed that agricultural extension is essential to agricultural development, why are agricultural extension organizations sometimes criticized? People have a wide range of views about the relative value of agricultural extension, because in different situations it has been organized in different ways to pursue different objectives. These views range from very positive to negative, depending, upon each observer's knowledge of, and experience with, agricultural extension.

For example, agricultural extension has been criticized because it has neglected certain categories of agricultural producers, such as women and small farmers. These omissions are, in fact, a reflection of many factors, among them the agricultural development objectives being pursued, the unusually high proportion of small farmers to extension personnel, the way in which extension is organized, the difficulty of reaching women in some cultures and the extension strategy being followed. Thus agricultural extension can have positive and/or negative consequences, depending upon the goals pursued, the clients served and how success is measured.

Extension has also been criticized because it has been ineffective in persuading farmers to adopt a particular recommendation, when, in fact, the technology being promoted may not have been appropriate or was poorly suited to farmer conditions. Finally, extension has been ineffective in certain situations because of inadequate resources, poorly trained field staff, mobility problems, few teaching resources, or the field staff having too many non-extension responsibilities that can result in role conflicts.

The past experience of agricultural extension (as will be more fully documented in the next section) demonstrates that the function of extension is essential to the agricultural development process. Farmers cannot successfully adopt a new technology unless they are aware of it and learn how to incorporate it into their farming systems. Where the extension concept has been inappropriately applied and has resulted in an increased gap between rich and poor, men and women, or young and old, the consequences are unfortunate. But it is essential to differentiate between the extension concept itself, the inappropriate technology that may have been disseminated, or the inappropriate objectives that may have been pursued, whether intentional or not. The extension process can be applied to help bring about broad-based agricultural development.

Given this introduction to agricultural extension and its role in the agricultural development process, it is interesting as well as useful to understand how the concept of agricultural extension started and how it has evolved over the past 100 or so years. Therefore, the next section will provide a brief overview of the historical development of agricultural extension.

HISTORICAL DEVELOPMENT OF AGRICULTURAL EXTENSION

The Beginning of Agricultural Education

The historical roots of agricultural extension go back to the Renaissance when there was a movement to relate education to the needs of human life and to the application of science to practical affairs. According to True (1929, p. 2), "with the beginnings of the modern science in the sixteenth and seventeenth centuries, a desire to use the new knowledge in education soon appeared. Among those who influenced this movement was Rabelais (1483-1553), who would have pupils study nature as well as books and use their knowledge in their daily occupations."

"In England, Samuel Hartlib (c. 1600-70) published a book in 1651 entitled An Essay for Advancement of Husbandry-Learning. In the Tractate of Education, published in 1644 and addressed to Hartlib, Milton followed very largely the plan of education suggested by Rabelais, which involved a very broad study of classical literature, including agriculture, as described by Cato, Columella, and Varro...Out of this came the academies in England, which often included in their curriculum studies having practical bearing. Jean Jacques Rousseau (1712-70) ...dwelt much on the importance of manual and industrial activities in education." (True, 1929, p. 3)

The Swiss educational reformer, Heinrich Pestalozzi (1746-1826), "being influenced by Rousseau, entered on an agricultural life, and from 1775 for several years he conducted a school for poor children in which part of their time was spent in raising farm products, spinning and weaving of cotton, etc."

Philip Emanuel Von Fellenberg (1771-1844) conducted very successfully from 1806-44, at Hofwyl, Switzerland, two manual-training schools which had considerable influence in the United States. They were located on an estate of 600 acres, and the boys in both schools had gardens and were expected to do farm work. There was also instruction in science related to agriculture. Subsequently there was a school for girls and a normal school." (True, 1929, p. 3)

Some of the earliest agricultural schools in Europe were established in Hungary, including one at Zarvas which started in 1779, another at Nagy-Michlos in 1786, and the Georgicon Academy at Kezthely, which was founded in 1797 and was for 50 years, "the model agricultural college of Europe" (True, 1929, p. 3).

Early Agricultural Publications

During the seventeenth and eighteenth centuries, considerable literature on agricultural subjects was developed in a number of European countries. In France the publication of works on agriculture was greatly stimulated by the large series of volumes commonly called the Encyclopedia (1751-70). In Great Britain there existed agricultural works by about 200 different authors before 1800. The Annals of Agriculture and Other Useful Arts, a periodical begun in London in 1784 by Arthur Young (1741-1820), widely promoted the advancement of agriculture in Europe and America (True, 1929).

Early Agricultural Societies

According to True (1928), the forerunner of agricultural extension in both Europe and North America was the agricultural society. The first such organization in Scotland (and perhaps anywhere) was known as the Society of Improvers in the Knowledge of Agriculture, which was begun in 1723. The American Philosophical Society, which was founded in 1744 under the leadership of Benjamin Franklin, published many articles on agricultural subjects in its early years. However, because it developed more as a scientific society, this led to the establishment in 1785 of the Philadelphia Society for Promoting Agriculture.

"The first agricultural society in Germany was established in 1764. In France there was an early Society of Agriculturalists that was succeeded by the Academy of Agriculture of France, which began the publication of proceedings as early as 1761. In Russia, the Free Economical Society was established in 1765, with a large experiment farm near St. Petersburg." (True, 1929, p. 6-7)

These societies were formed to acquaint their members with what was being done to improve agriculture, to establish local agricultural organizations and to disseminate agricultural information through their publications, newspaper articles and lectures. For example, in 1812, the Massachusetts Society for Promoting Agriculture sent out 1000 copies of a letter to stimulate farmers to improve agriculture. Town clerks were asked to read this letter in town meetings. The next year the society reported that numerous town societies were in operation (True, 1928).

The agricultural societies in North America were responsible for holding fairs, not merely for the sale of animals or farm products, but for educational purposes. A notable early example was the lecture by John Lowell at a fair held by the Massachusetts Society at Brighton in 1818.

The Beginning of Extension-Type Programmes

The use of itinerate teachers to improve agriculture was first started in North America in 1843 when the Committee on Agriculture, New York Assembly, suggested that "the legislature might authorize the State Agricultural Society to employ a practical and scientific farmer to give public lectures throughout the state upon practical and scientific knowledge" (True, 1928, p. 4). That year itinerate lectures were begun by the Society.

In Ohio, in 1845, N.S. Townshend, afterwards Dean of the College of Agriculture, suggested that the State Agricultural Society "might select a sufficient number of competent individuals to lecture...on all the sciences having relations with agriculture" (True, 1928, p. 4). According to True, he also advocated the formation of farmers' clubs in every township to hold meetings, at least monthly, at which there should be lectures on the sciences and their application to agriculture, reports from committees on their visits to member's farms, and so forth.

In 1848, when the office of state agricultural chemist was created in Maryland, "the act required him to deliver one public lecture in each elective district and a course of lectures at each county town and some central place in Baltimore County" (True, 1928, p. 5).

In 1853, President Edward Hitchcock of Amherst College, and a member of the Massachusetts State Board of Agriculture, recommended the establishment of farmers' institutes that became the primary educational forerunner of agricultural extension in the United States. In 1914, when agricultural extension was formally established in the United States, 8861 farmer institutes were held that year, with a total attendance of 3050150 (True, 1928).

The Rise of Extension

"The first, modern, agricultural advisory and instructional service was established in Ireland during the great potato famine of the mid-nineteenth century. This service operated from 1847-1851. It was begun in the autumn of 1847, initially as a small scale and temporary scheme, as a result of detailed proposals contained in a letter from the Earl of Clarendon, the Lord Lieutenant of Ireland, to the President of the Royal Agricultural Improvement Society of Ireland. This led to the institution of itinerant practical instructors to work among small-scale, peasant farmers in the areas worst affected by the famine in the south and west of Ireland". Initially 10 itinerant lecturers or "instructors", as they were popularly known, were appointed to carry out this work. The number increased to 33 instructors at the peak of this extension-type activity. "Lord Clarendon's letter...must be regarded as a classic document in the early history of agricultural extension." (Jones, 1982, p. 11)

The actual use of the term "extension" originated in England in 1866 with a system of university extension which was taken up first by Cambridge and Oxford Universities and later by other educational institutions in England and in other countries. According to Farquhar (1973) the term "extension education" was first used in 1873 by Cambridge University to describe this particular educational innovation. The objective of university extension was to take the educational advantages of universities to ordinary people.

Extension Work in the United States

In the United States this system of university extension was introduced through city libraries, especially in Buffalo, Chicago, and St. Louis. By 1890 the American Society for the Extension of University Teaching was established. In 1891 the state of New York appropriated $10000 for university extension, and in 1892 the universities of Chicago and Wisconsin began organizing university extension programmes.

The Land Grant Colleges in the United States were influenced by this university extension movement and other related extension-type activities that were also expanding in scope during this period. Therefore, the formal establishment of agricultural extension work in the United States was really the integration of these different extension-type thrusts.

The United States Department of Agriculture (USDA) had fostered several of these thrusts, including the Farmers' Institutes, which were supported by the Office of Experiment Stations, and Farmers' Cooperative Demonstration work, which was an outgrowth of the work started in 1902 by Seaman A. Knapp (1833-1911) in several Southern states. This latter programme had gained considerable momentum by the time of Knapp's death in 1911. Finally, the office of farm management at the USDA was also assigning agents to districts during this period to study farm management problems, the prevailing systems of farming, and to conduct on-farm trials of new crop varieties (True, 1928).

There was also strong interest in extension-type activities within the private sector of the United States to support the improvement of agriculture. Boards of Trade, grain associations, bankers, railroads and other commercial concerns were also directly funding extension-type activities, such as agricultural trains, to inform farmers about improved methods of farming.

As mentioned above, the Land Grant Colleges were involved in many of these early agricultural extension-type activities, including working closely with the Farmers' Institutes, conducting local experiments as a means of teaching, preparing and distributing extension bulletins and conducting correspondence courses. By 1907, 42 colleges in 39 states were involved in extension-type activities, and many were in the process of establishing departments of agricultural extension, with a superintendent in charge. By 1910, 35 colleges had such agricultural extension departments, and over the next four years these programmes grew rapidly in extent and complexity.

All of these efforts culminated in the passage of the Smith-Lever Cooperative Extension Act in 1914 that provided for a combination of federal, state, and local funding of agricultural and home economics extension work to be carried out under USDA approval. It is interesting to note that there was strong competition in the legislative process to institute vocational agricultural education in secondary schools before, or instead of, the passage of the agricultural extension legislation. Because the Association of American Agricultural Colleges and Experiment Stations was closely involved in supporting the agricultural extension legislation, it was passed first. However, in 1917 the Smith-Hughes Act was passed, and it provided strong support for vocational agricultural and home economics education in secondary schools (True, 1928).

The spread of agricultural extension-type activities in Europe, Australia, New Zealand and Canada tended to parallel events in the United States, but their organizations developed somewhat differently. The demand for extension-type services came largely from the agricultural societies and, in some cases, were organized by them. In other cases, these activities were institutionalized as part of the national ministry of agriculture. However, several of the European extension systems included a co-operative dimension that provided support at both the national and local levels, particularly through these farm organizations. In most North American and European countries, local-level farmer participation in programme planning, and even in the hiring and firing of staff, remains high today.

Development of Agricultural Extension in the Third World

The development of agricultural extension organizations in Third World countries was, to a very great extent, a post-independence phenomenon, occurring mainly after the second world war. In Latin America and the Caribbean, the majority of national agricultural extension organizations were started in the mid-1950s, with a few established in the late-1940s and others initiated in the early-1960s. The experience of Asia and Oceania was similar to Latin America and the Caribbean, except that the midpoint was around 1960, with some of these organizations not starting until the 1970s. The introduction of agricultural extension organizations in African nations was somewhat later, with most extension organizations starting in the 1960s and 1970s (Swanson and Rassi, 1981).

In most Third World countries, the introduction of general agricultural extension organizations was brought about by assistance, particularly from the United States. The lack of local or popular demand for extension-type services has been characteristic of the experience in most Third World countries; this is a major difference from the North American and European experience. Also, because few Third World countries had well-established colleges of agriculture or an agricultural university when they became independent, in nearly every country agricultural extension was attached to the ministry of agriculture, rather than to an agricultural college as was the case in the United States.

It must be noted, however, that extension-type activities were carried out in many Third World countries earlier in the century. These activities were usually associated with commodity improvement schemes. Colonial governments sponsored research and extension-type activities for export crops such as sugar, rubber, oil palm, groundnuts and tea, because they were interested in increasing the export of these crops. In most countries, these commodity improvement programmes continue today. Research and extension of the traditional food crops, however, was seldom given any attention in these countries until after independence. In some countries these trends continue, in large part, today.

In most Third World countries, there was, and still is, a severe shortage of trained agricultural personnel. Therefore, when agricultural extension organizations were established, ministry personnel were often assigned to operate at the local level, and generally became involved in many administrative and regulatory activities, in addition to their so-called extension activities. In fact, in most countries there was little improved agricultural technology to extend because a productive agricultural research system was absent.

Today, most Third World countries have some type of extension organization, although the experience with these organizations has not been very satisfactory. Many donors report frustration in trying to improve these systems. This frustration is revealed in a recent statement by Clifford Wharton, Jr. (1983, p. 11), President of the State University of New York and former Chairman of the Board for International Food and Agricultural Development, the entity that provides policy direction for the United States Agency for International Development:

"If there is an area where we have been most unsuccessful, it has been the development of cost-effective and program-efficient models for the delivery of new scientific and technical knowledge to the millions upon millions of farm producers of the Third World. We know how to harness the creative and inventive forces of science and technology in the war on hunger, but I submit that we still have not been fully successful in technology diffusion."

ALTERNATIVE MODELS AND RECENT TRENDS IN ORGANIZING AGRICULTURAL EXTENSION

Over the past three decades, many different approaches have been used to develop or improve the different national agricultural extension systems, which can be considered as alternative models in organizing agricultural extension. Familiarity with these models is important in order to understand extension as well as to recognize the advantages and disadvantages of each approach. Therefore, this section will outline the major characteristics of six different models.

Before beginning the discussion, however, it must be pointed out that in this manual, a functional approach to extension has been assumed; thus, no single extension model is promoted as being more appropriate than another. Each has certain advantages and disadvantages depending on the problem to be solved.

Common Forms of Agricultural Extension

The first three models to be examined fall under the category of common forms of agricultural extension systems. The first, a conventional approach, is an attempt to summarize some of the basic features and problems of agricultural extension organizations in many Third World countries. Most of these systems were established long before there were productive agricultural research systems; therefore, in most cases, these institutions did not have an extension message to transmit. In addition, these organizations were generally established under the ministry of agriculture; therefore, they became increasingly involved in carrying out all types of governmental activities at the local level.

The World Bank has introduced and substantially supported a new system to improve existing agricultural extension systems in countries where the agricultural extension organization is very weak and does little or no extension work. This approach is known as the Training and Visit System (T & V System) of agricultural extension and focuses on specific weaknesses of these national systems.

A third common form of agricultural extension is one organized by an agricultural college or university. The primary example of this approach is in the United States where agricultural extension was first formally established in 1914. Because extension was organized under higher agricultural educational institutions, it has been easier to preserve the autonomy and educational nature of extension, than in other countries where extension is part of a governmental agency.

From these three general systems, we can examine how extension is handled within Commodity Development and Production Systems, Integrated Agricultural Development Programmes, and under Integrated Rural Development Programmes.

The commodity systems approach is essentially a continuation of the technology development and transfer system that colonial governments initiated to increase the export of specific commodities, such as rubber, coffee, groundnuts and sugar.

Integrated agricultural development programmes were a product of the 1970s and a recognition that all of the institutional components of an agricultural system must be successfully co-ordinated and made available to farmers if agricultural development is to occur.

Integrated rural development programmes trace their roots to the community development efforts (particularly in South Asia), and the Animation rurale approach that was undertaken in a number of Francophone countries of Africa. As the name implies, these participatory rural development programmes are much broader than agricultural extension, per se. However, the high level of client participation in planning, implementing, and evaluating programmes makes this an important approach to consider when improving agricultural extension systems.

1. Conventional Agricultural Extension Approach

This is a broad categorization that covers many general agricultural extension organizations in the Third World. It is somewhat difficult to generalize accurately about these organizations as there is a wide range of national extension systems included in this category. Also there are the additional disparities between theory and practice. Nevertheless this section attempts to outline some of the general features of these national systems and to identify possible areas for improvement.

Objectives. Generally, the main objectives centre around increasing national agricultural production, including food crops, export crops and animal production. Additional objectives frequently stated or implied are an increase in farm incomes and quality of life of the rural population.

In analysing these different goals and objectives, it is very important to consider the policy context in which extension operates, to determine which objectives are given priority: farmer objectives and national objectives can be in conflict. For example, if a government is interested in increasing national food production, particularly for an urban population, while pursuing a cheap food policy, it may be in direct conflict with the objective of increasing farm income. Policymakers must recognize, however, that farmers will not adopt new technology to increase production unless they can recover the initial cost of the inputs, and make a sufficient profit. Therefore, it is essential that the goals of national agricultural development be consistent with, and supportive of, extension's objective to increase farm production and incomes, as well as to improve the quality of life of rural people.

Clientele. The clientele for general agricultural extension organizations should be all farmers. Because few Third World nations have enough well-trained agricultural extension workers to reach all of the farm population, target groups are frequently identified, or diffusion multipliers such as contact farmers, demonstration farmers, etc. are used.

If a country is interested in increasing total food production and/or export crop production to increase foreign exchange earnings, it may tend to focus on larger, more productive farmers. By doing so, it can maximize the impact of a limited extension input on these macro-economic objectives. In such cases, small farmers, women farmers and young farmers may receive little or no attention, both in terms of research and extension activities. On the other hand, if a government is interested in more broad-based agricultural development, and extension wants to maximize its impact on all human resources in rural areas, then all major categories of farmers must be included as primary clientele of extension. This decision will affect how extension is organized and staffed and will determine which extension strategies shall be pursued. More will be said about this issue in Chapter 6.

Organization and Scope. As indicated earlier in this chapter, most Third World nations have established general agricultural extension organizations within the ministry of agriculture. In large countries, such as India, these are organized and operated at the state or provincial level.

In some countries, extension is organized by different ministries representing sub-sectors of the agricultural economy. Therefore there may be separate extension organizations for forestry, livestock production, crop production or youth. This often means that workers have a limited focus, and contributes to duplication of organization and services. In other words, rather than one extension worker serving the needs of the farm population in a particular area, there may be two or more workers travelling and working in the same area serving the same clientele. But in practice this seldom happens because with limited budget only the national and provincial offices use up the funds with limited amounts left to hire the needed extension workers.

In addition to educational responsibilities in these general agricultural extension organizations, extension personnel may be responsible for carrying out most ministry programmes and activities at the local level. Therefore, they may sell and distribute inputs, perform regulatory functions, arbitrate disputes, collect agricultural data and handle subsidy programmes. They, in fact, become the local agricultural representative of government rather than a full time agricultural extension worker. This type of assignment directly influences the extension worker's ability and capacity to perform his or her extension assignment, generally in a negative manner.

Approach. In most countries, agricultural extension workers concentrate on individual methods of extension, particularly home and farm visits, as well as office calls by farmers. Some countries have small agricultural information units that produce radio programmes and other materials for use by the mass media, in addition to extension bulletins and teaching aids. However, the lack of improved technology and appropriate teaching material to use in group meetings typically results in little educational and communications work being carried out by field extension personnel.

Role of Extension Personnel. By definition the role of the agricultural extension worker is that of an educator and communicator. Extension workers should identify farmer problems and production constraints. Working closely with subject matter specialists and research workers, extension workers should disseminate useful information about new technology and teach farmers how to use it successfully to increase production and incomes. Clearly, this set of responsibilities is educational in nature and, given the number of farmers to be served in most Third World countries, is a full time job.

On the other hand, where extension workers serve as the agricultural representatives of government at the local level, they will largely be low level administrators rather than educators. These types of non-extension assignments usually result in serious role conflicts for field-level personnel to the detriment of the extension function. Therefore, the extension organization effectively fails to carry out the essential education and communication functions that are necessary for agricultural development.

2. Training and Visit System

The Training and Visit System (T & V System) is not a new extension model. It is an attempt to reform and improve upon the effectiveness of conventional agricultural extension organizations. However, because this approach to improve national agricultural extension systems is now being widely introduced and used in many Third World countries, largely through the encouragement and support of the World Bank, it merits inclusion here.

Objectives. The extension objectives of the T & V System are similar to the conventional model described above, although the primary objective is to increase individual farm production and income. The implicit assumption is that if farmers increase their production and income, then national agricultural production will also increase.

The objectives of the T & V System as a reform movement of conventional agricultural extension organizations should also be described. Some of the more important problems the T & V System attempts to solve are

(a) to improve the organization of extension by introducing a single, direct line of technical support and administrative control,

(b) to change the multi-purpose role of many extension workers to a clearly defined, single-purpose role involving only education and communication activities,

(c) to improve coverage by limiting the number of farm families or households one extension worker is expected to visit,

(d) to improve mobility by providing appropriate transport so each worker can regularly visit his or her contact farmers,

(e) to improve each extension worker's technical skills and knowledge about improved agricultural technology by providing regular in-service training sessions,

(f) to improve extension's ties with agricultural research through the addition of more subject-matter specialists, who are expected to maintain regular contact with their research counterparts and to ensure a continuing flow of information that transmits technology to farmers and farmer problems back to research personnel,

(g) to improve the status of extension personnel by giving them a relatively clear-cut extension job with reasonable expectations that they can successfully carry it out; this will increase their level of respect in the community and begin to build their self-confidence, and

(h) to reduce the duplication of services that occurs when extension is fragmented among different ministries (for example, agriculture, livestock or forestry) or is added to new area or commodity development schemes in a country or province that already has a general agricultural extension system. (Benor & Harrison, 1977)

Clientele. The clientele of the T & V System are all farmers in each extension worker's area of responsibility. The contact farmers in each community represent about 10 percent of the farmers in each village. Initially, the T & V System selected the more progressive farmers in each community as contact farmers. Because these farmers had greater access to resources and were not typical of the majority of small farmers, this limited the diffusion of new technology in the village. Now extension workers are instructed to select contact farmers who are representative of all the major groups of farmers in each village. This approach is expected to encourage the spread of technology to all categories of farmers in each community.

Organization and Scope. The organization of the T & V System is based on the total number of farm families or households that one extension worker can reasonably expect to cover. This number is affected by population density, availability of roads, and cropping intensity and diversity. The extension worker-farmer ratio may vary from 1:1200 to 1:300, with an average of 1:800.

All farm families or households under the jurisdiction of one extension worker are divided into eight groups of about equal size. Contact farmers represent about 10 percent of this group, therefore, in an average situation, an extension worker would work with 10 contact farmers in each of eight villages or areas who would themselves represent 100 farm families or households. In other words, there would be a total of 80 contact farmers who would represent 800 farm families or households.

Given the appropriate extension worker/farm family or household ratio, it becomes easy to determine the number of extension workers needed in an area by dividing the total number of farm families or households by the denominator of this ratio. The rest of the extension organization is based on the total number of extension workers needed. First, it would require one agricultural extension officer (AEO) to supervise, train, guide and administratively backstop each group of six to eight extension workers. Six to eight

AEOs can, in turn, be supervised, guided, and backstopped by a single, sub-divisional extension officer (SDEO), and so forth.

In addition to AEOs, who would provide a single, direct line of administrative control, there would need to be a team of at least three subject matter specialists (SMSs) assigned to each SDEO. According to Benor & Harrison (1977), it is recommended that this team consist of one agronomist, one plant protection specialist and one training officer. These SMSs would split their time among training, field/farm visits and working with their research counterparts.

Approach and Role of Extension Personnel. The T & V System revolves around an intensive series of fortnightly visits on a fixed schedule known to farmers, supervisory and technical staff. The extension worker receives one day of training each week; during the first week the SMSs conduct the training, while during the second week the AEOs conduct a more informal training session that deals with problem solving.

The extension worker visits four groups of contact farmers the first week and the other four groups during the second week. Because each extension worker's schedule is printed and made available to the supervisory staff, it is simple to make random supervisory visits to each extension worker. The purpose of these visits is to see how he or she is doing and to provide whatever support is needed.

Initially, the focus of extension is to transfer low-cost and low-risk technology to farmers (such as improved varieties and cultural practices) that will increase output and income. Later, when the farmer has confidence in the new technological recommendations, further improvements can be demonstrated that involve the use of inputs. At this point the farmer should be in a position to try these new technological components that will further increase productivity and income.

The only report or activity record an extension worker is expected to keep is his or her daily diary. This diary provides a daily record of contact farmers visited, training imparted, problems encountered, recommendations made and any other comments or observations that seem important. This diary provides an excellent record of work accomplished and problems encountered that can be discussed with the AEO and/or SMS during the weekly training sessions. Solutions to problems and answers obtained in the training sessions can then be given to farmers during the next fortnightly visit.

In addition to regular supervision, the T & V approach also includes a well-defined monitoring and evaluation (M & E) system. This M & E function is usually carried out by an autonomous unit in the ministry and is designed to make information available on a regular basis. This information is used for formative evaluations (the monitoring function), while the overall data base can be used to measure extension's accomplishments (summative evaluation). This particular approach is expanded in Chapter 13.

General Observations. The T & V approach attempts to reform a conventional extension organization that may be carrying out many non-extension activities; therefore the process of organizational change may not be smooth or readily accepted. The World Bank, in introducing this particular extension approach into a country, generally tries to improve the policy environment towards the agricultural sector in general (so farmers have some incentive to adopt new technology), and toward the extension organization in particular. Then it makes resources available to affect mobility, housing, job descriptions, training, technical backstopping, supervision, monitoring and evaluation, which are all potential major problems. Most of these investments are made at the field level or have direct implications for field-level extension personnel.

In general, the T & V System is designed to give each extension worker a well-defined job with timely training, technical backstopping, and adequate supervision. He or she is expected to live within his or her area of responsibility, and is provided with adequate transportation and teaching resources to conduct his or her work in an effective manner.

Some of the major criticisms of the T & V System include the following (a) it is too top-down oriented and does not allow enough farmer participation in programme planning, (b) it is too rigid in terms of the fortnightly schedule, particularly during the slack seasons, (c) it is too labour intensive, requiring a large number of extension workers which a country may not be able to afford, (d) it does not make effective use of mass media methods of communications, (e) due to the serious lack of SMSs in nearly all Third World nations, extension's linkages to research are weak, resulting in poor technical training and backstopping, and (f) because many of these extension organizations are run in an authoritarian manner, extension worker supervision is often not sufficiently positive and supportive to improve extension worker morale.

3.   Agricultural Extension Organized by Universities

The most comprehensive example of this system is found in the United States. The programme is carried out under federal and state legislation that sets up a co-operative programme among federal, state and local governments to fund and carry out extension work on a matching basis through one or more land-grant universities in each state. Local level extension offices and workers are normally deployed in each county (or in multi-county units) through a hierarchial administrative arrangement. These systems are characterized by co-operative relationships, the scope of subject matter taught, the broad nature of the clientele, and the focus on human development.

Objectives. The primary goal of this approach is to conduct educational programmes in selected subject matter areas to help clientele solve problems in a way that is socially desirable and personally satisfying. Some of the specific objectives pursued by these extension organizations include: improving the efficiency of agricultural production, farm incomes, and rural welfare; producing an adequate volume of high-quality farm and ranch output for consumers at acceptable prices; strengthening the family and home; aiding young people to learn and develop through "learning by doing" projects; enhancing the environment and the use of natural resources; and working with communities to improve the community as a place to live and work.

Clientele. The clientele normally includes all people who are interested in the subject matter. Audiences are targeted for each type of programme. For example, significant programmes are carried out in large cities, especially in home economics, youth work and home horticulture. Extension also works with a great many organizations and firms which are involved with providing services to agriculture.

Organization and Scope. The extension system is organized at the state level under one or more of the state land grant universities. Therefore, all extension personnel are staff members of the university. There are essentially four types of extension personnel: extension workers and extension assistants (para-professionals) at the local level, subject matter specialists at the state level, and administrative and supervisory personnel at the state and regional levels.

Subject matter specialists generally have joint appointments as regular members of the university faculty with academic rank. Their joint appointments with extension may be in either teaching or research; in a few cases they may have a three-way appointment in teaching, research, and extension. It is important to note that about 20 percent of the total extension staff time commitment is to subject matter specialists. In a state like Illinois, this actually involves 125 faculty members who have from 10 percent to 100 percent of their time allocated to extension work.

The United States extension organization is somewhat unique in three ways; local and county clientele are included in the financial and operational processes of the extension programme, there is concurrence of representatives of the county clientele in selecting personnel, and an evaluation of personnel performance is submitted directly to the supervisor. The scope of this programme includes agriculture, natural resources, home economics, rural development and youth programmes.

Parts of this concept of agricultural extension have been applied in other countries. India and the Philippines, for example, have developed an effective system of land grant type universities that are responsible for applied and adaptive research. These institutions make systematic inputs into extension, although the agricultural extension system continues to be operated by the respective ministries of agriculture.

It is suggested that one of the primary reasons that more countries have not adopted this model of organizing agricultural extension is that few Third World nations had well-organized agricultural universities in place at the time extension systems were introduced. Therefore, it was logical to attach them to the ministry of agriculture. In the United States the land grant universities had been established 50 years earlier and had active research programmes for 25 years before extension work was formally organized.

As discussed above, conventional agricultural extension systems in the Third World have been largely devoted to carrying out government programmes, rather than conducting extension (educational) work full time. In the United States, state extension organizations are attached to state universities; thus it has been relatively easy for them to maintain their autonomy from political forces, and to concentrate on locally-relevant extension activities.

Approach. The approach is educational with an orientation to using research to help people identify and solve problems. It uses representatives of the various clientele groups in each county to form an advisory extension council to help determine the major problems to be given priority each year. Requests are then made by the county staff for instructional assistance from the subject matter specialists on the topics selected.

Multi-media techniques, county meetings, demonstrations, and tours are typically used, with regional and state-wide meetings for the more specialized topics. Thus, audiences are roughly stratified both by subject-matter and by level of current knowledge, so that the instruction can be made more relevant. Teaching methods involve lectures, discussion groups, computers, distance-learning systems, mass media and visual aids. This approach is very clientele-oriented, stressing bottom-up rather than top-down programming. Evaluation of programmes and personnel by local people is built into the approach.

Role of Personnel. The normal role of extension personnel is to disseminate information and encourage the application of this information to solve specific problems. Extension, in this format, also plays a co-ordinating and developmental role. For example, by working in co-operation with input suppliers, they are able to make agri-service companies and their products more appropriate to farmer needs. When problems are observed for which there are no solutions, extension frequently takes the lead in helping solve the problems. Thus, if industry is supplying the wrong type of fertilizer, or if an insecticide is failing to produce results, extension takes it up with the industry to encourage a solution. If a service is needed that is not being provided in a satisfactory manner, a co-operative may be formed. Thus, extension is an action-oriented force in development.

4. Commodity Development and Production System

The extension function within commodity development and production systems is generally well-integrated with other aspects of technology transfer such as input supply and other agri-services, as well as having good links with both researchers and with farmers. Because these systems have a very narrow focus (an individual commodity) and fully organize each phase of the technology development and transfer process, and, generally, the marketing phase, they can be characterized as vertically integrated production systems.

Objectives. The objective of the commodity development and production systems is to produce and market relatively high value commodities efficiently and effectively. These commodities are generally produced for export (such as tea, oil palm, cashews, rubber and sugar), but can also be grown for domestic consumption (e.g., milk). Commodities grown primarily for export are very important to a country because they generate foreign exchange. Successful systems encourage farmer participation by profit-sharing which increases farm income, and by continuing to re-invest in technology development and transfer, thus ensuring continued improvement of these production systems.

Some governments use marketing boards for export crops simply to capture both the foreign exchange and surplus income, without sharing much of the profit with farmers and/or without re-investing in the overall system. These exploitive systems generally stagnate in the long run because they choke off the incentive for farmers to participate and/or because the technology of production is not maintained or improved, which can result in relatively higher costs and reduced output.

Clientele. Commodity development and production systems are frequently defined by particular ecological features in a country. For example, beverage crops such as coffee and tea are generally grown at a particular altitude in the tropics, which affects product quality. Cocoa, rubber and other crops also have a particular ecological niche in the humid tropics where they are most productive. Within these particular ecological zones, these production systems usually involve most farmers in a contiguous area.

In the past, many of these commodities were grown under a plantation system of production. However, in recent years (usually post-independence), many of these large estates have been wholly or partially broken up into small farm holdings; thus small-farm agriculture characterizes the primary system of production for these commodities. This situation increases the importance of strengthening the technology transfer system, particularly the extension function.

Organization and Scope. As mentioned above, commodity development and production systems are generally limited to a single commodity; therefore, technology development and transfer, as well as the marketing function, are frequently handled by a single parastatal body (semi-autonomous public sector firm). By controlling the marketing of the commodity (because it is an export crop or a highly perishable product such as milk), it is possible for the parastatal body to provide research, extension, input supply, credit and marketing services to farmers, and then to recover the cost of these services from profits.

Approach. Quality control is frequently the primary factor that dictates the technology of production. Thus farmers have little choice but to use the inputs and to follow the technical recommendations given. Failure to do so may result in a product that will not be purchased by the parastatal organization. However, in countries where these commodity systems are well-organized, net farm income generally far exceeds returns on traditional crops.

The technology of production is usually well-established for these commodities; therefore, the farmer must follow the advice of the extension worker or technician to participate in this programme and to sell his or her crop or product. This approach is sometimes called contract extension. The farmer has a contract to produce the crop using the recommended practices, and the parastatal organization has a contract to buy the crop at the specified price.

Role of Extension Personnel. Technical recommendations (the extension message), input supply, and other agri-services are all closely related in these commodity systems. Therefore, it is common for one individual, the technical agent, to handle all aspects of technology transfer. The technical agent monitors the crop in his or her assigned zone, and when problems arise, or when a particular input is needed, the extension worker provides both the technical advice and the input at the same time. Because these commodity systems generally deal with high-value products, it is possible (and perhaps even more efficient) to have the technical extension worker specialize by commodity rather than by function.

5. Integrated Agricultural Development Programmes

In the early 1970s it was recognized that for agricultural development to occur, all of the institutional components that affect this process must be co-ordinated and applied to achieve increased agricultural output. Usually these efforts revolved around donor-assisted projects in a particular geographical area, and focused on a common set of production problems.

Objectives. The objectives of these projects were generally production-oriented, aimed at increasing food or agricultural output in the project area. In addition, these projects were expected to demonstrate that agricultural development could occur if an intregrated approach was used.

Clientele. The clientele included all farmers in the project area. However, where resources were insufficient, generally the larger, more progressive farmers tended to take advantage of the new inputs, credit and marketing services in the project area.

Organization and Scope. Generally these projects created their own management structure and technical support system, because it was considered too difficult to change the existing institutions that were supposed to be assisting farmers. Thus the project tended to create an artificial environment by recruiting some of the more competent technical personnel in the country. It did so by increasing salaries of project personnel and by providing transportation, inputs, and other factors considered essential to a well-co-ordinated, agricultural development project.

Approach. In many cases, these projects did not introduce any significantly improved technological components to farming systems. Essentially these projects assumed that the existing technology of production was adequate and that the major limiting factors were a lack of co-ordination or of inputs. Therefore, special input supply, credit, extension, marketing and other agri-services were made available in a well-co-ordinated approach. The mechanism at the local level was sometimes a farm service centre, where a farmer could come, obtain credit and purchase inputs, as well as receive advice about how to use these inputs effectively.

Role of Extension Personnel. In integrated agricultural development programmes, extension workers can play different roles. In theory they are expected to help farmers learn about new technological alternatives, and how to gain access to inputs, credit and marketing services, so that farm output and incomes can be increased. In situations where there is a shortage of trained agricultural personnel, extension personnel frequently become directly involved in supplying inputs and services. It appears that dispensing inputs, credit and other services is a more clear-cut, rewarding job than conducting extension programmes. Therefore, the latter may be neglected in integrated agricultural development programmes, unless more priority and more direction is given to the extension function.

6.   Integrated Rural Development Programmes

Integrated or participatory rural development programmes are, in some respects, a blend between the community development projects of the 1950s and early 1960s and the Animation rurale approach in Francophone Africa. These approaches continue to reflect a broader concept of rural development, including both social and economic factors. In doing so, there is a concern that these projects should include an income-producing component, probably involving new agricultural technology. At the same time, there continues to be a strong emphasis on the broad-based participation of the rural poor in planning, implementing, and evaluating programmes. These efforts are also clearly designed to enable rural people to strengthen their indigenous institutions.

Objectives. These programmes generally reflect both economic and social objectives. Therefore, the introduction of appropriate new technology, particularly to increase agricultural output, is expected to produce the new income that will support and enhance social objectives. Increased participation is a central concern of these programmes, particularly to increase self-reliance and local initiative. These rural development programmes also pursue objectives such as improved health, nutrition, and basic education.

Clientele. Most rural development programmes are aimed at the rural poor. Community development programmes of two and three decades ago tended to work within the existing power structure of rural communities; therefore, many of the benefits actually accrued to the traditional village elite (Holdcroft, 1982). By aiming these programmes at the rural poor, there is a direct attempt to achieve increased equity in terms of the new or expanded rural services.

Organization and Scope. As Holdcroft (1982) points out, integrated rural development programmes will face the same inevitable conflicts between the generalists and specialists as was the case with the community development programmes of the recent past. The problem is how to co-ordinate and bring technical departments (particularly agriculture) under unified control. He suggests that rather than imposing a super-department from above, a more appropriate approach might be to create autonomous institutions at or near the village level.

Approach. One approach that can be used in initiating an integrated rural development programme is to establish a pilot project in the target area. The purpose of the pilot project is to work out the methodology of establishing a rural development programme. One of the first areas of concern will be identifying a profitable package of technology that can increase agricultural production and income. In addition, the rural poor should be directly involved as a needs assessment is made in the pilot area to identify their rural development priorities.

It is likely that the project team for the pilot area will include both generalists, who will be more concerned with the process itself, and technical specialists, who will be directly concerned with identifying and introducing new agricultural technology, as well as other innnovations and/or services. Once the project team has identified areas that can be improved, then the project area can be expanded, using the pilot project as a training site for new project personnel.

Role of Extension Personnel. The generalists will be primarily facilitators and catalysts to involve the rural poor in the participatory process of programme planning, implementation and evaluation. Technical specialists would work directly with small farmers to develop, test and then demonstrate improved agricultural technology. How different service organizations should co-ordinate their work with extension personnel needs to be worked out during the pilot project stage.

COMMON DIMENSIONS

Overall, the expectations placed on extension are complex and diverse. Different groups in a country may have different goals for extension. Sometimes these goals are in conflict. Farmers may hope to improve farm income and family welfare, while urban people may want an adequate amount of quality food at affordable prices. Government may hope for both of these, plus an improvement in the balance of payments, through increased agricultural exports or reduced food imports (import substitution). In spite of these diverse expectations, there are some common elements which merit attention.

First and foremost, extension must be guided by client needs and objectives. Generally, the farm family or farm household is considered to be the primary clientele to be served. However, as we consider the many cultures and ethnic groups being served by extension across the Third World, these terms may need to be further defined. Also, the gap that exists in some countries between a small number of large-farm operators and the vast majority of small, subsistence farmers requires careful consideration in identifying the primary client groups for extension and the appropriate goals and objectives for each group. This dimension will be considered in more detail below.

Another characteristic of agricultural extension is that it is an organized, non-formal educational activity, usually supported and/or operated by government, to improve the productivity and welfare of rural people who engage in all types of agricultural production, including livestock husbandry, fisheries and forestry. This educational effort is generally tied to the problems of farmers and involves the use of problem-solving skills and information. The overall objective of these educational programmes is to develop people as better decision makers and as better managers of their own resources, so they can achieve their own goals.

Extension, then, is developmental in concept. Extension workers must start with what farmers know and their current farm practices. Furthermore, they must relate change to the particular concerns and values of farmers. Research stresses that direct contact and a farmer-centred approach is very important, especially in reaching the rural poor (Rogers, 1983) .

While there may be widespread agreement that extension must be client centred, it is also a fact that most rural communities are heterogeneous in nature. New technology may be quite appropriate and useful to one group of farmers but quite inappropriate for others. There is a growing recognition that to achieve broad-based agricultural development, the problems of each major client group must be solved. To understand the complexity of this problem better, it is useful to examine the major client groups to be served by extension.

CLIENTELE TO BE SERVED

Large and Small Farmers

Agricultural extension, as a public sector institution, has an obligation to serve the needs of all agricultural producers, either directly or indirectly. In some cases, agricultural extension has purposely focused its primary attention on the larger, more progressive farmers who can have the greatest short-term impact on overall agricultural production. Röling (1982) calls these individuals high access farmers. These farmers frequently have somewhat better education, they may have greater access to land, capital (credit) and inputs, and they generally are tied more closely to information networks. Therefore, these high access farmers are frequently in a better position to try new forms of technology and handle the concomitant risks.

The Third World, however, is primarily made up of low access farmers (Röling, 1982) who are generally operating at or near the subsistence level. The sheer number of farmers who fall into this category, with their very limited access to resources, particularly land and capital, and their minimal capacity to handle risk, make this dominant group of agricultural producers a major area of concern for a country and its development objectives.

The policy question to be answered in each country concerns what development strategy will be pursued and what potential consequences can result from following dif-

15

ferent strategies. More specifically, what are the competing interests between the rural and urban areas, and, within the rural areas, what are the common and/or competing interests between different groups of farmers, such as between large and small farmers, male and female farmers, and established and young farmers?

Too often the policy decision, about what types of technology should be developed and extended to farmers in less developed countries is either ignored or made by research workers who do not carefully consider the implications of different technological alternatives. Too often the decision is made to favour urban rather than rural areas. This decision tends to imply that larger, more efficient farmers will gain access to most agricultural resources, including extension services. In other cases, policy directives may be ambiguous, and a de facto decision is made by extension personnel who tend to work more easily and comfortably within the local village power structure and with progressive farmers, because they are more receptive to change.

The consequences of agricultural extension and the other agricultural service agencies favouring larger, more progressive farmers must be analyzed. A very brief examination of the agricultural development experience of the United States over the last few decades is instructive. In this case, extension personnel were explicitly taught how to understand, and then work within, the power structure of local communities. Extension workers generally focused their efforts on the larger, more progressive farmers because that is where progress could be made more rapidly. The assumption was that the new technology would trickle down to the smaller, lower access farmers.

The results of this strategy, particularly since the second world war, has been a rapid decline in the number of farm operators, as the larger, more efficient farmers adopted new technology (particularly labour-saving technology) and expanded their farming operations. The smaller, less efficient farmers were unable to compete in the marketplace and were forced out of agriculture production. However, during this period in North America, there was an expanding industrial sector that effectively absorbed the excess labour from the agricultural sector.

The situation in many Third World countries is markedly different. The majority of people still draw their living from agriculture. Close to two-thirds of the developing world's population gains its livelihood from the land either as farmers or farm workers (World Bank, 1983, pp. 188-189). On the other hand, the industrial sector in many of these countries is relatively small and is expanding slowly. Too much rural-urban migration has already occurred, and there are growing masses of unemployed and underemployed people in urban areas. Approaches that focus on the technological needs of larger, more progressive farmers will tend to exacerbate this situation in many rural areas.

In recognition of the need for a different strategy of agricultural and rural development in many less developed countries, the World Conference on Agrarian Reform and Rural Development (WCARRD) in 1979 recommended that rural development be based on a strategy of growth with equity. In particular, attention is to be focused on small producers, of which many are rural women. In addition, there are large numbers of rural youth in the 15-24 year age group who are employed in, or are just becoming established in, agricultural production. This group of young men and women also merit special attention, as their educational needs are different from those of established farmers.

To better appreciate the extent of involvement and contribution of women and youth to the agricultural sector, it is useful to review briefly the magnitude and role of these particular client groups to agricultural production in the Third World.

Women as Farmers and Farm Labourers

A significant proportion of small farmers and farm workers in the Third World are women. Estimates of women's contribution to agricultural production vary widely, but all estimates suggest that women constitute a significant proportion of the agricultural labour force in developing countries (see Chapter 6). Women's active role in subsistence farming and in production for market exchange has been well-documented (Beneria, 1982; Boserup, 1970; Buvinic, Lycette, & McGreevey, 1983; Lewis, 1981), and their input into agriculturally related decision-making appears to be substantial. Empirically based case studies in South America, Africa, and India suggest women's input varies among and within geographical and cropping regions and in relation to farm size (Bagchi, 1982; Mickelewait, Riegleman & Sweet, 1976). These studies tentatively conclude that women do influence men's decisions in specific areas, such as the purchase and use of certain modern inputs and the joining of co-operatives. United Nations figures show that women serve as head of over one-third of the rural households in the developing world (Blumberg, 1981, p. 41). Where these are primarily farming households, women's overall role in decision-making and production would be considerable.

The situation of women farmers in the extension process has been similar to that of small farmers. In most cases, where large farmers are the primary beneficiaries of extension efforts, both men and women small farmers receive less attention than their overall contribution merits. For African women who manage their own farms, this situation is exacerbated (Bond, 1974; Fortmann, 1981; Staudt, 1975). For women who work along with their husbands, it is generally assumed that extension efforts reach them through their husbands. This assumption is being questioned (Ashby, 1981). Research in Tanzania suggests extension information is infrequently transmitted from husbands to wives; however, the percentage distribution of extension workers visits and discussion of recommendations is higher for husbands and wives together than for either group alone (Fortmann, 1977).

Rural Young People

Rural young people are another major client group that have received too little attention in previous agricultural extension programmes. The millions of young people living in rural areas are a significant and untapped resource available to assist in the rural development process.

The United Nations defines youth as individuals from 15 to 24 years of age. It is estimated that about 20 percent of the world's population falls in the youth category, and that in 1985, the International Youth Year, there will be approximately 742 million young people living in the less developed countries. Of this population, over 70 percent, or some 520 million, live in rural areas and the majority are victims of rural poverty.

The opportunities for formal education and technical training for youth living in rural areas are very limited in many countries, especially for girls. The number of school drop-outs is high, and too many rural children never attend school.

The consequences of this situation for the youth of low-income farm families is particularly serious. These young people tend to have a very low level of functional literacy. There is frequently an obligation to work as family or casual labour from a very early age, often for long hours during the busy season, but they are virtually unemployed during slack seasons; they are expected to contribute most, if not all, earnings in support of the family; they tend to marry early; they are unwilling and generally unable to mix with more literate youth, even of the same age group; they have a fear of formal learning situations; they want to leave the rural areas, and farming in particular, and they are confronted with diminishing farm size and the prospect of living near or below the poverty level.

Rural youth has a widespread need for practical training in agriculture, home economics, group leadership and progressive rural living. Non-formal education programmes offered through agricultural extension provide some opportunities for training, but the low ratio of extension workers to farm families, and the general lack of preparation to organize and support youth programmes severely limit these opportunities. There is an enormous potential for agricultural extension to improve the future of rural youth through the development of community-based rural youth and young farmer organizations by providing training in improved methods of agricultural production, as well as income-earning skills, and by organizing extension programmes that would contribute to better family life, including nutrition improvement and population or family planning education.

Special efforts are needed in agricultural education, extension, and training programmes to include a much higher proportion of rural young women. In addition, traditional home economics programmes need to be broadened to include agricultural and income-earning skill training. Agricultural schools, colleges, and universities that prepare extension personnel need to make a greater effort to attract female students who are sorely needed, especially to work with women farmers.

CONCLUSION

The purpose of this chapter has been to define extension and to indicate its role in agricultural development. The historical development of agricultural extension was traced from its earliest roots through the spread of these programmes in Third World countries.

Agricultural extension has been organized in different ways to achieve different objectives. These different extension models were examined briefly, including their main objectives, client groups served, organizational features, approach and the role of extension personnel.

A basic premise of extension has been that agricultural technology will diffuse from more progressive farmers to most other farmers in a rural community. This premise is based on the assumption that rural communities are relatively homogeneous; therefore, new technology should be more or less appropriate for most farmers in these communities. This assumption is now being seriously questioned, and the concept of high access and low access farmers is proposed to explain why some types of so-called improved agricultural technology is not being adopted by small farmers. This new perspective suggests that the resource endowment of small farmers may be sufficiently different to make some types of agricultural technology inappropriate for their operating conditions.

This on-going discussion suggests that not only should agricultural extension be client-oriented, but also that there may be multiple client groups in rural communities who have different needs. These groups include not only large and small farmers, but also women and young farmers. Therefore, if the objectives of "growth with equity", which was adopted by the World Conference on Agrarian Reform and Rural Development (WCARRD) in 1979 is to be achieved, then the technological and related needs of these different client groups must be identified and solved.

REFERENCES CITED

Ashby, J. A. (1981). New models for agricultural research and extension: The need to integrate women. In B. C. Lewis (Ed.), Invisible farmers: Women and the crisis in agriculture pp. 144-195). Washington, D. C.: Office of Women in Development, Agency for International Development.

Bagchi, D. (1982). Female roles in agricultural modernization: An Indian case study (WID working paper 10). East Lansing, Mich: Michigan State University.

Beneria, L. (Ed.) (1982). Women and development: The sexual division of labor in rural societies. New York: Praeger.

Benor, D., & Harrison, J. Q. (1977). Agricultural extension: The training and visit system. Washington, D.C.: The World Bank.

Blumberg, R. L. (1981). Females, farming and food: Rural development and women's participation in agricultural production systems. In B. C. Lewis (Ed.), Invisible farmers: Women and the crisis in agriculture (pp. 23-102). Washington, D. C.: Office of Women in Development, Agency for International Development.

Bond, C. A. (1974). Women's involvement in agriculture in Botswana. Unpublished manuscript.

Boserup, E. (1970). Women's role in economic development. New York: St. Martin's Press.

Buvinic, M., Lycette, M. A., & McGreevey, W. R. (Eds.) (1983). Women and poverty in the third world. Baltimore, Maryland: The Johns Hopkins University Press.

Coombs, P. H., & Ahmed, M. (1980). Attacking rural poverty. Baltimore, Md.: Johns Hopkins University Press.

Farquhar, R. N. (1962). Reviews, papers and reports, Australian Extension Conference.

Fortmann, L. (1981). Women's agriculture in a cattle economy. Gaborone, Botswana: Rural Sociology Unit, Ministry of Agriculture. Ithaca, N.Y.: Center for International Studies, Cornell University.

Fortmann, L. (1977). Women and Tanzania agricultural development (Economic Research Bureau paper 77.4). Dar es Salaam: Economic Research Bureau.

Holdcroft, L. E. (1982). The rise and fall of community development in developing countries, 1950-1965: A critical analysis and implications. In G. E. Jones & M. J. Rolls (Eds.), Progress in rural extension and community development. Vol. 1, Extension and relative advantage in rural development (pp. 207-231).

Jones, G. E. (1982). The Clarendon Letter. In G. E. Jones & M. J. Rolls (Eds.), Progress in rural extension and community development. Vol. 1, Extension and relative advantage in rural development (pp. 11-19). Chichester, UK: John Wiley.

Judd, P. (1983, May). The training and visit extension system in South East Asia. Paper presented at the Regional Seminar on Extension and Rural Development Strategies, Universiti Pertanian Malaysia, Serdang, Selangor, Malaysia.

Lewis, B. C. (Ed.) (1981). Invisible farmers: Women and the crisis in agriculture. Development.

Maunder, A. H. (1973). Agricultural extension: A reference manual (Abridged edition). Rome: Food and Agricultural Organization of the United Nations.

Mickelwait, D. R., Riegelman, M. A., & Sweet, C. F. (1976). Women in rural development. Boulder, Colo.: Westview Press.

Prodipto, R., Fliegel, F. C., Kinder, J. E., & Sen, L. K. (1968). Agricultural innovation among Indian farmers. Hyderabad, India: National Institute of Community Development.

Rogers, E. (1983). Diffusion of innovations. 3rd ed. New York: The Free Press.

Röling, N. (1982). Alternative approaches in extension. In G. E. Jones & M. J. Rolls (Eds.), Progress in rural extension and community development. Vol. 1, Extension and relative advantage in rural development (pp. 87-115). Chichester, UK: John Wiley.

Sigman, V. (1981). Selected problems affecting agricultural extension as perceived by directors of national extension institutions in the developing world. Unpublished master's thesis, University of Illinois at Urbana-Champaign, Urbana.

Stevens, H. (1981). Problems of agricultural extension in Africa. Training for Agriculture and Rural Development, 1980, 23-37.

Swanson, B. E., & Rassi, J. (1981). International directory of national extension systems. Urbana, Ill.: Bureau of Educational Research, College of Education, University of Illinois at Urbana-Champaign.

True, A. C. (1928). A history of agricultural extension work in the United States 1785-1923 (U. S. Dept. of Agriculture. Miscellaneous publication no. 15). Washington, D. C.: Government Printing Office.

True, A. C. (1929). A history of agricultural education in the United States 1785-1925 (U. S. Dept. of Agriculture. Miscellaneous publication no. 36). Washington, D. C.: Government Printing Office.

Wharton, C., Jr. (1983). BIFAD's sixth birthday: A personal exaugural. BIFAD Briefs, 6 (4), Appendix pp. 1-11.

World Bank (1983). World development report 1983. New York: Oxford University Press.

Chapter 2
The Organizational Setting for Agricultural Extension

L. H. Watts

AGRICULTURAL EXTENSION: AN ESSENTIAL COMPONENT FOR AGRICULTURAL DEVELOPMENT

Regardless of its degree of urbanization, no nation is completely immune from concerns regarding the management of its agricultural base and the welfare of its rural population.

In many nations of the world, especially the developing countries, agricultural development ranks as one of the most important goals of government. Governmental interest is expressed in the nature and functions of the various ministries and agencies created to deal with agricultural development. Although ministries or departments of agriculture are usually considered the primary agencies for governmental initiatives pertaining to agricultural development, those agricultural ministries are by no means exclusive. In many countries closely allied ministries or departments, for water and power development or irrigation, for forestry and natural resources, for rural development or land reform, and for the organization of rural communities may all have assignments of one type or another dealing specifically or peripherally with agricultural development.

In the more developed countries there may be a myriad of agencies that are not exclusively agricultural or rural in orientation or purpose, but that provide governmental support of one type or another which has impact upon rural citizens.

It is common to think of highly developed nations as having the most complicated governmental bureaucracies, but the same general complexities can be found in even the least developed countries.

Moris comments as follows (1981, p. 121): "Under-developed countries often have overdeveloped institutions. A low level of material development does not imply a simple administrative system. Some of the poorest nations have the most elaborate public service structures. It is this highly differentiated and complex bundle of agencies that one must come to terms with in sponsoring rural development."

It is in such a milieu that ministries or departments of agriculture are located. Agricultural extension is, therefore, only one function among many and is influenced in a variety of ways by both governmental policy and organizational structure. Moris (1981, p. 121) observes that "At some point in the rural development process, clients must come in contact with the inter-agency matrix of official programs...it is inevitable that the task of establishing linkages to farmers should have been elaborated into a specialized function, "extension" as it has come to be known. "Extending" is a logical requisite of any program requiring farmers involvement."

Extension is basically an educational function. Its job may vary considerably country by country, but without exception it will be expected to inform, advise, and educate in a practical manner. Maunder (1973, p. 23) included an even broader role of inter-institutional relationships. "Agricultural extension services are established for the purpose of changing the knowledge, skills, practices and attitudes of masses of rural people. Schools, health services, regulatory agencies, churches, buyers of agricultural products, suppliers of production requisites and many other institutions and services are also involved in activities effecting rural people. It is the function of extension service organizations not only to establish a system of harmonious internal relationships, but also to establish complementary rather than competitive relationships with all other institutions, services and organizations contributing to progress in the rural community.

The work to be done is the basic consideration in determining the structure of any organization. The broader the scope of responsibilities and program of an extension service, the more involved become relationships with other institutions, services and agencies."

It is unrealistic in today's world to expect extension to be charged with responsibility for inter-institutional harmony, because it has neither the authority nor the ability to do so at agency levels. Extension can, however, have a significant impact at the local level and should encourage collaboration.

Agricultural development should be viewed as a dynamic developmental process that uses government to reach and to improve the quality of life for the rural family.

Although extension is one of the components supporting development, it is also supported and affected by the quality of agricultural research, the degree to which policy and prices support the use of technological adoption, and the effectiveness of a support-ing infra-structure. Considering this from an extension perspective, failure of any of those supporting functions can cause a collapse or a weakening of an otherwise effective extension organization.

In an aggregate sense extension can be illustrated as the link between research and farmers (Figure 2.1).

Figure 2.1   Extension linkage with research and farmers

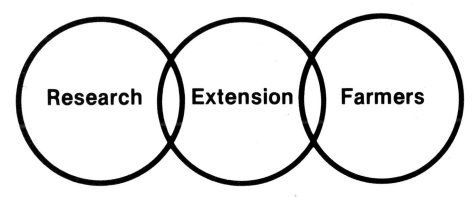

Figure 2.2 illustrates the flow of technology to farmers from research through exten-sion. Ideally there should be a return flow of research needs from farmers plus some direct farmer-to-researcher feedback. Technology is the product of research. Extension is the "diffusion-adoption" system. Farmers are the users.

An agency or function-type diagram is used to illustrate the linkages between various activities and their focus on the farm family (see Figure 2.3). The relationships between

Figure 2.2   Flow of technology to farmer from research through extension

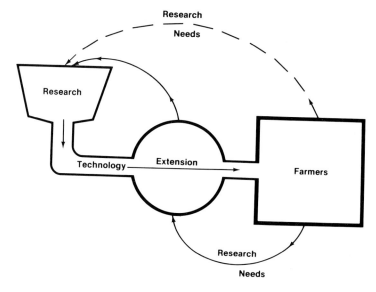

agencies can vary with governmental organization, personnel attitudes and agency assignments. It is important, however, to keep in mind the fact that the various functions are inter-dependent and should be mutually reinforcing. They become increasingly ineffective as they compete as separate bureaucratic functions. For example, research without extension is seldom used; extension without technology, markets or inputs cannot achieve effective farmer response; good markets without production are useless, and so on.

THE CENTRAL FOCUS OF AGRICULTURAL DEVELOPMENT, THE FARM FAMILY

Governmental agencies, including extension services, often view themselves as the centre of agricultural development. Each agency in its own way has a contribution to make, but the central focus for all agricultural development must be the farm or ranch family. The term "family" is used in this context in recognition of the fact that women, as well as men, may be farm heads-of-households and/or decision makers. Additionally, in male-headed households, women may influence decisions or be decision makers and they, along with children in the household, contribute a significant proportion of the human resources involved in agricultural production.

Figure 2.3 Linkages supporting the farm family

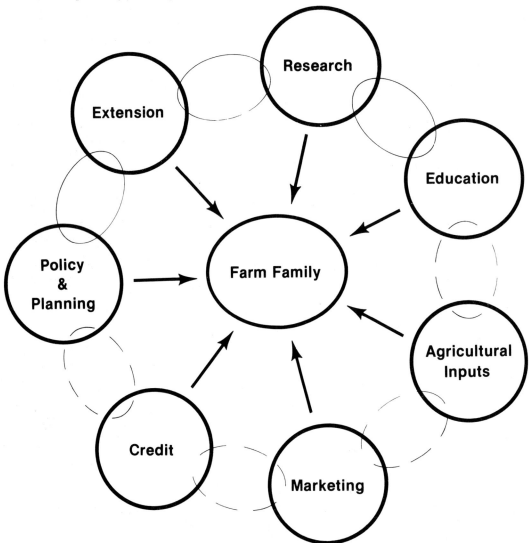

The focus of agricultural development on the rural family can be directed from a variety of perspectives and be delivered through many agencies or organizations. If extension is effective, it is an important ingredient but by no means the whole. The private sector may, and usually does, contribute as a supplier of inputs, often as a marketing agent, processor or source of credit. In other situations some or all of these functions may be provided through government parastatal bodies or agencies. The development support structures include research units, various ministries or agencies that may include irrigation, water and power development, rural development, health or community service agencies, and many other functions of the government. In many situations, religion provides a focus for developmental services which can be either peripheral or primary in influence.

Agricultural development is often thought of as the degree to which farm and ranch people adopt technology to increase crop production. Although the rate of technology adoption is an important aspect of development, it is by no means the only factor in the process.

Successful agricultural development programmes must improve the quality of life of rural people. Quality of life may mean different things in different societies. It will almost always require adequate diets, a reasonable, even if minimal, level of financial security, and an environment in which the entire family can function in accepted norms. The level of expectations will relate directly to the standard of living found in the country in question. Agricultural development can appropriately be viewed in the context of the rural family, its economic and cultural environment, its needs, desires, and expectations, and not simply in terms of yields per hectare. Development should be designed to assist people. People are the reason development is sought in democratic societies.

The underlying philosophy of extension as it evolved in North America and Europe was based upon offering assistance through an extension service and encouraging voluntary response. The concept was based upon an assumption that governmental assistance could generate greater incentive for the people to respond to national or local goals if they were enticed to voluntary action in their own self-interest rather than forced into compliance.

Various terms have been used to describe farmers: large scale, small scale, renters, share-croppers, hired labour, migrant labour, nomads, pastoralists, extensive, intensive, subsistence, dry-land or rain-fed, mixed farmers, specialized farmers, mechanized, non-mechanized, capital intensive, labour intensive, and so forth. All of these terms describe different situations in which the operations take place. For extension, it is important to think of the farm family and its relation to the rural community, as well as to the type of farm or agricultural system in which it is located.

A primary challenge for extension services is to concentrate their available resources in a manner most effective in terms of promoting the welfare of rural families, who are the key to agricultural development. Limited extension resources have frequently resulted in a concentration upon the larger, more innovative farmers, usually men, who seek information and are easier to reach. It is widely recognized, however, that in most nations the host of small and even subsistence farmers, both men and women, represents the basic agricultural resource of the nation and is, therefore, the appropriate target and concern for extension. The landless pose a special problem because they may not be decision makers, and therefore, may not be in a position to implement new ideas even if they were aware of them.

Although cultural variations will significantly affect the gap between generations, there is a pervasive problem in educational efforts which deal with a population composed of all age groups. A special challenge occurs for extension, therefore, because of its need to reach not only the mature and established farmers, but to influence and educate the younger generations who often may be the most amenable to new ideas and concepts.

Women, as farm managers in their own right and as contributors to agricultural production, have been identified as a target group of extension. And appropriately so, even though the role of women in agricultural production varies significantly between countries and world regions.

Typically, in farm families in the United States and Western Europe, women participate as co-managers of farms along with their husbands. They perform numerous activities, such as tractor-driving during peak periods and record-keeping and analysis, which are essential to a successful farm operation. Additionally, men, women and children make up the migrant labour force found working in the fields.

Generally, in developing countries, women constitute a larger proportion of the agricultural labour force, as farmers and farm workers, than in the industrialized countries (see Table 6.1). In Lesotho, for example, women farm family members carry a particular responsibility for agricultural production because significant numbers of male family members find employment in the South African mines. Likewise, in some Middle Eastern countries, men seek work in the oil fields of the Gulf States and thus, women family members are responsible for farming. Further, in many societies, women are the

23

primary producers of food crops. In fact, on a world-wide basis, it has been estimated that women contribute a full 44 percent of subsistence food production, largely through subsistence agricultural production (Aronoff & Crano, 1975). Extension has grown to recognize the importance of men small farmers to agricultural productivity and recognizes them as a target group. Likewise, recognition of women small farmers and workers has begun. Basically, this means sensitizing extension personnel to the agricultural contribution made by women in their respective countries and then directing extension efforts to include women.

In those societies where clear cut divisions exist between men and women, there may be special problems associated with reaching women through extension efforts. This situation has led to efforts to recruit women extension personnel so that women may extend agricultural information to women. This change, however, is moving slowly.

In the past, there were few trained women agriculturalists to recruit. This is gradually being corrected as access to intermediate-level training and higher education in agriculture is increasing for women. Still, it has been noted that some countries are encouraging employment of women extension workers, but placing them in non-extension jobs or in work that is not relevant to their training. This later situation is exemplified in cases where production agriculturalists are assigned to family nutrition programmes. The stereotyped view of women as home economics extension workers is being altered as competent women are proving their value in agricultural extension work as well.

Nomads have posed a special problem for extension because of their mobility and because many of the research-based recommendations which are carried by extension services are not well-adapted to nomadic life-styles or to the agricultural situations faced by nomadic peoples.

Successful agricultural development requires that an accurate assessment be made of the various types of farmers in the country, their capabilities, their limitations, and the policies which support, or fail to support, their operations. A successful extension service is different from many agricultural development activities because it deals with the social and community structures as well as the agricultural aspects of development.

An encouraging trend in recent years has been the emergence of Farming Systems Research and Development (FSRD) or Farming Systems Research and Extension (FSR/E) that places heavy emphasis on the farmer's situation or system in order to orient research priorities toward in situ needs and problems rather than on the research perception of them. Factors basic to FSRD as identified by Hildebrand and Waugh (1983) are: (1.) a concern with small-scale family farmers who generally reap a disproportionately small share of the benefits of organized research, extension, and other developmental activities; (2.) recognition that thorough understanding of the farmers' situation gained first-hand is critical to increase their productivity and to forming a basis for improving their welfare; (3.) the use of scientists and technicians from more than one discipline as a means of understanding the farm as an entire system rather than the isolation of components within the system.

FSRD is currently seeking to develop stronger research-extension linkages. The effort in concept as well as implementation indicates a growing awareness of the fact that farm families are the most critical factor to successful agricultural development, and correct interpretation of the farmers' view of agriculture is necessary for appropriate research and effective extension.

ESSENTIAL COMPONENTS AND LINKAGES

Agricultural Research for Technological Development

Agricultural research in the twentieth century has changed the nature of, and the outlook for, agricultural development. Research is the fuel that powers the engine of new technology and the development derived from it. The truly significant changes in agriculture that have taken place in the last fifty years owe their genesis to systematic analysis and study by agricultural researchers.

The evolution of extension services actually began as sufficient research knowledge was accumulated to permit effective, production-enhancing, educational programmes to keep farmers abreast of the new technology.

A majority of nations created research departments within their ministries of agriculture. In some of these nations the emergence of special problems such as on-farm water management and irrigation led to the creation of additional research organizations outside as well as inside agricultural ministries. In the United States, research was organized both at the national Department of Agriculture level and at the state level, where it was made a part of the land grant university system.

The rationale for publicly supported agricultural research is based on the assumption that farmers cannot afford risks associated with experimentation. The nearer agriculture is to a subsistence level, the greater is the risk to individual farmers and the greater their reluctance to experiment "on their own."

As societies develop their agricultural infra-structure, they also tend to provide opportunities, in free market economies, for private enterprise. Privately sponsored research may exceed that conducted by government or universities. It is estimated that at the present time more than half of applied research in the United States is carried out in the private sector. Today a very high percentage of agricultural research is conducted by commercial firms as they seek to develop new products for agriculture. Research is, therefore, most accurately viewed as a function in which both private enterprise and government participate.

The "Green Revolution" has captured the fancy of many in key leadership positions. It has, unfortunately, developed in the minds of some a belief that reliance upon huge genetic, chemical, or other technological breakthroughs can provide such significant increases that other aspects of agricultural development, including extension, can be dispensed with. Significant as some new research will continue to be, the reliance on research alone is insufficient for appropriate progress. McDermott (1982, p. 200) has pointed out that the farm sector is..."a dynamic social system actively seeking better ways of doing things." He adds that we often view the farmer as much more passive than is the actual situation and have not fully appreciated the desire to innovate or the "experimental" nature of the farmer's social system. The task of providing information, as well as developing the community organizational structures in which to utilize new research, is critical and may well become even more so in the future. An assumption that an extension programme is no longer needed because there was success with a specific variety of rice or wheat is over-simplistic. The means, however, by which McDermott's dynamic social system is energized, informed and set in motion may require modification in organization and techniques.

Many of the problems of the developing nations, and a huge proportion of the global problem related to food production, require increased research attention to basic food crops. Additionally, research efforts must consider who grows the food crops, because in many countries, women are the primary producers of food crops. These observations highlight the need for an expansion of research on food crops, with special attention to crops which have important nutritional value. Various reports on global hunger such as those by FAO, the United States President's Commission on World Hunger (1980), and others lend a sense of urgency to food crops, malnutrition, and starvation.

One of the risks to farmers and to agriculture in general from the Green Revolution package is the temptation to standardize or concentrate on only one or two varieties of a specific crop. Over-reliance on a few varieties of a crop increases the risk of genetic vulnerability and the potential for total crop failure if a particular disease or other problem proves catastrophic in its impact. Concentration on single crops also runs counter to the practice common in many countries, such as China, of multiple cropping or inter-planting that may not yield the responses in the field equal to significant single-crop successes on the experimental farm.

A significant difficulty in many Third World research projects has been the low level of involvement of indigenous and well-trained scientists in actual field experimentation. While many agricultural scientists trained at the Ph.D. level have returned to countries badly needing their expertise, many of them can be found behind an administrator's desk running a bureaucracy rather than an experiment.

Research organization has many forms. Almost without exception, one can find the central staff operating out of a metropolitan centre, usually the capital, with outlying research centres scattered throughout the country. It is at the regional or district research centre where the research personnel have the greatest opportunity to interact with extension staff and with practising farmers. Emergence of the farming systems research (FSR) approach is now calling upon researchers to begin their efforts by identifying problems of the farmer rather than relying upon those perceived from outside through their own eyes.

The technical or subject matter specialist support for field extension staff is basic to ensure valid research information is used in farmer contacts. Without it, extension workers tend to use their best (and often unreliable) judgement. This is especially true where field personnel are poorly trained. It is not sufficient to rely on an information service or written communication.

Research today is supported by a globally oriented effort which has arisen within the past three decades. The evolution of International Agricultural Research Centers (IARCs) has provided a basis for effective and comprehensive research keyed to specific problems or crops. For example, the International Rice Research Institute (IRRI) works on rice; the Centro Internacional de Mejoramiento de Maiz y Trigo (CIMMYT) works on maize, wheat, barley and triticale; the International Livestock Center for Africa (ILCA) works on livestock production problems in Africa and the International Institute of Tropical Agriculture (IITA) works on food crops in the humid tropics. These centres have the advantage of an international base. They are also serving as an excellent training ground for junior scientists. They are clearly a significant development in agricultural research and can be expected to provide strong technological influence in coming years. They have already demonstrated significant influence in their contributions to the Green Revolution.

From an extension perspective, the evolution and development of more effective linkages, between the international research institutes and national research organizations, and between them and extension, should receive increased attention. Some efforts of the IARCs to disseminate their findings have been viewed as an opportunity to improve effectiveness of extension services. Serious examination of the function of the IARCs, however, indicates that while such efforts are admirable, the function and staffing of the institutes cannot provide for the intensive, farmer-oriented, site-specific and demonstration-centred educational programmes needed in various countries. Further efforts are needed to ensure that new technology developed by IARCs is tested and/or modified, and is then made available to extension services. However, national research agencies should carry the primary linkage responsibility (see Figure 2.4).

A central concern of any book which concentrates on extension is the research-extension linkage. The complexity and importance of this problem is perhaps best illustrated in the International Service for National Agricultural Research (ISNAR) 1982 annual report (p. 31). ISNAR is an IARC which has as its mission the strengthening of national agricultural research systems. Less than two years after its establishment, it found... "One of the most consistent recommendations of the ISNAR teams concerned the upgrading and further development of the relationship between research and extension organizations...ISNAR teams have found in many developing countries that extension workers without close links to research pass on improvised technical messages; and some research is carried out in isolation, building up technical solutions--some of which are unusable or unused. Research systems that cannot transmit findings to the extension service and to farmers make little practical contribution. Extension work that is not sustained by results obtained through research has little value, and may even be detrimental."

Agricultural Extension for Technological Diffusion

Agricultural extension is widely accepted as the governmental function that is responsible for providing information and educational services to farmers about new technology. Role and mission may vary widely, however. Debates are common among extension practitioners as to whether a system of information transfer can be applied to a wide range of rural and sometimes even urban problems, or whether such broad definitions of extension tend to blur the focus and deny effective targeting of impact. Perhaps the most relevant perspective is to consider the focus as educational versus regulatory, or as a provider of agricultural inputs. Johnson (1983, p. 5) raises an appropriate question: "Given whatever organization exists for agricultural development, where does the extension or knowledge/technology transfer function fit in and relate to all else? There are many options to be considered and many decisions to be made if optimal possibilities are to be realized. In the world as a whole, U.S. domestic experience is atypical. The more common pattern of extension organization relies on the regular civil service structure (i.e., on the ministries of agriculture and of related functions), while the U.S. relies on a variety of mechanisms to produce relative autonomy, intergovernmental checks and rela-

Figure 2.4 Primary research linkages

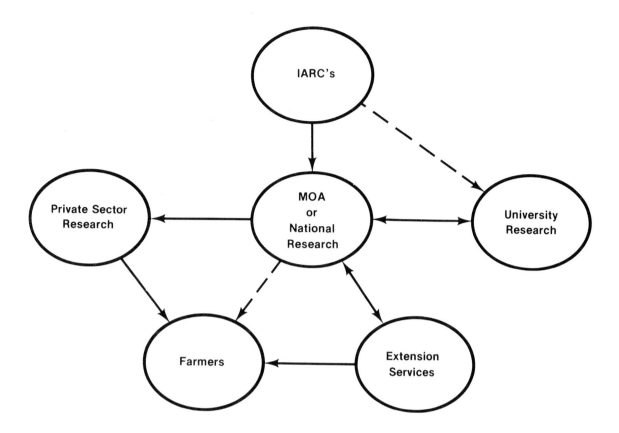

IARC = International Agricultural Research Center
MOA = Ministry of Agriculture

tions, and primary responsibility through the land grant universities with integrated research instruction and extension.

Organization is a way of ordering necessary relationships among the components of the agricultural knowledge/technology transfer process--the officers and their clients, the headquarters and the field, the knowledge sources and their application, the producers and consumers of information, the subject matter specialists and the farm agents, the handed-down policy and the feedback response."

The greater the degree of "top down" or autocratic operation of the central government, the less will extension carry out an effective educational role and the greater will be its use as a conveyor of government policy. The more extension is assigned jobs of supplying inputs, representing government policy, or regulating, the less will be its time for, or acceptance as, an objective educational activity.

For the purposes of this discussion, the extension role will be considered in terms of its primary mission of carrying new knowledge to rural people and assisting them to use it effectively. The technology may not always be production oriented. It may deal with home and family needs as well as farming and their interaction. A secondary, often ignored, but important role is advising researchers of farmers' problems which require research attention. This need has led to Farming Systems Research (see Figure 6.2).

Extension personnel are in close contact with farmers and should be more able to understand their problems, their inhibitions and their needs than other developmental agency representatives. However, extension personnel interact primarily with men farmers and, therefore lack knowledge about women farmers. Substantial effort to redress this bias is required. Extension personnel also must depend on good research to bring new knowledge to farmers. Systems are needed which provide a dependable flow of information on farmer's needs to researchers, through extension.

27

The basic philosophy of government can affect the content as well as the organization of extension's educational programmes. The degree to which extension may be free from requirements to promote government policies or enforce regulations may hinge largely upon the willingness of central government to provide for voluntary rather than coercive influence. Those governmental policies and attitudes may also be expected to influence the basic statutory or legislative authority upon which agricultural extension functions are based.

Successful extension programmes depend upon a continuous and reasonably stable financial and policy foundation. These aspects should be provided for in legislation which provides a specific legal basis for extension in each country. If extension is expected to operate on the basis of ministerial guidelines established by each different minister of agriculture, it will be subject to the vagaries of administrative change and can become an undependable source of objective information. An extension law is needed to define the basic purpose or mission, identify the means of financing, and spell out the broad scope of responsibilities and expectations of the extension service.

A 1962 survey indicated that approximately one quarter of the countries surveyed at that time had no legal provision for extension work, and over one-third were operating under general legislation of a ministry of agriculture rather than specific legislation for extension (Maunder, 1973). In the United States specific legislation and various amendments have provided a legal foundation for co-operative extension work. However, there is a wide variation in state legislation pertaining to extension (Hoffert, 1979).

Although the ability of various societies to absorb and utilize technology can vary significantly because of geography, climate, national policy, population density, soils and a host of other factors, progress in meeting global food needs and in enhancing agricultural development can be expected to require an effective and continuous flow of technology to agricultural producers. Under these conditions continued significant investment in extension will be required in the future.

These continued investments, however, should be placed into a context of balance and national capacity. Woods (1983) points out that extension services are expensive, that factors contributing to costs apply in varying degree to virtually all extension services, and that there is little cause to expect their efforts to diminish in the foreseeable future. He identifies several factors which tend to push salary costs to higher levels, with increasing costs of travel, materials, allowances and so forth as reasons for his concern.

Precise measures of extension performance and results are sketchy at best. Extension has too often reported its level of activity rather than results. Factors other than extension also are involved. Variations in market systems, weather, and infra-structure support cloud the issue. Aspects often overlooked and most difficult to measure are the social values or community improvement aspects of extension which may appear subtly in a variety of ways and be described best as a "quality of life" influence upon families benefitting from extension work. These factors, too, can vary widely from country to country, depending upon the nature and competence of the extension personnel employed. Extension, as only one of many components in agricultural development, must be viewed as effective in direct proportion to the degree that it provides a collaborative and supportive function in the total agricultural development activity.

One of the most significant limitations to successful extension is the absence of a supportive infra-structure and a national policy which provides incentives to agricultural producers to respond to the educational offerings of an extension service. Extension cannot be productive in the absence of incentives that make it worthwhile for farmers or livestock producers to use new techniques and new knowledge. Farmers cannot be expected to improve productivity in the absence of a supporting infra-structure for inputs and for marketing (Berg et al., 1983).

In those nations where a majority of farmers are operating on a subsistence basis, the risk element for any type of change is so high as to discourage experimentation and innovation. Under these conditions new technology must be relevant to the situation in which farmers find themselves, and the extension activities must rely heavily upon demonstrating feasible ways by which farmers can improve their situation without undue risk. Even relevant, new research can carry risk. For example, a new crop variety must have the foundation seed stock given to a reliable multiplication system; otherwise an insufficient supply can result, or even a supply of an impure variety.

28

Perceptions of extension vary greatly within the donor community. Total investments in extension represent a significant expenditure, especially in developing countries. There are critics who accurately point out that very little real growth has taken place in many nations. These critics indicate that extension is no longer needed in industrial nations with sophisticated communications capabilities. They also point to the slow and often unimpressive gains made in agricultural development in Third World countries, even though extension services exist. What they may overlook is the fact that extension, to be effective, must function in situations where other components of agricultural development also are working effectively and where an appropriate incentive structure exists (Berg et al., 1983). Again, the partnerships of extension must be stressed as only one component of agricultural development. In addition, extension must have effective linkages to research and the ability to provide continuous in-service training. But even when development projects have attempted to eliminate infrastructure weaknesses, expectations have been only partially realized (Feder et al. 1982, p. 1). No serious consideration can be given to extension without recognizing the total web of development agencies, governmental policies, supporting infrastructure and variability in human response due to wide variation in literary, educational levels, social organization, customs and levels of economic development.

World Bank officials identify (a) appropriate technology, (b) attractive markets, and (c) available inputs as the critical ingredients for successful development (Russell, 1981). Extension effectiveness will be dependent upon those ingredients, which are outside its direct control.

Many so-called action agencies seek to develop their own extension services because they distrust a broader-based and truly objective educational effort and are seeking promotion critical to understanding the nature of agricultural extension. The most effective extension work is usually done by extension personnel whose behind-the-scenes, personalized, helpful, objective, educational effort is less visible than the role of other agencies but whose credibility is unquestioned by farmers.

Kearl (nd., p. 5) refers to the importance of commitment to farmers in extension programmes as "the requirement of the agents to hold primary loyalty to the welfare of the rural family in preference to rigid commitment to governmental goals and targets". If the extension role is basically one of technology transfer or delivering useful information to rural families, what then is its optimal linkage within the governmental system?

To be effective, extension must have adequate knowledge of relationships with, and ability to refer farm families to other agencies or organizations for appropriate assistance. At the same time, extension's credibility with farmers will be based on solid research information appropriate to farmers needs. It will also require extension to develop consistent objectivity in its recommendations and to qualify them when information is incomplete, unavailable, or not site specific for local needs. Hard as it may be for the extension worker, it is far better to admit that he or she does not have data needed to answer a question than to give out erroneous information.

Extension will be weakened to the extent that it is included within small sub-units of government, or to which it is splintered as a function with similar roles among a variety of rural development agencies. A single agricultural extension agency is able to maintain a stronger central staff, hold higher visibility within government and operate with a less complicated and more visible local mission.

Thus, characteristics of a strong extension system are

1. effective linkages with research organizations,
2. statutory basis and mission,
3. stable financial support,
4. on-going and effective in-service training,
5. adequate field offices, transportation, communications systems,
6. an orientation to understand and to serve the farm family,
7. freedom from regulatory or input supply responsibilities,
8. continuing farmer input to guide programme priorities,
9. a competitive salary system with incentives for professional development and advancement,
10. an effective communications (information) staff.

Agricultural Education for Knowledge and Skill Development

The concept is well-accepted that universal free education for students played a major part in the progress made by the United States in application of technology. This concept was accelerated by passage in 1862 of the Morrill Act, more commonly known as the Land Grant College Act. This act, which provided for college opportunities for the sons and daughters of the working classes, established a collegiate-level educational opportunity that had not been available for young people planning careers in agriculture or the mechanic arts.

The educational level of society at large and rural residents in particular carries a significant influence for any extension programme. First of all, it affects the educational level of the pool of young recruits coming into the extension system. Secondly, it influences the technical level required for extension to assist farm families. As might be expected, there is a tremendous variation among the different countries in the type of education available to rural youth.

A tenth-grade or ten-year educational programme is common in many developing nations. At this level, depending upon the agricultural emphasis in the school curriculum, students may be exposed to rudimentary scientific and technological education. It is obvious that such efforts are minimal at best. In many situations, the education may be conducted on a rote-learning method with no actual practical experience. Obviously, people educated in this system, who may be employed as para-professional extension workers, require considerable practical experience as well as pre-service and in-service training. This aspect is covered in Chapters 14 and 15.

Vocational agriculture at the high school level has long been a part of training in the more developed countries and can frequently provide an excellent basic education for beginning extension workers or those engaged in the private sector. Properly conducted vocational agricultural studies can expose students to practical field or project experience, farm shop and machinery repair, basic science, and a good foundation for university-level agricultural studies, if available.

Most nations prefer agricultural officers in extension to have at least a bachelor's degree in some agricultural field. These degrees are obtained through colleges and universities which are accredited for granting degrees. Colleges may give only a baccalaureate degree. Most universities will offer advanced degrees at least to the master's level, and many will provide doctorates in certain specialized areas of competence. Third World educational entities continue to need highly trained scientists. As a result, many of the donors as well as developing nations have sought to provide out-of-country, graduate level educational opportunities. This is providing an increased pool of highly trained personnel for professional agriculture, scientific research and extension leadership. Unfortunately, in many countries which lack trained administrators, there is still the tendency for the well-trained graduates with technical and agricultural skills to seek and obtain administrative positions where their technical skills are not used in a direct sense. Another concern arises where many newly trained scientists elect to remain in a more developed nation where salaries are more lucrative than in their home country.

Graduates in agriculture find professional job opportunities not only in research or extension, but often at higher salary levels working for financial institutions, for agricultural input supply firms, for marketing concerns or for parastatal companies. It has been noted in several of the African countries that parastatal bodies frequently pay higher salaries to individuals who might otherwise be working in extension or research. Private industry frequently hires extension and research personnel. This process lowers the pool of trained labour available for extension and research. However, some private research has been effective, and is often made available directly to farmers through the technical inputs resulting from that research.

In summary, agricultural education can be viewed as the training provided through vocational or technical institutions, generally at the diploma level; education at the college or university level which may be at any stage from the B.Sc. up to the Ph.D. degree; and in terms of less formal training conducted for practical rather than academic use, which may be made available to farmers through farmer training centres or similar techniques. Some colleges may provide non-credit and non-degree training by using faculties for off-campus as well as on-campus training of farmers, which may significantly add to efforts made by an extension service. In other situations the only such training available may be that provided by extension or, as is frequently the case, by agricultural training institutes. These institutes are most often operated by the ministry of agriculture of the country concerned.

OTHER AGRICULTURAL SUPPORT SYSTEMS

The World Conference on Agrarian Reform and Rural Development (WCARRD) concluded that all farmers should have equitable access to agricultural inputs, credit services, markets and all other agricultural institutions that serve rural communities. The following section is a brief examination of some of these important agri-services.

Marketing

Farm production at all except subsistence levels is significantly affected by the marketing system, and women are integrally involved in the marketing system throughout the world. Often, the marketing system can be as significant to the welfare of the farm family as the efficiency or inefficiency of the production activities themselves.

Many farm products are sold in a barter system or as cash in local village markets. In these situations, price control or crop value is determined primarily by local supply and demand. Marketing beyond the village immediately brings into play a variety of other factors such as transportation, marketing organizations, processing capacity and, if crops are intended for export, the ability to transport them to the points of embarkation, and deliver in a satisfactory condition for export.

Significant marketing problems for small farmers in some countries with certain crops are the availability of trucks and organizations able and willing to make roadside collection, with delivery to a central collection point. Local processing for certain crops can also be critical where bulk is a factor and transport vehicles are limited. Marketing is affected by the ability of a country to use standards of weights and measures, and quality upon which price variations may be determined for certain types of crops. This latter gradation may not apply in many small farm operations, but can be important in areas producing for export, or for certain quality-related commodities. Marketing of livestock is often complicated by distances, as the stock must be transported to a central market place, and by the quality and availability of abattoirs.

Simply stated, marketing represents the process by which an individual farm family's produce is transferred from the farm to become a value-laden commodity to be used by others. For extension personnel, knowledge of pricing policies, transportation difficulties, marketing firms and related aspects represent an important source of knowledge needed by farmers. These are all too often overlooked by extension organizations that concentrate too heavily and too specifically upon production itself.

Policy and Planning

Policy within any country is a critical factor in extension effectiveness. Policy can be examined from two perspectives, the first, in terms of the importance given to extension, and the second, in terms of agricultural policy and its impact upon the extension's educational effort.

If a nation's agricultural development policy is strongly supportive of extension and agriculture research, extension can be expected to enjoy a reasonable share of that government's financial resources allocated to agriculture. Obviously, such a policy enhances the ability of extension to be effective, and improves its status as an agency.

Agricultural policy which does not support recommendations from extension on input availability, pricing, transportation and related factors can quickly negate the most attractive extension recommendations. What value can be ascribed to increases in production if price policy, poor markets and lack of transportation combine to negate the financial rewards to the farm family which has followed otherwise good production recommendations?

Good, long-range agricultural planning, designed to develop and maintain effective policies in support of agricultural development, obviously goes hand-in-hand with supportive policies. A nation's ability to provide a level of confidence to its farmers, based upon rational forward planning, improves not only individual confidence, but enables farmers to plan their production enterprises with some assurance of policy and marketing stability. Farming, at best, is subject to the vagaries of weather, insects, diseases and related problems. Farmers who are unable to make plans prior to planting because of a lack of coherent agricultural policies or planned priorities for agriculture suffer a serious disadvantage, regardless of the size of their operation. Under such conditions, good extension efforts yield little, if any, benefit.

Input Supply and Credit Services

Input Supply

The supply of seeds, fertilizers, agricultural chemicals, machinery, repair parts and other inputs is critical to the success of any extension programme. A response in 1980 in an African nation indicated that if extension recommended use of a certain fertilizer, it would probably take 11 months before the request could be processed and the fertilizer delivered, if it was still available. This type of delay, and inability to provide the input is a serious disincentive for any extension programme. The extension worker who encourages a new seed to improve crop yields is the first to be discredited if seed stocks are not available, or if the variety has different characteristics or poor germination, or if they reach the farmers too late for planting. If a recommended pesticide proves unavailable or of less than standard quality, the resultant poor results discredit the extension worker who made the recommendation.

Information by the extension service about limitations in the total input supply system can provide much-needed help for farmers, and must be considered by extension personnel as they make recommendations to farmers. Admitting there are such limitations, however, may not be supported by the governments. In the absence of dependable information to farmers on inputs, on credit, and on marketing problems the credibility of extension will be seriously eroded.

One factor in the success of agriculture in the United States has been the availability of high-quality seeds, agricultural chemicals to meet specific needs, and the ability of machinery companies to provide a wide variety of machines and, just as important, spare parts. Significant changes in price or availability of inputs can make major differences in options available to farmers and in recommendations projected by extension.

Credit Services

An almost universal requirement in agriculture is financing. Credit for farming operations is provided in a variety of ways. However, in general, women farmers have experienced more credit constraints then men. Credit can be provided directly by private money lenders, or by financial institutions such as banks; it can come from government-subsidized credit services; or can, as in the Unites States, come through commodity credit provided as a part of government farm policy. Some agricultural co-operatives provide credit. In many countries, the private money lender is preferred by small farmers even though the cost of money borrowed may be exorbitant. Many under-educated farmers are of the opinion that they are borrowing interest-free, simply because they do not understand how the monetary system works, or fail to recognize the with-holding of part of their crop as interest. The availability of the local money lender is much less complicated than the more formal bureaucracy involved in servicing institutional loans. While costs of money borrowed from them may be high, money lenders do provide a unique service as a handy, flexible, personalized source of credit and may be more tolerant of local hardships from weather extremes than large and distant banks.

Donald (1976) categorized farmers by their credit needs. He identifies as type one, those farmers with larger or medium operations and ready access to bank credit; type two, farmers with small-scale but reasonably profitable operations, eligible for bank credit (assuming physical access to banks); type three, farmers with the potential for establishing profitable operations if greater access to technology, inputs, and markets at fair prices were possible but who are not now eligible for bank credit; type four, farmers who could attain profitability, given access to technology, inputs and markets, but only when provided with special incentives (this implies a need for subsidization for an undefined period); and type five, farmers with such poor resources that even with improved access to the factors listed, they could not develop a viable farm enterprise capable of supporting a family without a permanent subsidy or substantial off-farm income (this group includes landless labourers and part-time or garden-plot farmers).

Donald points out that the so-called small farmers would be those of types two to five, but the ones with particular credit problems are types three, four and five. In view of these factors, Donald places a priority upon the importance of the small farmer and his or her need for credit to maintain operations. He says that, "It must be recognized that the normal flows of credit through banks and related institutions have been subject to a number of inhibitions that have worked to exclude most small farmers" (Donald, 1976, p. 17). The two principal objectives of small-farmer credit programmes identified by Donald are an increase in agricultural production, and an increase in social equity. Thus it is obvious that attention to small-farmer credit, particularly for women, represents an important aspect of development credit that is critical to the extension function, even though the extension service is not responsible for it as an organizational activity.

Material and Capital Transfer

A serious limitation in rural development is lack of available capital on which to develop a base of economic activity. In the preceding paragraph, credit was discussed as it applies to the individual farm family. Capital transfer in this section is in a larger context, in which the difficulty of low-income rural residents is considered as a community problem in generating sufficient economic activity for viable development.

Various means have been explored to develop methods by which capital may be transferred to the rural sector. For the low-income small farmer, the concepts have included donor infusions of capital, and efforts by group action involving sufficient numbers of rural residents to obtain credit, or accumulate sufficient leverage to entice additional capital to be transferred into the rural community. In subsistence situations, even small amounts of money for seeds, hand tools, a water pump or insecticides can make a significant difference. An extension role in such situations can be assisting farmers to organize access to sources of capital from government or private sources.

ENVIRONMENTAL FACTORS WHICH AFFECT AGRICULTURAL EXTENSION AND DEVELOPMENT

Political Environment

The political environment in any nation is critically important to the success or failure of any extension effort. The extension model may suggest that the organization "stay out of politics". In developing countries, food policy is of paramount concern. "It is too important to escape politics. It may, in fact, be the essence of politics" (Johnson, 1983, p. 9). Because much of extension is directed towards food production, food policy alone dictates a significant influence upon extension.

If extension is an agency of change, it can also be viewed as a threat to those for whom change is not desirable. The power elite can be threatened by a strong and active extension service. Effective extension can, for example, create a stronger and more politically articulate agricultural population through the process of encouraging progress which inevitably involves change. For extension leaders to fail to understand the "politics of progress" is to operate with tunnel vision in a complex and dynamic social/political system. Realistic evaluation of extension requires identification of accepted and achievable goals. Those goals will be encased in a political as well as technological framework.

Effective extension requires a linkage to governmental planning, strategy, and policy, all of which have an impact upon the mission for, and support of, extension. In spite of the need for extension to be neutral and objective in its handling of information, the realities also require that it be compatible with and aware of the political policies of its government.

In several African nations, sharp shifts in governmental policy following independence, including expropriation and land reform, had significant impact upon extension work. The priority given by national governments to agriculture can also be critical. If agriculture is perceived as a primary industry and has strong support by the government, policies are most likely to be developed which support rural development.

Policies which are critical to agricultural extension work include the following

1.   price incentives or pricing policies,

2.   export and marketing policies,

3.   agrarian and land tenure policies,

4.   policies which provide the rural infra-structure, including farm-to-market roads, distribution or collection centres and fertilizer and pesticide supplies,

5.   policies which support a processing and distribution system for agricultural products,

6.   support of agricultural research and extension,

7.   policies which support serving families headed by women,

8.   policies which support the development of agricultural institutions, such as universities and credit systems,

9.   dependable financial, organizational and policy framework for extension,

10.   governmental policy specifically concerning the extension mission.

Four illustrations may serve to expand this point. (a) Where governmental policy has used profits from corn and rice to subsidize industrial development, it has worked to the disadvantage of agriculture and negated extension efforts to encourage increased production of those crops. (b) In one instance a major government programme to increase rice production required all extension personnel including home extension workers, to concentrate on rice production. A glut in production under these conditions produced a negative attitude by extension workers, and an imbalance in extension programming. (c) In another example, a shortage of edible oilseeds led to importation of those products. The government support of sugar but not oilseed crops resulted in an over-supply of sugar and under-supply, and continued importing, of edible oilseeds. The best extension efforts cannot result in a major shift from sugar cane to oilseeds unless governmental policy provides higher profit motives for such a change. (d) In many countries, extension suffers from poor salaries and a weak professional image. Under such circumstances a policy permitting extension salaries to rise would assist in employing more competent personnel and maintaining high staff morale.

Economic Environment

Two principal economic factors are important to both the scope of activity and perception of extension assistance to farmers and ranchers. The first, of course, which has been discussed earlier, is the profit opportunity available to the farmers who improve productivity by applying new technologies brought to their attention by extension service personnel. The second factor is one which relates to the general economic environment in which a nation's agriculture finds itself.

Obviously, farmer response to new technological innovations is directly related to his or her perception of financial advantage from applying such recommendations. It is through demonstration of increased productivity and the convincing of farmers that such increases can mean greater income or family security that recommendations coming from extension personnel find their most positive response.

It is well to keep in mind that the more prosperous an agriculture is, the less interest there may be in seeking assistance from an extension service. When weather is good, prices are high, and other conditions are favourable, the opportunity for a farm family to make a profit is improved. Under such conditions even relatively inefficient farmers may appear to be successful. Even though they may be in position at such times to take a greater risk through innovative practices, farmers may feel that there is little urgent need to listen to extension recommendations. The converse of this situation is found in times of great stress in the agricultural industry. It is at these times when farmers recognize a need for assistance and may be willing to reach out for and accept help.

To illustrate the above point, two examples are given. In one example we might assume that very favourable weather conditions have resulted in minimal land erosion, adequate rainfall, and generally good growing conditions. During such a period, it may be more difficult to interest the average farmer in soil conservation practices than during periods of extreme wind or water erosion, or plant debilitation due to stress from lack of water.

The second example may relate to a period in which cash crop returns and basic food crop availability have been quite adequate due to a favourable combination of weather and prices. Even though adoption of new technology during such a period might increase profits exponentially, a poorly motivated farmer might consider the situation quite adequate since he is under no particular stress in his operations and sees little risk to his family in holding to the status quo. This feeling can be intensified in terms of basic food crops if storage facilities are not available and the farm family sees no advantage in producing excessive crops for which there may be no market and which exceed their own family demands. If the above examples hold true during a given situation, they also pose a challenge to an extension service to develop initiatives for farmer contacts and educationa programmes which are strong and visible.

Access to land is an important economic factor. Too frequently, women's access to land is not given the consideration given to men's access to land. Land may be limited by land tenure policies which lead to control by a few, or to control by men. Access may be limited, as it is in the mountains or near urban areas, by development pressure and resultant unreasonable prices; or it may be controlled by zoning regulations. Non-agricultural demand for any type of development can pull land away from agriculture if it causes prices to rise so high that there can be no profit from farming. A major concern for the landless and the poor arises from their lack of access to land, the primary source of production. So long as lack of the source of agricultural production exists, there will be poverty and lack of income earning potential.

Socio-cultural Environment

Within the past decade, there has been a significant increase in attention to the socio-cultural environment in which extension services function. The assumption that agricultural technology related to crop and livestock production is the only significant factor is no longer considered an adequate programmatic base for extension. Many examples could be given which illustrate in a variety of ways the manner in which the socio-cultural environment can affect the organization and operation of extension services.

Extension services to be effective must operate at the local community level. This point cannot be over-emphasized. A strength of extension is, and has been, the location of the extension worker or field assistant in a local community where his or her access to farmers, and the farmers' access to him or her is easy and, where the work is as visible as possible.

Accessibility to offices by farmers is important from a sociological as well as geographic perspective. In Islamic societies, location of an extension office near a mosque can prove advantageous, not only to the extension worker personally, but in providing the opportunity for easy access after midday prayers. In Spain, some extension offices are located on the second floor, with a popular local bar on the first floor. Social access between extension workers and local farmers in both cases is much easier and more frequent than would be the case if the office were in a different location. Conversely, these examples clearly show that what improves access for one client group may exclude another group. This complexity requires careful attention. Women seldom have ease of access to bars. Where the client group is or includes women farmers, an extension office situated close to the market place or to the primary school may facilitate their access to extension.

The principle to stress is that ease and frequency of contact for extension workers will increase in direct relationship to the ease of access by local farmers in a social context. Use of local gathering spots can provide a means of access to farmers that can be used to partially offset situations where travel capabilities are restricted by budgets or lack of vehicles.

Levels of literacy and education affect extension in direct ways. Illiterate farmers or those with very little education require more simple information that is easily understood. Simple audio-visuals, radio, and personal contact are mandatory techniques, as is demonstration.

The socio-cultural environment will dictate to a significant degree the manner in which females are considered to be a target of extension's educational programme, or from which they are excluded. In many Islamic societies, for example, access to women by male extension workers is simply not tolerated. The obvious response has been to encourage employment of female extension workers. Even when female extension workers are employed, there may be other cultural barriers preventing effective extension work. In some rural villages the phenomenon of a woman working outside the house is still the exception and may be resented. The denial of community meeting facilities for women where audio-visual materials are readily available may effectively deter effective extension programmes. Conversely, women's organizations, such as informal saving groups, formal co-operatives, religious groups abound in rural life. They may be successfully used as channels through which to extend information to women.

Many examples have been reported which stress the barriers to adoption of technology based upon cultural patterns, sociological considerations, and general community attitudes. Sallam, et al. (n.d., p. 21), states that: "Developing a sense of social responsibility cannot be achieved through a simple training course, but that does not diminish the importance of a strong social technology for the success of rural development or of any other form of advancement in a society. A community with a strong social sense of cooperative problem-solving, with effective leaders and organizations, is the best basis for beneficial change. And in its own way, such communities can coerce and convince reluctant traditionalists as no outside could."

A World Bank survey (Feder, et al., 1982, p. 49) considers the adoption of innovation in developing countries. The paper states that..."conflicting conclusions (as to rate of adoption) which are sometimes indicated by studies from different regions or countries may, in many cases, be the result of differing social, cultural and institutional environments (aside from "pure" economic factors)".

A specific example may be noted in the sociological arena of rural development. In one African country, for example, a policy of relocating rural families in new communities posed major constraints and changes in the manner in which services were designed and in the attitude of the local residents toward those services. Extension inevitably is caught in this type of situation and often finds itself in a position of some difficulty if the policies of relocation are unpopular.

An example could be illustrated by land reform in situations such as have been found in Latin America. Peasant farmers can see that land reform supported by government and discussed by extension programmes is a rewarding activity. Conversely, if such policies are perceived as working to the disadvantage of the peasants, the extension support will be associated with government policy, and affected by it.

Agro-ecological Environment

One of the most pervasive limitations in crop production research is development of crop varieties which appear to be universally advantageous. This is seldom the case. As a result, most nations have developed or are developing to the maximum extent possible site-specific research stations in order to adapt varieties and recommendations to various ecological conditions. Extension services must develop and maintain effective linkages to those location-specific research activities which can provide them with information of particular relevance to specific areas.

The farming systems approach provides one means by which site-specific research as well as extension can be conducted. Under this concept, the system of agriculture and the problems of the farmers are analyzed to develop priorities for the research programme. To the degree that the extension personnel in the area are involved with the research staff in farming systems, the programme will be enhanced. Farming systems approaches do not provide a ready knowledge base for extension on a national basis in most cases, but they do provide an opportunity for more relevant research for a specific area. In addition, a proper mix of extension and research people in the initial as well as final stages of farming systems activity provides an opportunity for effective inter-action between researchers and extension personnel, which is a significant achievement in its own right.

The Administrative/Organizational Environment for Extension

Many different administrative and organizational arrangements may be employed for extension (see Chapter 12). The most common include, in the United States, the formal tie between extension and the Department of Agriculture (and, in some cases, county governments) but with the primary administrative linkage being to the land grant university. In most nations of the world, administrative arrangements locate extension within the ministry of agriculture. Within these two frameworks, however, there can be significant differences in the access of extension administrators to the policy-making levels of government, and a significant difference in the representation which extension may or may not enjoy at the decision making tables.

As pointed out above, locating extension in ministries of agriculture is a predominant organizational mode. Extension may be represented in the policy councils of the ministry of agriculture by a director or director general who is equal with other members of the minister's executive council. Extension in this arrangement would normally be considered as the "educational arm of the Ministry of Agriculture". It would also enjoy strong representation during decisions on budgets, staffing, and programme leadership.

A slightly different organizational arrangement might find the extension service represented by a director reporting to an intermediary on the cabinet of the minister. This arrangement is perhaps more common than the earlier example. Extension's representation will be somewhat less direct and its influence perhaps somewhat weaker within the administrative hierarchy than under the earlier example. In both cases, however, the image of extension and its mission will be affected by other extension functions permitted or charged to other agencies of government as well as the director's position within the administrative hierarchy.

In those nations which have elected to develop separate extension structures for different agricultural commodities or activities, a proliferation of the extension function is inevitable with resultant bureaucratic competition.

The greater the number of separate extension organizations found within a nation, the more complex will be the administrative structures, the more wasteful will be the appropriations system, and the more competitive will be the various extension functions at the farm level. In almost every nation, the fact that there are finite and limited resources to support extension dictates that efficiencies are needed, not only in the basic appropriations but in the management of the programmes. Permitting duplicate functions and double or triple staffing not only requires additional organizational overheads but may result in unfortunate competition and confusion, for example, an extension worker interested in one aspect of agriculture may tend to compete with another whose perceptions are based upon a different crop, livestock or crop priority.

At the land grant university level, the placement of extension within the university administrative hierarchy is also important. In most cases the director of extension reports to a dean or vice president for agriculture, who reports to the office of the president of the university. In some land grant institutions the director may hold a vice presidential title and report directly to the president, but this is less common.

A major difference between the ministry of agriculture and the land grant system is in the straight-line authority held by the agricultural ministry as opposed to the somewhat diffused authority of the land grant system, which involves federal, state, university, and county influences in the United States. Ministries of agriculture may, by administrative edict, organize the administrative heads of various agencies in a manner which can permit or demand co-ordination, or deny it.

From the standpoint of effectiveness, if the extension service is to be the educational arm of the ministry of agriculture, the head of that activity should be placed in a position to participate in policy and planning so that he or she is fully aware of, and attuned to, the policies which extension education programmes are related to. If an effective extension service is assumed, the director should be in a position to provide relevant observations from the field to the senior officials of the ministry in any policy-planning.

Maunder (1973) identified six common deficiencies in extension organization as the following

1. lack of general understanding and appreciation of the role of extension education in rural development,

2. failure to establish a national policy for the scope of extension service responsibility and programme,

3. lack of continuity of extension programmes, due to political instability and attendant changes in agricultural policy, personnel and priorities for economic development,

4. weaknesses in the organizational structure of government which inhibit the development of co-operation between agricultural extension and other government services and institutions,

5. failure to provide an effective balance in the allocation of limited resources among the necessary elements of rural development such as extension education, agri-

cultural research, credit, agrarian reform and other elements of agriculture modernization, and

6.   failure to provide a proper balance between technical and educational competence in the staffing of the extension service.

In the 1980s, additional deficiencies including the following might be added

1.   failure of the extension organization to provide for continuous effective in-service training,

2.   failure to recognize the need for extension initiatives in maintaining linkages to research, and

3.   lack of a definitive reporting and evaluation process (for greatest effectiveness, such reporting and evaluation should be linked closely with the research organization).

A particular problem for extension in international development is the need to limit the scope of activity to a reasonable cost and a manageable size. At the same time, it is critical that extension programmes be built into an effective relationship with the provincial or national extension reward system. Many development projects are constructed on a site-specific or limited geographic basis in order to develop a manageable and effective effort at the local level. All too often, such projects have been mounted without adequate attention to the national policies which must be put in place if the project is to truly serve as an effective pilot concept. Designing extension projects should incorporate these administrative concerns as well as technical factors.

Development projects involving extension must by their very nature reach out to and involve farmers and farm families at the local village level. At the same time, such projects should be undertaken only after there is understanding and agreement at the national, state, province, or district level of the goals, objectives, and the possibilities for successful concepts to be incorporated within a larger system.

REFERENCES CITED

Aronoff, J. & Crano, W.D.   (1975).   A re-examination of the cross-cultural principles of task segregation and sex role differentiation in the family. American Sociological Review, 40, 12-20.

Berg, E., Ameoko, K. Y., Gusten, R., Meerman, J., & Tidreck, G. (1983).   Accelerated development in Sub-Saharan Africa.   Washington, D.C.:   The World Bank.

Donald, G. (1976).   Credit for small farmers in developing countries.   Boulder, Colo.: Westview Press.

Feder, G., Just, R., & Silberman, D. (1983).   Adoption of agricultural innovation in developing countries (World Bank staff working paper no. 542).   Washington, D.C.: The World Bank.

Hoffert, R. W. (1979).   American states and the Cooperative Extension Service.   Fort Collins, Colo.:   Colorado State University Printing Service.

Hildebrand, P. E. & Waugh, R. K. (1983).   Farming systems research and development. Farming Systems Support Project (FSSP) Newsletter, 1 (1), 4-5.

International Service for National Agricultural Research (ISNAR) (1982).   1982 annual report.   The Hague, ISNAR.

Johnson, E. (July, 1983). Summary and comment. In J. B. Claar & L. H. Watts (eds.), Proceedings of a Conference on International Extension at Steam Boat Springs, Colorado.   Urbana, Illinois: University of Illinois at Urbana-Champaign and Fort Collins: Colorado State University.

Kearl, B. (n.d.).   An analysis of extension models for developing countries.   Unpublished manuscript.

Maunder, A. (1973). Agricultural extension: A reference manual (Abridged Edition). Rome: Food and Agriculture Organization.

McDermott, K. (1982). FSR project evaluation. In C. B. Flora (Ed.), Farming systems in the field: Proceedings of Kansas State University's 1982 Farming Systems Research Symposium (pp. 198-205) (Farming systems research paper series no. 5). Manhattan, Kan.: Kansas State University.

Moris, J. (1981). Managing induced rural development. Bloomington, Ind.: International Development Institute.

Presidential Commission on World Hunger (1980). Overcoming world hunger: The challenge ahead. Washington, D.C.: U.S. Government Printing Office.

Russell, J. (1981). Adapting extension work. Finance and Development, 18 (2), 30-33.

Sallam, J., Knop, E., & Knop, S. (n.d.). Rural development and effective extension strategies: Farmers' and officials views (Technical report no. 19). Tucson, Ariz.: Consortium for International Development.

Woods, B. (1983). Altering the present paradigm: A different path to sustainable development in the rural sector. Washington, D.C.: The World Bank. Unpublished manuscript.

Chapter 3
Extension's Role in Adapting and Evaluating New Technology for Farmers

S. H. Johnson III and E. D. Kellogg

One of the most important means of accelerating national development in economies with large agricultural sectors is the development, adaptation and evaluation of new agricultural technology that can be adopted by small farmers. This adoption can result in higher incomes for small farmers, lower real prices of agricultural products for consumers, and greater economic efficiency and growth in the national economy. Therefore, the identification, development, adaptation, verification and farmer adoption of new agricultural technology has become an important part of the economic development strategies in many countries.

Extension services can play a substantial part in this process (Kellogg, et al., 1984), although the success of past efforts has often been limited. In order to overcome these limitations, new approaches to the adaptation and evaluation of technology have been developed. The purpose of this chapter is to specify extension's role in the new approaches. First, problems which inhibit successful technology adaptation and evaluation will be discussed. Then, a general approach will be outlined to alleviate many of these problems. Finally, specific extension activities to implement the approach will be identified.

PROBLEMS FACING TECHNOLOGY DEVELOPMENT, ADAPTATION AND TRANSFER

In many countries there are too few resources devoted to the technology development and transfer process. Additionally, there may be problems such as the lack of trained personnel, too little basic and/or component research, or agricultural policies that discourage farmers from adopting new technology. These problems are difficult to solve within the extension-research system. However, there are other important problems that have limited the development, adaptation, and adoption of agricultural technology by small farmers that can be addressed by extension personnel. For example, researchers do not fully understand that agricultural production in less developed countries takes place within a complex farming system that has evolved over a number of years, based on experience and sound reasoning. In many cases, researchers have limited knowledge about important problems confronting farmers. In addition, there is often little information about the physical, economic, and socio-cultural factors which create the environments within which farmers work. Because these environments are spatially different and change over time, agricultural technology development and transfer activities must deal with these differences to be effective. Also, new technology, as adopted and managed by farmers, often does not perform as well as expected, and additional adaptation may be necessary. Many of these problems can be alleviated by developing and/or strengthening extension activities in their inter-action with researchers and farmers.

These problems fall into four major groups (a) lack of knowledge and understanding of farming systems, (b) insufficient feedback from farmers to research programmes, (c) insufficient understanding of the environment within which farmers work, and (d) lack of mechanisms for testing and adapting technology on farmers' fields (Gilbert, Norman and Winch, 1980).

The first problem area entails having insufficient knowledge and understanding of the farming systems. These farming systems are often characterized by activities related to crop and animal production, family and household consumption, production, labour and leisure time usage, and off-farm household tasks. Technology development and transfer activities that do not consider these farming systems might attempt to extend inappropriate technology which will not be accepted by the farmer. For example, new crop varieties often have higher yields than traditional varieties but require a longer or different growing season; therefore, they may not be readily adopted in well-developed cropping sequences. Another example is the rejection of new crops which need attention during seasons where off-farm labour demands are high.

A second problem area is the lack of communication and information feedback from farmers to research programmes. In some cases, there are few, if any, mechanisms established to facilitate this process. In other cases, researchers and extension personnel do not appreciate the need for eliciting information and evaluation from farmers. Technological developments, adaptation and transfer based on problems identified by farmers themselves will be more effective in producing new technology acceptable to farmers. In adapting and transferring new technology, it is important to understand farmers' own evaluations of its performance during early adoption. Farmers' evaluations can also assist research personnel in their efforts to adapt technologies to fit a greater number of situations.

A third problem area involves the need to identify the environments farmers must operate within more precisely so that technological development and transfer can be tailored to specific situations. Agricultural and family activities are greatly influenced by physical, economic, and socio-cultural factors which tend to change with time, and are different among geographic areas within the country. Technological development and transfer must accommodate these different environments. For example, topography, climate, and water availability in different zones where crops exhibit roughly the same biological expression (agro-climatic zones) dictate options for the adaptation of relevant technology. Off-farm labour opportunities, input and output market accessibility and government price policy also determine the environment, as do family structure, kinship ties and habits of consumption.

The fourth problem area is the lack of well-developed mechanisms for testing and adapting technology on the farmers' fields. It is clear that adaptation of technology cannot be accomplished entirely on experimental fields or by tightly-controlled experiments on farmers' fields. For successful technology transfer, farm-level testing, adaptation and verification must be done in co-operation with farmers and extension personnel.

A NEW APPROACH TO ADAPTING AND EXTENDING
NEW TECHNOLOGY TO FARMERS

Over the past decade, new approaches have evolved for technological development and transfer that are oriented to alleviating many of the problems discussed above. Although specific approaches vary among locations, the general approach involves the following steps (see Figure 3.1) (Kellogg, 1977a).

I.   Diagnosis of farmers' circumstances and actions in target area.

II. Planning and design of technological adaptation.

III. On-farm testing and verification.

IV. Multi-locational field trials and dissemination.

Step I

This new approach is usually initiated by defining target geographical areas which have similar characteristics. These similar characteristics define a relatively homogeneous set of agricultural activities throughout a particular agroclimatic zone; it is distinguishable from other zones. Extension personnel can contribute to this identification of target areas because they often know conditions at the farm level better than research personnel. While physical data, such as topography, soil type, temperature and rainfall can usually be developed from secondary sources, extension personnel can be useful in helping to define other environmental variables which help to define homogeneous farming areas such as off-farm labour opportunities, market accessibility, actual water availability, and kinship or tribal ties. Agricultural economics extension specialists, if available, can be particularly helpful in determining these variables (Kellogg, 1977a).

Once these target areas are defined, the next task is to diagnose the situation farmers must work within and important problems that need attention. Extension person-

Figure 3.1   Steps in a general approach to adapting and extending new agricultural technology

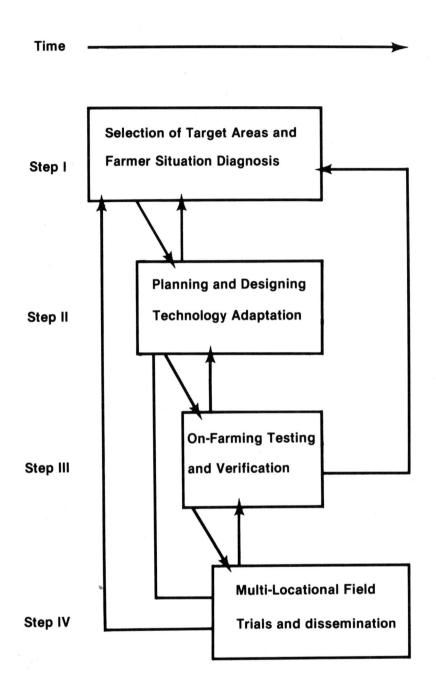

Note: The arrows indicate that steps overlap in time and results in later steps may require going back to previous steps for further analysis.

Note: From Role of social scientists in cropping systems research, by E. Kellogg, 1977, in Proceedings of the first national conference on cropping systems research in Thailand, pp. 500-547, Chiang Mai, Thailand: Department of Agriculture Economics, Chiang Mai University.

nel are particularly suited to helping researchers understand farmers' problems and existing farming systems with which new technology must fit. This information is usually developed through informal target area assessment surveys by observing farmers' fields and actions and discussing farming with farmers. In many cases, major impressions gained from these quick assessment surveys are further analysed through implementation of formal surveys of farmers and members of farm households. Again, extension personnel can take a leadership role in these activities. General extension field workers are probably more suited to contribute to this activity because they know farmers and general agricultural conditions better than other extension personnel.

Step II

By using information developed in Step I planning and design of technological adaptation can begin (Byerlee, et al., 1982). Because of their knowledge of research techniques and farmers' situations, the subject matter specialists play a critical role in determining how productive the research directions chosen by the researcher may be. Certain research projects may need to be implemented on experimental stations before being moved to farm locations, while other projects can be started directly on farmers' fields.

Step III

Once these research directions are chosen and preliminary results of experiments obtained, on-farm field testing and verification can be begun. In general, this involves planning on-farm trials, choosing method(s) for implementing these trials, and adjusting trials as results occur (Kellogg, 1977b). The role of extension personnel can be significant in this step.

Step IV

During the on-farm testing and verification process, certain trials will yield results that indicate that the technological options can be managed by farmers and are a significant improvement over current farming practices. Then, similar trials should be implemented in a number of locations within the target area for further verification. These multi-locational field trials can be primarily operated by extension personnel and utilized in the dissemination process.

The steps, as illustrated in Figure 3.1, overlap in time. This means that as sufficient information is gained in one step, the next step can be initiated without waiting for the completion of the previous step. For example, planning and designing technological adaptation may be initiated as soon as certain problems confronting farmers are identified. These steps are also iterative, in that results from one step may require reworking parts of the previous step. If, for example, on-farm trials are not producing results superior to current farming practices, new planning and design of technological adaptation must occur.

As illustrated in Figure 3.2, the role of extension personnel as intermediaries between farmers and researchers varies from step to step. The following section expands on these rules and provides a description of the tasks of extension personnel in ensuring that more appropriate technology is developed and extended to farmers.

Extension's Role in Selection of Target Areas and Farmer Situation Diagnosis

Effective selection of target geographical areas is a critical activity in the development and extension of new production technologies. Selection generally begins with high-level decision makers in government deciding on one or more areas needing increased attention. National and regional development goals are usually important criteria for selecting an area. For example, in one African country, decision makers selected an area that contained a large number of very poor families, while in another instance, an area that had a high potential for increased production of export crops was selected. These alternatives are illustrated in Table 3.1. In another example, decision makers in a different African country selected a sparsely populated region with improvement potential as a target area.

This target area selection decision was based on the national policy to alleviate land pressure on the fertile river plains. As another alternative, decision makers may select an area on the basis of specific physical limitations or problems such as flooding, soil salinity, steep slopes or inadequate rainfall. In Southeast Asia, decision makers in Thailand have selected target tracts based on criteria of poverty and erratic rainfall,

Figure 3.2 Division of effort in adapting and extending new technology

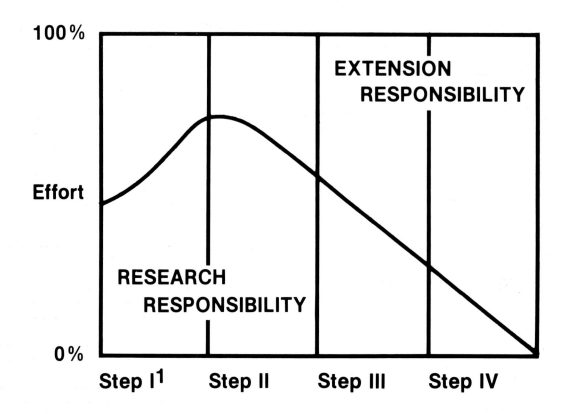

[1]See Figure 3.1 for definition of steps

Table 3.1
Comparison of Characteristics of Two Potential Target Areas

Characteristics	Target Area I	Target Area II
Total number of families	150,000	125,000
Number of families below poverty level	85,000	25,000
Production of export crops ($)	5,000	1,750,000
Average family's cash income as percent of national average	55%	135%

Note: From Understanding small farmers by M. C. Collinson, 1979, December, Paper presented at Conference on Rapid Rural Appraisal, University of Sussex, Brighton, England.

while in Sumatra areas that have large areas of land infested with Imperata have been chosen (Shaner, Philipp and Schmehl, 1982).

Once target areas are selected, it is necessary to subdivide them into recommendation domains that include a group of roughly homogenous farmers of similar circumstances. These similar circumstances define a relatively homogenous set of agricultural activities throughout the area that distinguishes it from other areas. These activities are usually

Table 3.2
Farm System Zoning Questionnaire, Central Province, Zambia

District_____ Ward No._____ Farmer Group_____

A.	Animals kept by most farmers	1.	Three main types of animals kept	1 _____ 2 _____ 3 _____
		2.	If cattle, main purposes for keeping	1 _____ 2 _____ 3 _____
B.	Foods Grown (G) or Bought (B) by Most Farmers	1.	Starch Staples	1 _____ 2 _____ 3 _____
		2.	Relish crops to flavour staples	1 _____ 2 _____ 3 _____
		3.	Animal products for food	1 _____ 2 _____
C.	Main Cash Sources for Most Farmers	1.	New cash crops and % growing	1 _____ 2 _____
		2.	Crop sales as a cash source	1 _____ 2 _____
		3.	Livestock as a cash source	1 _____ 2 _____
		4.	Off-farm cash source	1 _____ 2 _____
D.	Land Use and Preparation Methods of Most Farmers	1.	Years cultivated	1 _____
		2.	Typical area (ha.) preparation	1 _____ 2 _____
		3.	Main methods of land preparation	1 _____
		4.	Main months of land preparation	1 _____ 2 _____
E.	Hire and Purchase of Resources by Most Farmers	1.	Types of hired labour & payment	1 _____ 2 _____
		2.	Work done by hired labour	1 _____ 2 _____
		3.	Main inputs purchased and crops using	1 _____ 2 _____ 3 _____

Note: From Understanding small farmers by M. C. Collinson, 1979, December, Paper presented at Conference on Rapid Rural Appraisal, University of Sussex, Brighton, England.

defined by natural features, such as soil, climate and topography, and agro-economic features, such as access to resources and marketing opportunities. Site selection may also be concerned with certain subsets of the area, such as small farmers following a particular cropping rotation or those with mixed livestock farming systems (Tripp, 1982).

Table 3.2 illustrates a simple questionnaire that was developed and administered by agricultural extension staff to collect descriptive information about a target area in Zambia. The information was used to define specific recommendation domains within the target area. This exercise involved some 20-30 professional work days to survey 100 farmers. Using such a questionnaire, it is possible to develop tables, such as Table 3.3, which detail a tabulation of farmer practices by recommendation domain in a tropical maize-producing area.

Table 3.3

Tabulation of Farmer Practices by Recommendation Domain - Tropical Maize

| | Recommendation Domain | |
General Farm Data	Flat Land	Steep Land
Average Farm Size (ha.)	11.1	10.2
Area in Maize in August (ha.)	4.6	2.6
Area in Tree Crops (ha.)	3.5	2.7
Annual Cropping Pattern in Selected Field		
Percent Maize-Maize	31	37
Percent Maize-Maize-Beans	33	33
Percent Maize-Squash-Maize	12	10
Percent Other Systems	24	20
Land Prepration		
Percent Plough-Harrow (with tractor)	38	0
Percent Harrow Only (with tractor)	24	0
Percent Hand Hoe	34	68
Percent Chop Only or Chop and Burn	3	27
Planting		
Percent Plant "Improved" Variety	18	3
Distance Between Rows (cm.)	103	102
Distance Between Hills (cm.)	92	94
Average Seeds per Hill	3.7	3.9
Percent Replant	26	35
Weeding		
Percent Weed with Horse or Tractor	15	3
Percent Weed with Hoe	85	97
Percent Weed Twice	83	80
Average Time of First Weeding (weeks after planting)	4.5	3.8
Other Inputs		
Percent Apply Insecticides	86	82
Percent Use Fertilizer	2	0
Production		
Average Yield (ton/ha.)	1.2	1.1
Percent Maize Sold	63	56

Note: From Planning technologies appropriate to farmers concepts and procedures by D. Byerlee, M. P. Collinson, R. K. Perrin, D. L. Winkelmann, S. Briggs, E. R. Moscardi, J. C. Martinez, L. Harrington and A. Benjamin, 1980, Mexico City: International Maize and Wheat Improvement Center.

Assessment Survey

Once recommendation domains are defined, the next task is to analyse the situation of farmers in the area. Extension personnel are well qualified to help researchers under0 stand (a) the problems farmers see as important, (b) existing farming systems within which new technology must fit, and (c) research areas that may yield potentially useful technologies. This kind of information is usually developed through assessment surveys by examining farmers' fields and discussing farming with them. In many cases major impressions gained from these quick assessment surveys are supplemented by working closely with key informants and by implementation of formal targeted surveys of farmers and members of farm households. These survey activities, which should be led by extension personnel, are critical in obtaining a complete understanding of farming practices and systems in the selected areas.

Initially, farmers are interviewed very informally at their farms, preferably in their fields. Farmers are asked about their agricultural production levels. As farmers respond, they are asked if they would like to increase their production. Assuming the answer is positive, the field staff then ask farmers questions such as, "What is limiting your production?" At the end of the day the responses are summarized to try to improve the structure of future interviews. This daily summary might look something like Table 3.4.

Table 3.4
Assessment Survey--Daily Summary

What Keeps Farmers From Obtaining Higher Rice Yields?

(rank 1 = most important   6 = least important)

Reason	Importance
Water supply	3
Lack of fertilizer	4
Poor drainage	2
Insects	6
Weeds	1
Other	5

By applying this approach to the informal survey, extension staff will have a better basis for developing a formal survey document.

Formal Targeted Surveys

Formal targeted surveys are used to gain a better understanding of farming systems in an area and to identify the most relevant problems and opportunities for improvement. Before developing the questionnaire, it is necessary to determine what information the survey is designed to produce. Depending upon the situation, a variety of information might be required. For example, information may be obtained about:

1.   Farmers' practices with a particular crop,

2.   Land preparation,

3.   Post-harvest operations,

4.   Harvest, including distribution of the produce and use of crop residue,

5.   Farmers' knowledge of plants, soils, pests, weather and plant damage,

6.   Those factors of a total farming system that bear on a particular crop, for example, labour bottlenecks, crop sequences, family food preference, cash flows and water availability.

The length of the questionnaire is important. The longer the questionnaire, the more difficulty the interviewer will have holding the farmer's attention. Ideally, administering a questionnaire should take no more than an hour. This is important, because to provide timely information, it is necessary to process the data quickly. Early knowledge of analysis procedures significantly reduces the time required to process data. It must be emphasized that the purpose of the survey is to be able to diagnose more accurately existing farmer situations in the area and not to obtain information concerning all facets of rural life. Thus, the questionnaire can be shorter and more focused than a general census or rural survey (Bernsten, 1979). Appendix 1 (see Appendix at end of manual) contains a sample of such a questionnaire that was used to interview farmers in Turkey regarding their maize production and general farming system.

Extension personnel can play a major role in the development of the formal survey document as their intimate knowledge of the area combined with their involvement in the informal surveys allows them to prepare well-focused, relevant questions. As a result of the survey process, new technology options that fit local requirements can be identified. In addition, participation in these surveys provides extension personnel with better insights into farmers' needs and problems. Again, extension personnel's knowledge and understanding of farmers' situations and actions can be invaluable in reviewing the direction of technological development proposed by research scientists.

Extension's Role in Planning and Design of Technology Adaptation

The planning effort is designed to produce a work plan for on-farm trials leading to improvements for an identified group of farmers within the target area. Potential for improvement becomes the basis for setting objectives, selecting field methods, co-ordinating the efforts of experimental stations and other supporting organizations, and outlining the tasks and responsibilities of field staff (Gilbert, et al., 1980). Planning activities involve

1.   identifying the target group of farmers,

2.   laying the groundwork for field activities,

3.   considering alternative activities and methods,

4.   finalizing plans for field trials.

In many instances, particularly in situations where there are weak links between research and extension organizations, planning for on-farm trials is viewed as the role of researchers. Yet, when these trials are to be located on a large number of sites, extension staff involvement is critical to the success of the project. In addition, when conditions within a target area are not homogeneous, it may be necessary to divide the target area into smaller sub-areas which eventually will be used as recommendation domains for more or less the same recommendations. Here again, extension staff with their intimate knowledge of the area can facilitate this process. Field and regional extension staff are familiar with local conditions and can work closely with researchers to select target area and sub-area locations. As the selection process requires a knowledge of physical, biological, and socio-economic conditions, extension staff are in a position to accelerate this process and to minimize the costs. Figure 3.3 illustrates the selection of recommendation domains and the identification of on-farm sites within these areas.

An analytical framework can aid in laying the groundwork for on-farm trials. This framework provides guidance in making preliminary estimates of overall feasibility and includes consideration of physical, biological, economic, financial and socio-cultural factors. While the first two factors are in the domain of biological researchers, the last three are clearly in the domain of the social scientists and extension staff. Thus, extension can contribute to the analytic framework. In designing trials, it is best to work towards technically viable designs rather than toward optimal designs. Optimality usually does not have much operational meaning within the complexity of farmers' circumstances, while technical viability implies designs that are responsive to conditions likely to prevail when the technologies are introduced to farmers (Harwood, 1979).

Figure 3.3 A hypothetical target area divided into four sub-areas showing trial locations

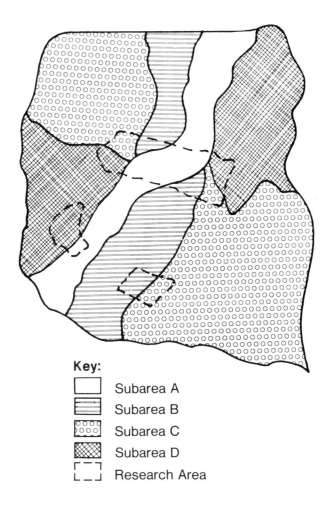

Key:

☐ Subarea A
▤ Subarea B
⊡ Subarea C
▨ Subarea D
⌐ ¬ Research Area
L _ ⌐

Note: From Planning technologies appropriate to farmers concepts and procedures by D. Byerlee, et al., 1980, Mexico City: International Maize and Wheat Improvement Center.

Before deciding on a particular approach, it is necessary to consider alternative crops, research trials, and design. Failure to consider alternatives will produce inferior results no matter how well the trial is designed. Searching for alternatives takes an open mind, imagination, and considerable judgement and experience. The field knowledge of extension staff members can be invaluable at this point. Often the substitution of traditional farming techniques for research station-oriented practices can mean the difference between success and failure (Rogers, 1971). Alternatives under consideration should always attempt to reduce farmers' risk and increase farmers' satisfaction.

In developing alternative designs, it is necessary to establish standards for on-farm trials. General standards include (a) setting up trials that will yield useful results, (b) avoiding unnecessary detail and complexity, and (c) gaining uniformity among trials across areas and over time. Other standards are specific to different countries and to the objectives of the programme. These standards are related to types of farmers, location and number of trials, design complexity and methods of evaluating results (Shaner, et al., 1982).

Before finalizing field trial design, it is necessary to determine the characteristics of farmers who are to be chosen as co-operators. It is important to decide whether to select more progressive farmers or farmers who are representative of the areas' average farmers. More progressive farmers are usually easier to work with and ensure a higher degree of success with trials. However, the results from such trials are probably not representative and may provide biased answers to the question of the appropriateness of the technology (Kirkby, Gallegos, and Cornick, 1981).

Farmers who meet the requirements must be approached to see if they are willing to participate in a trial. Some small farmers are themselves experimenters, and the concept of trying a different variety of seed or a new technique is acceptable to them. Nevertheless, in many places, farmers have limited experience in working with government agencies and none at all in cooperative on-farm trials. Therefore, a careful explanation of the farmer's duties and expectations must be made prior to initiating the trial. Extension personnel can help select and contact co-operating farmers, unless it is felt their choice is biased. Collaboration between on-farm researchers and extension workers is important at this stage if a good set of representative farmers is to be selected (Tripp, 1982).

Another consideration in choosing sites is a logistic one. The ideal distribution of sites would have on-farm trials scattered throughout the recommendation domain. Problems of transportation, resources and time often make this impracticable. If trials are placed too far from one another, it is not possible to visit all of them frequently enough. One compromise is to choose several areas within a domain to concentrate on during a given crop season and cluster trials in this area. Trial areas can, of course, be shifted from year to year. As will be emphasized later, the more responsibility is shifted to extension personnel the less any scatter of trial areas is a major problem (Moscardi, 1982).

Selection of trial sites requires a great deal of thought and effort and cannot be left to chance. The planners must have definite criteria for site selection. Some of the more important criteria include the degree to which the

1.   site conforms to the basic characteristics of the target area and target group;

2.   trial site fulfills the requirements for the type of on-farm trial that is proposed;

3.   trial sites can be visited and managed throughout the year, and can provide opportunities for farmer participation;

4.   trials are frequently seen by non-participating farmers, if they are to be used as demonstrations;

5.   site selection is reviewed to correct previous biases or to reflect changes in emphasis;

6.   sites are spread within the target area.

Once locations and farmers have been selected, it is necessary to decide how to choose among alternative fields, and locations within fields. Replicated experiments conducted within the same farmers' fields provide the greatest uniformity of conditions, but where farmers' fields are small, it is often necessary to spread the trials throughout the target area. In general, if trials are managed by research staff, the number must be limited to less than 20. Farmer-managed, extension-supervised trials allow for a much larger number of sites, perhaps double the number supervised by research staff alone.

The level of complexity of on-farm trials is critical. If trials are very complex and involve a large number of treatments, they are difficult to manage under actual farm conditions. In general it is better to have less complex trials in order to meet the limited resources of co-operating farmers and to facilitate eventual diffusion by extension field staff (Zandstra, et al., 1981).

Finally, it is critical to identify a methodology for evaluating the degree of success or failure of a trial. Where possible, it is best to use some type of total-farm budget analysis to determine the success of the trial within the context of the entire farm operation. It is particularly important to learn why trials fail and to be alert to the implications of this failure for design of future trials (Perrin, et al., 1976).

Extension's Role in On-Farm Testing and Verification

Once on-farm trials have been planned and designed, they are ready to be implemented. The most critical question at this stage is that of management. Which group shall actually be responsible for management of the trial sites? Often research staff attempt to manage all trials, but due to their limited numbers in rural areas, the number of trial sites is severely curtailed. Management by extension staff, or even by farmers themselves, offers the opportunity for a greater number of sites. Of course, this does not allow for the same level of research sophistication. Real-world illustrations of these various field trial management techniques can be found in Limpinuntana, et al., (1982) and Kellogg (1977b).

Researcher-Managed Trials

Researchers manage trials on farmers' fields to develop appropriate technologies for specific groups of farmers. These trials tend to use more sophisticated methods, as used on experimental stations, although non-experimental variables are generally set to represent farmers' conditions. Researchers will often pay farmers for their labour and the use of their land, so that farmers do not suffer losses from poor experimental results. Because researchers manage the trials the designs can be more complex. Small plots are generally chosen, although larger plots may be needed to study constraints on water or soil resources. Replications can be contiguous or dispersed on different farms, depending on the experiment and field sizes. The necessity of researchers to visit the trials at least on a weekly basis severely restricts the number of trials managed by any individual researcher.

Extension-Managed Trials

Extension staff manage trials on farmers' fields to identify and verify appropriate technologies for target groups of farmers. In contrast to researcher-managed trials, extension-managed trials do not use experimental techniques typical of the experimental station, nor do they generally pay farmers for their labour or the use of their land. Extension-managed trials attempt only to control a limited number of experimental variables, although their level of control is usually higher than that of farmer-managed trials. By focusing on farmer conditions, the results of extension-managed trials provide a much better indicator about how farmers will respond to new technologies than do researcher-managed trials. Even though the trials are usually less complex than researcher-managed trials, extension-managed trials can be operated at a level of complexity higher than farmer-managed trials. As the extension worker is stationed and working in the area, it is possible for extension staff to manage a larger number of trials than can be managed by researchers alone.

Farmer-Managed Trials

Farmer-managed trials are particularly important because they allow farmers to participate in testing new technologies. In doing so, farmers reveal their reactions to these technoloiges. Selection of farmer co-operators requires knowledge, not only of individual abilities, but also about circumstances of the entire farm household. This is important, as field-level staff have often found that the limitations of farmers to manage on-farm trials properly are related directly to the household's resources and the uncertainties of the farmer's environment. Usually, farmers are not provided with financial encouragement, such as outright payments, subsidies for input costs or reimbursement for losses. The reason for this policy is that on-farm, farmer-managed trials are intended to show how farmers will react to the new technologies when applied to their actual conditions.

A critical factor in farmer-managed trials is the relationship between outside staff and farmer's. This relationship hinges on the degree to which staff attempt to direct a farmers activities. Fundamental to the effectiveness of farmer-managed trials is recognition that these tests belong to the farmers. Farmers must be allowed freedom so that the trials reflect their responses to the new technologies.

Staff must realize that the results of farmer-managed trials will be different from those of researcher- and extension-managed trials. Because farmers carry out their activities to fit their available time, they may omit some operations. These adjustments allow the trials to become fully integrated into the farmers' systems. These adjustments are important as they provide the means for identifying and measuring farmers' constraints in adapting the technology to their conditions, and they help to identify problems and opportunities for additional trials.

While it may be possible for researchers to supervise farmer-managed trials, given the shortage of research staff and the necessity of having a large number of location-specific sites, it is usually best to have extension staff responsible for these trials. This can be done in conjunction with other farmer-oriented activities and fits well with the role of an extension worker as an educator. Although extension workers are usually not well-trained in field plot layout and design, the planning exercise and the simplicity of the trial design developed during this exercise should overcome this problem. Extension workers are exceptionally well prepared to determine acceptability of farmer-managed trials, as this is usually based on farmers' own actions and comments. Farmers will accept a new practice or technology when they perceive that the benefits are great enough to outweigh the costs.

It is easy to plant a trial, make observations, harvest the crop, and take yield data, all without talking to the farmer. This is a great loss and defeats the purpose of on-farm trials. Some of the most valuable information obtained by on-farm research are the observations and opinions of the farmer. Recommendation domains are defined as "groups of farmers." Collecting data from fields fulfills only a part of the goals of on-farm trials. As the season progresses, and as the research or extension personnel and the farmer participate in agricultural activities together, they come to know each other. This development of mutual respect is important as it facilitates honest exchange and encourages the farmer to provide a frank opinion of what he observes in the trial. It is a good idea to visit farmers with specific questions in mind, although farmers should be encouraged to talk about whatever is of particular importance (Tripp, 1982).

Every on-farm trial has to be closely monitored in order to provide the maximum amount of information concerning that production technology. Data collection is separated into data sets that include information on climate, field characteristics, crop performance, field operations and management, farmer characteristics and harvest data. An example of a field book used for on-farm research is provided in Appendix 2.

Analysis of production practices covering the agronomic and economic performance of the practice should be collected at the end of every season. Because of year-to-year differences in weather, input costs, and product prices, it may be necessary to evaluate the practice over a number of seasons. By working closely with extension staff familiar with the area, it is possible to weight the evaluation of test results to obtain a better estimate of the performance of the practice under the more common conditions of the different test sites. Important criteria for evaluation of performance of the trial technologies are the size of variation in yield and economic measures of the returns from the new practices. An experimental practice that offers returns above variable costs (RAVC) more than 30 percent greater than that of prevalent farmers' practices may be recommended for introduction to farmers (Zandstra, et al., 1981).

Once a new technology has proved itself in the field verification phase, it must be tested in a wider variety of locations. This is done by implementing a larger series of multi-locational trials in other areas that appear to have soil and climatic characteristics similar to those of the original trial area.

Extension's Role in Multi-Locational Field Trials and Dissemination

The target area and sub-areas contain some variability in farmers' conditions. In multi-locational testing, successful technologies identified within specific test areas are evaluated at many sites representative of the conditions for which the patterns were designed. Specification of the land type is an important aspect of multi-locational testing because it allows for a clearer delineation of the domain of adaption of the recommended practice. Procedures for multi-locational testing include

1.   identification of extrapolation area by using rainfall records and soil, irrigation and land-use classification maps, where they exist. Extrapolation areas must be sufficiently large to merit future production programmes;

2.   within the selected extrapolation area, identify the location and approximate extent of the area that coincides with the land types that were identified at the original site;

3.   develop a set of designs for the extrapolation area that allow comparison of the technology's performance with that in the original sites;

4.   establish and manage trials;

5.   evaluate performance of the technology from production data.

Based on the results of the multi-locational trials, staff should plot the results of the trials on a map and attempt to associate performance with environmental and/or social factors. Through this process it is possible to determine how much, if any, adaptation is required to adapt the technology to the varying conditions in the extrapolation areas. In the process of multi-conditional trials, staff are able to

1.   associate differences in performance of new technologies with the factors causing the differences;

2.   discover what adjustments are needed to adapt the new technologies to conditions somewhat different from those encountered in the original on-farm sites;

3.   verify, or revise if necessary, boundaries for various recommendations associated with each new technology.

As a result of knowledge developed from multi-locational trials, it is possible to define the conditions for which the technologies are suitable and to formulate recommendations for their extension. This requires defining the domain of adaptation in site specific terms (soil types, rainfall patterns, or drainage characteristics) that are easily identified by extension workers on the basis of simple field observations.

Multi-locational trials should be done by extension or production agencies with the majority of the management provided by farmers. However, in order to ensure timely application of critical inputs such as fertilizer, pesticides and irrigation water, a limited number of trials are usually managed by extension staff. Management and supervision of farmer-managed trials by extension personnel serves to familiarize a large number of extension staff with the new technologies, which helps to facilitate broad-spread extension and to generate the interest and support of local and regional supervisors (Haws and Dilag, 1980).

Once a technology has been proposed, tested, retested at a variety of sites, and proven acceptable, it is ready for regional diffusion. At this stage diffusion responsibilities for education and promotion at the farm level become that of the extension organization. The appropriate approach to use depends upon a number of factors which are discussed further in Chapters 9 and 10. Often it is felt that a pilot production programme is required in order to test how support systems--suppliers of inputs and markets react when the new technologies are introduced on a relatively large scale. Such a pilot programme provides valuable information about local factors like commodity markets, credit labour, agricultural chemicals, transportation and information systems.

ORGANIZING FOR ADAPTING AND EVALUATING NEW TECHNOLOGIES

As indicated in the previous section, on-farm trials can be organized and supervised in a number of different ways. Partly this is a function of the purpose of the trial, but it is also dependent upon the mandate and organizational structure of agricultural research and extension agencies implementing the trials. When trials are designed and supervised by researchers, it is important to make certain that trials are not simply research plots transplanted from experimental stations. There may also be problems feeding the results into the broader extension system in order to diffuse them to as many farmers as possible. Where trials are extension organized and supervised, there may be problems gaining access to new technologies from research institutions or stations. In addition, there are usually problems with the general lack of training in plot design for most extension personnel. The best organizational arrangement is one that mobilizes the specialized skills of the various entities. These skills should be focused on the problems of on-farm trials and should be organized in such a manner that they meet and come together in farmers' fields.

For the transition from research to extension to be effective, it is imperitive that both extension and research play an active role during the on-farm trial stage. Unfortunately, this is rarely the case. Too often extension is only involved at the point of diffusion. Field trials must be a shared responsibility of both agencies with joint budgetary, staffing and operational responsibility. In order to operate in this manner, an organizational structure such as that illustrated in Figure 3.4 is recommended. This organizational arrangement effectively links research and extension by creating a joint office responsible for field trials. This office, by having staff from both research and extension, especially subject matter specialists, creates the equivalent of a joint research-extension appointment and ensures that both research and extension are represented in on-farm field trials. Responsibility, and credit, for successful farm trials can be shared equally by the two organizations.

Figure 3.4   Proposed organizational arrangement linking research and extension for adaptation of on-farm technology

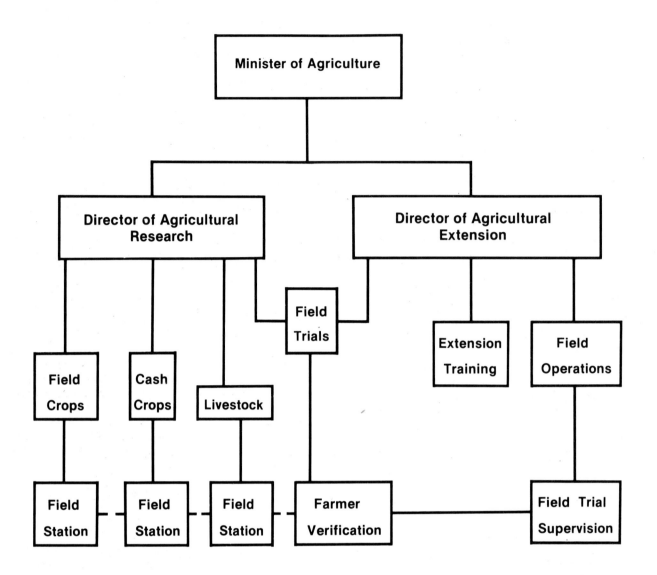

REFERENCES CITED

Bernsten, R. (1979). Design and management of survey research: A guide for agricultural researchers. Los Banos, Philippines: CRIA/IRRI Cooperative Program, International Rice Research Institute.

Byerlee, D., Collinson, M. P., Perrin, R. K., Winkelmann, D. L., Briggs, S., Moscardi, E. R., Martinez, J. C., Harrington, L., & Benjamin, A. (1980). Planning technologies appropriate to farmers concepts and procedures. Mexico City: International Maize and Wheat Improvement Center.

Byerlee, D., Harrington, L., & Winkelmann, D. L. (1982). Farming systems research: Issues in research strategy and technology design. American Journal of Agricultural Economics, 64, 897-906.

Collinson, M. P. (1979a). Farming systems research. Nairobi, Kenya: CIMMYT/ Nairobi. Unpublished manuscript.

Collinson, M. P. (1979b, December). Understanding small farmers. Paper presented at Conference on Rapid Rural Appraisal, Institute of Development Studies, University of Sussex, Brighton, England.

Compton, J. L. (1983, May). Linking scientist and farmer: Rethinking extension's role. Paper presented at Regional Seminar on Extension and Rural Development Strategies, Universiti Pertanian Mayalsia, Serdang, Malaysia.

Gilbert, E. H., Norman, D. W., & Winch, F. E. (1980). Farming systems research: A critical appraisal (MSU rural development paper no. 6). East Lansing, Mich.: Department of Agricultural Economics, Michigan State University.

Harwood, R. R. (1979). Small farm development: Understanding and improving farming systems in the humid tropics. Boulder, Colo.: Westview Press.

Haws, L. D., & Dilag, R. T., Jr. (1980, March). Development and implementation of pilot production programs. Paper presented at Cropping Systems Conference, International Rice Research Institute, Los Banos, Philippines.

Kellogg, E. (1977a). Farm level testing of cropping systems - An economic analysis of multiple cropping project experience. (Agricultural Economics Research Report No. 2). Chiang Mai, Thailand: Department of Agricultural Economics, Chiang Mai University.

Kellogg, E. (1977b). Role of social scientists in cropping systems research. In Proceedings of the First National Conference on Cropping Systems Research in Thailand (pp. 500-547). Chiang Mai, Thailand: Chiang Mai University.

Kellogg, E., Compton, L., Butler, R., Johnson, S., Swisher, M., & Pribble, C. (in press). Role of extension in farming systems research programs. In C. B. Flora (Ed.), Proceedings of the Farming Systems Research Symposium - 1983. Manhattan, Kansas: Kansas State University.

Kirkby, R., Gallegos, P., & Cornick, T. (1981). "On-farm research methods: A comparative approach. Experience of the Quimiag-Penipe Project, Ecuador" (Cornell international agricultural mimeograph no. 91). Ithaca, N.Y.: Department of Agricultural Economics, Cornell University.

Limpinuntana, V., Patanothai, A., Charoenwatana, T., Conway, G,. Sectisarn, M., & Rerkasem, K. (1982). An agroecosystem analysis of northeast Thailand. Khon Kaen, Thailand: Faculty of Agriculture, Khon Kaen University.

Moscardi, E. (1982). The establishment of national on-farm research entity in Ecuador. Mexico City: International Maize and Wheat Improvement Center.

Perrin, R. K., Winkelmann, D. L., Moscardi, E. R., & Anderson, J. R. (1976). From agronomic data to farmer recommendations: An economic training manual (Information bulletin 27). Mexico City: International Maize and Wheat Improvement Center.

Rogers, E. M., & Shoemaker, F. F. (1971). Communication of innovations: A cross-cultural approach. New York: Free Press.

Shaner, W. W., Philipp, P. F., & Schmehl, W. R. (1982). Farming systems research and development: Guidelines for developing countries. Boulder, Colo.: Westview Press.

Tripp, R. (1982). Data collection, site selection and farmer participation in on-farm experimentation (CIMMYT economics working papers 82/1). Mexico City: International Maize and Wheat Improvement Center.

Zandstra, H. G., Price, E. C., Litsinger, U. A., & Morris, R. A. (1981). A methodology for on-farm cropping systems research. Los Banos, Philippines: International Rice Research Institute.

Chapter 4
Introducing Technological Change: The Social Setting

A. J. Sofranko

Modern agriculture "requires an innovative technology which systematically adapts scientific knowledge to farming" (Sanders, 1977). In many countries farm-level technologies superior to those currently in use are already available. The gap between the existing level of technological knowledge and what is in use in a particular farm setting is not easily closed, however, as experience has shown over the years. Technological change is a difficult, time-consuming process, made even more difficult because much of the technology being promoted is not suitable to a particular locality, complementary services and delivery systems are not available, or unexpected cultural resistance often emerges among the intended beneficiaries.

This chapter provides an understanding of what is required to promote technological change among farmers. The focus is on the social and cultural setting into which new technology is introduced. The point that will be stressed is that extension workers, who are the main link between villages and farm families and the new technologies of the outside world, should be knowledgeable about the cultural setting into which technology is introduced, have an understanding of potential obstacles and of ways to reduce resistance to change, and be sensitive to how technological change might affect peoples' living conditions.

The chapter starts with a discussion of concepts that are important to understanding what is meant by the word "change" as it is used by social scientists. It discusses several concepts intended to sensitize the reader to important features of a local culture. This is followed by a discussion of obstacles and resistance to change, and a presentation of alternative ways of introducing change. Finally the chapter discusses the impact of technology on farmers and examines the extension worker's role in technological change. This chapter is not meant to be a list of tried-and-true methods, applicable in each and every rural setting. Such lists do not exist. It tries instead to ensure that extension workers are made aware of social and cultural conditions, and thus can take them into account in their own activities.

THE NATURE OF CHANGE

Despite frequent observations to the contrary, norms and value systems do change. Even the most traditional cultures change over time. For example, peoples' expectations, aspirations and values change; groups form and dissolve; leadership changes; power and influence shift; standards of behaviour are modified; and roles of men, women, and children gradually alter. Change is normal and inevitable. What varies is the speed with which it occurs, whether it is deliberate or accidental, and the significance of the change.

Sources of Change

There are both internal and external sources of change in any rural area. The former occur as a result of normal forces acting within a culture or village entity; for example, people age and die and are replaced by new members through birth or migration. These normal population processes alone create demands on existing resources, and often require cultural adjustments or adaptations. Apart from demographically generated changes, other pressures, innovations, conflicts, or changes in leadership and influence in a village can also lead to significant and lasting local changes. An emerging source of change in rural areas is the women agriculturists who are requesting and being given extension assistance. Externally produced changes, on the other hand, occur as a result of forces originating outside the village. These occur through contacts with the outside world, especially communications, education and transportation which link villages with larger towns. In a Costa Rican village Barlett (1982) has shown how contacts with the external world have increased through construction of all-weather roads, inauguration of regular bus services, possession of radios, and the operation of several national-level institutions in the village itself, including schools, churches, political groups, police, health services, and three agricultural agencies. At the present time, these types of contact are an important source of change in the rural areas of developing nations.

Types of Change

Social science terminology distinguishes different types of change. Social change generally refers to changes that occur in the structure and functioning of a social system (Rogers, 1969). This term includes changes in the roles individuals perform, changes in the status of individuals in a village or other social system, alterations in social relationships among people, changes in patterns of social interaction, and changes in the functions carried out by different groups and institutions within a village community. The more significant social changes are those which endure, become permanent and affect a broad segment of the population.

Cultural change is a broader term used to describe all changes occurring in any part of a culture such as its art forms, its values and beliefs or its science and technology. Cultural change involves alterations in the way people perceive and relate to their environment, changes in their customs, their artifacts, inventions and laws, the common knowledge shared among people, and in their attitudes, beliefs, and values (Allen, 1971). The term "sociocultural" is usually used to encompass both of the above types of change, and is the concept which will be used in this chapter.

Socio-cultural change is viewed as a requirement for technological change and as a consequence of the introduction of new technologies into a village system. To use a common example, changes in farmers' values and beliefs are frequently viewed as a requirement for agricultural development. As a result, certain change strategies (education, indoctrination, and communication) specifically emphasize modification in values and beliefs. To illustrate that socio-cultural change is also a consequence of technological change, we might show that changes in beliefs and values often occur as a result of farmers using more modern technologies. For example, when farmers recognize the superiority of fertilizer over the use of totems in ensuring fertility, the importance they attach to other traditional beliefs begins to wane. Ultimately their beliefs and values change. Many of these points will be discussed in greater depth later.

Much of the socio-cultural change that occurs in any society is unplanned. It is a result of accidental discoveries, cultural drift, diffusion from neighbouring groups or cultures, infusion of new technologies, the influx of new members into a culture, or even the consequences of crises, conflicts, and natural disasters. Throughout history this is the way in which most change has taken place. Now, however, there is a growing realization that if the quality of peoples' lives is to be improved, and agriculture is to become more productive, change will have to be more deliberate, that is, planned.

Planned Change

There are several good reasons why planned change is being emphasized in agricultural development efforts.

1. A gap exists between the level of scientific knowledge and technology available in the world and what is actually being used. It is unlikely the gap will be narrowed without acceleration of the change process.

2. Within countries there are frequently varying rates and levels of change among certain regions, villages, ethnic groups or types of farming areas. Planned change is necessary to readdress any imbalances that have developed over time.

3. Planned change is necessary to correct some of the consequences of unintended change, or to reduce the severity of problems which could arise if nothing were done.

Technological Change

In many rural areas agriculture is still carried on with simple tools by traditional methods, using practices based on trial-and-error. As a result, farmers produce only slightly more food than is required by their households. There is little question that changes must be made in production methods, and new technologies are increasingly being viewed as the vehicle for solving agricultural problems. While the solution seems to be simple, in practice it is not. Even where new technologies exist they may be inappropriate for particular agricultural settings, they cannot be transferred easily, or they collide with traditional cultural practices and preferences.

Any definition of technology encompasses a wide range of phenomena. In the broadest sense, technology is defined as the translation of scientific laws into machines, tools, mechanical devices, instruments, innovations, procedures and techniques to accomplish tangible ends, attain specific needs, or manipulate the environment for practical purposes (Theodorson and Theodorson, 1969). Although the study of technology has most often focused on material objects, there are also lesser known social technologies. In agriculture these represent new modes of organizing production, as well as the development of techniques for regulating and controlling behaviour, delivering services and improving the efficiency of organizations. Industry has recognized that productivity and efficiency can be improved through changes in equipment and techniques, as well as through changes in the way production is organized. This type of thinking is reflected in many of the current attempts to organize and reorganize farming enterprises into co-operatives, communal farming systems, state farms and supervised production schemes.

For many years technological change has emphasized the introduction of machinery. Now, however, agricultural technology is viewed as representing much more than only mechanization. It includes

1.   introduction of new farm inputs, such as compounded fertilizers tailored to soils and plants, insecticides and herbicides, new irrigation systems, and new plant varieties that are immune to fungus and diseases, less sensitive to sunlight, responsive to new fertilizers, of shorter stock and faster maturing.

2.   introduction of new techniques or practices, such as new planting and cultivation techniques, different cropping sequences, new crop rotations, improved storage methods, and improved uses of animal power. Such improvements have the best potential for improving farm output. The effects of these non-mechanical improvements will be limited, however, if farmers do not have the skills and resources for utilizing the technologies, or if there is not a delivery system for extending them to farmers. In some respects, it is more difficult to introduce these types of improvements rather than machinery, because they are dependent on the institutions which must adapt, modify and deliver them to farmers.

CONCEPTS FOR UNDERSTANDING RURAL CULTURE

Extension workers are frequently criticized for not "taking cultural factors into consideration" and are constantly being reminded to "understand the local culture" of people and villages in which they are trying to introduce change. It is assumed that if they had a better understanding of local culture they would be more effective in persuading farmers to accept change. Extension workers frequently misunderstand the culture of rural people. It is also true, however, that extension workers can never be expected to know all or even most of what there is to know about a culture. They generally have little training in the social sciences; they are accountable for carrying out specific activities and not necessarily for developing cultural understanding; and they are responsible for working among villages and groups that may be culturally quite diverse.

Somewhere between ignoring the local culture and becoming totally immersed in it there is a middle position which suggests that an extension worker must have some knowledge and give some consideration to local cultural conditions. Which are the essential features of rural culture that are important to extension workers who want to be more effective in their change efforts is the question that then arises. These features can be captured in a set of concepts which focus attention on phenomena which might otherwise be overlooked. Some of the important social science concepts that can inform and sensitize extension workers about local conditions are values and beliefs, stratification, power and influence, and social organization.

Values and Beliefs

Values and beliefs have been singled out as important elements in the change process. In most cases, they are viewed as obstacles to change, thereby becoming legitimate targets of change efforts. Values have been defined as conceptions of the desirable, as standards of evaluation, as guides for decision-making behaviour, or simply as expressions of preference (Kahl, 1968). They are seen as having a central role in the change process because it is assumed they are crucial in influencing farmers' goals and behaviour. Technological change requires behavioural change on the part of farmers, whether it be using new inputs or developing extra-local ties with input suppliers and technical expertise. It is feared that these required behavioural changes will not occur until values change, that is, until traditional values are replaced by more modern values.

Values of Rural People

At one time or another, a variety of value orientations have been attributed to rural people in developing nations (cf. Foster, 1973; Kahl, 1968; Rogers, 1969; Sanders, 1977). Table 4.1 presents some of the more frequently mentioned values and their characteristics.

Table 4.1
Selected List of Value Orientations

Value		Characteristics
Familism	1.	Subordination of individual accomplishments and goals to those of the family.
	2.	High level of integration with family and relatives.
	3.	Unwillingness to engage in activities with persons outside one's family.
Fatalism	1.	Resignation.
	2.	Passivity.
	3.	Feeling that one lacks the ability to influence the future.
	4.	Feeling that the outside world is unpredictable and cannot be understood.
Low empathy	1.	Inability to envisage oneself in a different role.
	2.	Difficulty in viewing oneself in a relatively better off position.
Aversion to risk-taking	1.	Unwillingness to take chances.
	2.	Reluctance to experiment or venture out beyond one's immediate social environment.
Traditionalism	1.	Emphasis on the past, preserving it and continuing it.
	2.	Belief that one's ways of doing things is natural and best.
Immediate gratification	1.	Unwillingness to save or invest for the future.
	2.	Unwillingness to postpone present satisfaction in anticipation of future rewards.
Submission to nature	1.	Indifference to the passage of time which no one dreams of mastering, saving up, or using.

Beliefs

Beliefs, which are closely related to values, are the mental convictions one has about the truth or actuality of something. They refer to what people believe or accept to be true, what people can trust or place confidence in. Like values, they are viewed as an underlying support for behaviours: "men of all cultures support their actions by elaborate systems of beliefs. There are beliefs of what is right or wrong, what is proper or improper, what is lucky or unlucky. Logically, there is no cultural behavior for which men do not have supporting beliefs" (Niehoff, 1969, p. 45). Various beliefs serving as barrier to change have been identified. For example, beliefs in how one controls events in this world, in cause and effect relationships, in the possibility of self-improvement, and in the likely outcomes of individual actions. Foster (1973) points out that with so much of the world not subject to control, the peasant has a poorly developed critical sense and is able to believe improbable things. There may be beliefs that could seriously undermine one's efforts if they were not taken into account in an extension effort.

Foundation of Values and Beliefs

Extension workers often view traditional values as reflections of backwardness without attempting to understand why people hold particular values. They view traditional values as obstacles to change rather than as reflections of the conditions under which people live. It is easy to criticize people for being opposed to risk-taking, for example, without recognizing that taking risks can have serious consequences in a resource-poor environment. Similarly, it is easy to be critical of farmers who put family concerns over personal achievement without realizing the importance of family ties to one's security and welfare. Extension workers should be sensitive to the origins and functions of traditional values and beliefs.

Prevailing values and beliefs reflect the ways in which people have been taught to behave and view the world. They are either reinforced or modified in relation to the opportunities people have and their contact with individuals holding different values. The linkages between values and their causes and consequences are illustrated in Figure 4.1.

Figure 4.1 Underlying causes of individuals' beliefs, values, and behaviour

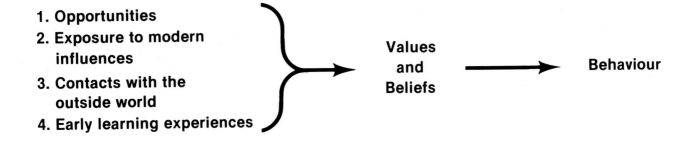

Determining Peoples' Values and Beliefs

Values and beliefs cannot be observed directly but have to be inferred. An observant extension worker can begin to infer values from:

1. <u>choices people make</u>; what do people choose to do with their free time; where do people choose to invest their scarce resources?

2. <u>behaviour</u>; what will people organize themselves, or attend a meeting for; what do people get angry about or readily approve?

3. <u>statements people make</u>; how do people justify a particular behaviour or choice of actions; what do they say they would prefer, given a set of options?

These are simple ways of establishing what people value highly and believe in. One must be cautious, however, because values and behaviour are not always consistent. People will not always reveal their true reasons for a particular type of behaviour, if they are aware of them.

It is rare these days to find a rural community which is completely isolated from contact with the influences of the modern world. With more such contacts and increased opportunities for participation in the larger society, one would expect changes in values and beliefs.

Changes in Values and Beliefs

Values and beliefs do change over time, depending on (a) how well values and beliefs serve people, that is, how well they continue to explain and predict what occurs in people's lives, (b) the amount of culture contact people have with others holding different and competing values and beliefs, and (c) the range of opportunities people are presented with which question existing values or force individuals to actually modify their beliefs and values. Technological change is a major source of value change. Use of modern agricultural inputs, for example, can demonstrate the superiority of science over other more traditional practices when it comes to increasing yields. The benefits farmers get from using modern technologies can convince them that planning and investing are worthwhile goals, in which case it is unlikely farmers will remain fatalistic or present-oriented for long. Modernization of agriculture through technological change also involves a certain amount of culture contact on the part of farmers, with experts, the mass media, and other people with new perspectives, different behaviour and contrasting lifestyles. Ultimately these types of contacts will have an influence on traditional values and beliefs.

It is easy to characterize all rural people as being traditional in outlook, but rural communities have never been completely homogeneous with respect to the types of values and beliefs their people hold. Even in the most traditional cultures some people, by virtue of their experiences, position in the community, age or level of education have different values and beliefs. Extension workers are aware of such differences, for they frequently identify and work with those individuals who are less traditional and more modern in their orientations. One of the concerns of extension workers should be that they should not devote most of their efforts to these individuals, even though they are frequently the most accessible and easiest to work with.

Extension workers should also recognize that there are "hierarchies of values." In other words, not all values are equally resistant to change or charged with the same emotional commitment. There are clearly some values which are relatively difficult to change, and extension workers would be wise to avoid clashes over them. Clever extension workers can also capitalize on traditional values to implement technological change. For example, one can emphasize increased production as a means of maintaining strong family ties. Familism and strong village ties can be utilized to encourage investments in agriculture, education, and community development efforts. This is a common practice in many rural areas of Africa.

Power and Influence

All communities have individuals whose opinions are taken more seriously than others, whose co-operation is essential if anything is to be accomplished, and who have a major role in making decisions and mobilizing people. These individuals are powerful and influential, and to ignore this aspect of village life is to increase the likelihood of failure in any effort for change. Knowing how power, influence, and leadership are exerted in a community is part of "understanding a local culture"; according to Niehoff (1969), leadership is the most important feature of local culture. If influential leaders cannot be persuaded to change their practices, lend approval to change or at least remain neutral, extension workers will have a difficult time making progress. It is important that extension workers learn who the influential leaders are in a particular cultural setting and enlist their support or participation. Barry (1971) employed this tactic when he convinced the town chief to plant a new rice variety one year, and the Iman of the local mosque to plant it in the following year. The enlistment and support of such locally influential persons is invaluable in the transfer of technology at the village level.

There are three methods of determining who is powerful and influential. The first, the reputational method, is based on asking people in a community to identify the influential person or persons. People might be asked who has high prestige or respect, whose opinion or word is highly regarded, or whose approval is essential for a particular activity in the community. The second method, called the positional approach, equates position or title with influence, and assumes that people in hereditary, appointed or elected positions are influential. It is likely they have some influence, but it is often limited. Finally, the decision method establishes power and influence by examining who was instrumental in recent major decisions in the community. For example, who was instrumental in getting water, a branch post office or a clinic into a village? None of these methods is perfect, but a combination of them should provide the extension worker with enough information to establish whose co-operation should be sought before proceeding.

A study by Niehoff (1969) of planned change efforts in developing nations points out some of the types of leaders and influential people that have to be recognized and acknowledged. These include (a) administrative or formal political leaders, (b) educators, (c) religious leaders, of the major and indigenous religions, (d) civic leaders who are active in various local groups, associations or clubs and (e) the non-institutional leaders, the people who are looked up to and emulated because of their wealth, personality or family background. This list may not be inclusive in some cases. In some villages, only a few of these listed leaders may be present, but the list is an adequate guide.

When examining power and influence in a village setting, extension workers ought to be aware of several possibilities;

1.   power and influence may lie not so much in the hands of individuals but in certain groups or organizations;

2.   power and influence are not always equated with wealth. There are many bases of power, for example, age, sex, expertise, control over information, or lineage;

3.   the structure of power and influence does change over time, and one should not assume that today's influential people were also influential in the past;

4.   there are limits to power and influence, and some things can be accomplished without the support of influential people. It may be considerably more difficult, but it is possible to circumvent some leaders;

5.   influentials in one area of community life are not necessarily influential in all others. Some persons or groups may be influential in religious matters or civic improvements, while others may be influential in agriculture.

Agricultural development efforts have a much better chance of success if influential persons are convinced of the utility of such efforts and are given the opportunity to provide some say in adapting programmes to existing local economic and social conditions.

Social Stratification

The arrangement of individuals or groups into a hierarchy represents a stratification system in a community. Even in what appear to be homogeneous villages, there are internal inequalities, divisions and distinctions. Over time these distinctions become patterned and stabilized, and different life-styles, beliefs, values and behaviour come to be associated with different status positions. The basis for status distinctions may be ownership of property, family wealth, amount of education, type of occupation, skills, personal qualities, or ascribed characteristics such as age, sex, race or ethnic group membership. In some communities there may be elaborate distinctions, while in others they may be few in number and blurred. Often there is also a spatial and interactional aspect of stratification. Where people live within a community may be influenced by their status, and social interaction may be limited among certain strata or groups. What is important for the extension worker to recognize is that such internal differences do exist, and that they are likely to vary from place to place.

There are several features of a local stratification system an extension worker should be able to identify

1.   how many strata there are, and in what order they are arranged,

2.   on what basis people are placed into different strata,

3.   how much interaction there is among strata,

4.   how much influence each strata has on village life,

5.   how the different strata perceive change, and

6.   how different strata would be affected by a particular development.

Social Organization

"Social organization" refers to the "network of social relationships within which people live, such as families, organized and informal groups within villages and neighborhoods, marketing and credit organizations, and the legal framework (group norms) within which they live" (Mosher, 1976, p. 296). An examination of social organization focuses our attention on the basic ways in which people are organized, the enduring social relations that exist, and the dominant social bonds and collective relationships that exist in a rural community. Many modern agricultural technologies create new forms of social organization based on vocational interests and problems. For example, the installation of a new well or irrigation system may require several unrelated farmers to make decisions jointly, or there may be instances where the use of new inputs requires new patterns of distribution and new linkages to the modern sector. Technological change can also have an effect on traditional forms of social organization, by displacing existing work groups, creating new bases for prestige in the local community, or by creating new associations which compete and conflict with established groups.

There are several elements of social organization which can affect the process of change; therefore the extension worker should have some understanding of their importance and role in villagers' behaviour to function effectively. For example, with respect to agriculture, what are the social arrangements governing this activity? How are the basic production activities carried out: on the basis of the nuclear family, the extended family, or some other groupings? On what basis are agricultural decisions made? For example, what role do family/kinship factors play in detemining what farmers plant or the size of holdings? One cannot simply assume that it is possible to get individuals to act alone, especially where family units or other forms of co-operative arrangements are the core units of organization.

In many villages kinship is the central social bond and the family is the basis of social organization. The culture establishes which family relationships are significant, what the rights and obligations of related individuals are, and what forms of organization exist among related persons. Where kinship is exceptionally strong, activities are initiated if and when they do not threaten the family and if the roles within the family unit are not disrupted. Emphasis on kinship frequently precludes either co-operation among families or action by individuals alone, and in this respect inhibits activities requiring the development of community consciousness and co-operation (Niehoff, 1969). Where kinship ties are paramount, the extension worker should work to introduce technological change at the family level. It might be possible, for example, to utilize the extended family system as a way to spread innovations.

There are other bases of social organization, although few are as important as family and kinship in most rural areas. For some purposes, groupings are formed on the basis of sex. Traditionally, some tasks are carried out by men and some by women. The extension worker should become aware of the basic work grouping in an area. For example, who does what type of work on the farm? At times, these work groups may consist of all males; at other times, they may be made up of husband and wife, or wives; and in some villages there is strong preference for kinsmen as work partners (Niehoff, 1966). In many rural areas women are the heads of households and, along with their children, constitute the primary work group. Yet, little attention has been given to the role and importance of women agriculturalists to change efforts, or to the role of children.

Finally, in some rural areas, farmers are organized into more formal "lobby groups" which directly petition and try to influence the government. More common are the traditional forms of co-operation that appear in rural areas, sometimes formed along caste or ethnic lines, and sometimes cutting across social boundaries. For example, in Nepal, Messerschmidt (1981) identified the presence of land tenure associations, credit associations and reciprocal labour exchanges, as well as temporary associations of youths engaged in agricultural field work on a seasonal or task-specific basis.

OBSTACLES TO AGRICULTURAL CHANGE

As a way of introducing the issue of obstacles, it might be useful to define what is meant by "agricultural development." It is a term that has been used to refer to the process of (a) getting farmers to produce more than what is required for immediate consumption or use by the farm household, (b) encouraging farmers to grow crops for which a demand exists but which they may be unaccustomed to growing, or (c) persuading farmers to improve the quality of the crop they produce to meet consumer demands or the demands of an export market. For the most part, extension workers have focused their efforts on the first goal, which basically involves changing the behaviour of farmers. What, then, are some of the obstacles to changing farmer behaviour?

Socio-cultural Obstacles

Perhaps the simplest classification of obstacles to agricultural development has been on the basis of whether they lie within the farmer or within the farm environment. Obstacles residing within farmers themselves and their immediate cultures have been identified as traditional value and beliefs, illiteracy, lack of motivation for achievement, insufficient resources to take advantage of opportunities, low-level skills and limited aspirations. Because of certain traditional beliefs, values, or cultural practices, farmers are felt to be unconcerned with improvement, unwilling to take risks, or unable to take advantage of existing opportunities and use existing production technologies. At other times it has been argued that peasant farmers have a culturally defined "target income" which, when attained, inhibits further productive efforts.

The underlying theme of this perspective is that the traditional socio-cultural milieu, consisting of beliefs and values, traditional explanations of natural phenomena, family and village dependencies, and traditional agricultural practices, is incompatible with the requirements of modern agriculture. In that respect, it represents obstacles that have to be overcome. According to this reasoning, the necessary ingredient for agricultural modernization is a general reorientaton of farmers' beliefs, values and behaviour through education, training in literacy, exposure to information through the mass media, or training via demonstrations of various sorts. This is the type of role extension has typically played in the agricultural change process.

Economic Obstacles

An alternative explanation for the failure of farmers to change centres not directly on farmers themselves but on obstacles in the farm environment. These have been identified as inadequate financial incentives and lack of inputs necessary for getting agriculture moving. The assumption is that farmers are economically motivated, not tradition-bound, and are responsive to prices and various infrastructural alterations. At one time or another, these alterations have included making available, in a timely manner, new technological inputs such as fertilizers and other chemicals, new seed varieties, tools and machines. Included also in the list of infrastructural alterations necessary for change are low-interest credit, improved transportation, marketing and distribution systems, storage facilities, improved extension activity and research facilities, changes in land tenure arrangements where necessary, and improvement in the productive capacity of the land through irrigation and erosion control.

An example of the positive changes that can result from providing farmers access to proper inputs and training took place in Gambia. Women in several Gambian villages had often asked extension workers for a chance to become involved in cash crop production, an area traditionally dominated by men. In response, the ministry of agriculture set up a project to train 30 women in the production of onion crops for export--each woman being responsible for 10 onion beds. The ministry and the Gambia Cooperative Union provide fertilizer and seed and purchase the onion crop at an agreed price. Within two years, the project grew to involve over 900 women in 32 different projects--providing the women with income and with the opportunity to join cooperatives (WCARRD, 1980). In this case, it seems clear that the farmers were economically motivated, not tradition bound, and were responsive to prices and infrastructural alterations. Other similar examples could be cited for various parts of the world.

Assessing the Situation

Both the socio-cultural and economic perspectives correctly identify some of the obstacles to agricultural change and sensitize extension workers to the types of obstacles they may have to deal with. There are some general comments that can be made about the way in which extension workers should view obstacles. There is no finite list of obstacles that, when drawn up, will provide a complete set of tasks to work on, and it is impossible in the abstract to rank obstacles in importance. It is fruitless to compare groups of farmers or villages with different levels of development and conclude that the differences represent obstacles.

Hirschman (1967) presents several insights into the lists of obstacles that have been identified as impediments to change. He claims that some of the so-called obstacles can be turned into assets. For example, a strong extended family orientation may enable family members to undertake co-operative tasks without hiring labour, or family members may pool their resources for investment purposes. There are many obstacles whose elimination turns out to be unnecessary. For example, literacy may not have to precede agricultural change, nor should education necessarily be a prerequisite for being able to absorb new agricultural information. He also claims that there are obstacles whose elimination can be postponed. For example, attitudes, beliefs and values may not have to change prior to behavioural change; they may change as a result of behavioural change or, more specifically, as a result of having used new technologies.

There is no simple solution to deciding which perspective on obstacles is correct, or which is more correct than another. In some village settings the main obstacles may lie within the farmers themselves, and in others the obstacles may not be in the farmers but in the general socio-economic environment that discourages change on the part of farmers.

64

The extension worker must attempt to establish what the real obstacles are from observing, talking to farmers, or even undertaking small-scale farming to better understand what farmers are experiencing.

Incentives to Change

Most of the discussion has centred on obstacles and resistance to change. Undoubtedly there are many forces promoting stability in rural areas, and there are many obstacles to introducing the types of behavioural change necessary for agricultural development. We generally pay more attention, however, to those forces which inhibit change than we do to the forces for change in a particular rural setting. What we fail to make explicit is that behavioural change is a result of the interaction of two sets of opposing forces--change obstacles and change incentives. These forces act on individuals in the way shown in Figure 4.2.

Figure 4.2   Illustration of obstacles and incentives affecting farmer behaviour in developing nations

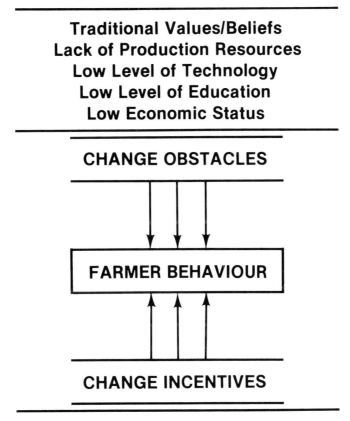

Traditional Values/Beliefs
Lack of Production Resources
Low Level of Technology
Low Level of Education
Low Economic Status

CHANGE OBSTACLES

FARMER BEHAVIOUR

CHANGE INCENTIVES

Desire for Economic Improvement
Availability of Modern Technology
Improved Extension Services
Desire for Social Improvement
Expanded Opportunities

The "change obstacles" have been discussed throughout this section. In the figure they represent a partial list of the type of forces that influence people not to change their behaviour. There are also, however, positive or "push" forces which motivate people to change by creating dissatisfaction with the status quo, and a favourable judgement of potential future situations (Leagans, 1971). Some of the more general push forces disposing people toward behavioural change are shown in the lower portion of the figure. This list recognizes that most people have a desire for economic and social improvement, whether for the purpose of purchasing more consumer goods, educating their children or improving health and nutrition. In most rural areas these types of motivations already exist, and they are becoming even stronger as rural people become more exposed to the modern world. There is also a recognition that changes in the environment in which farmers live and work, and changes in the opportunities available to them, can act as incentives to further change. The availability of new technologies, access to extension services and external markets, availability of credit and so forth, represent the types of changes in opportunity that act as incentives to further change.

The extension worker should have an understanding of both the change incentives and the change obstacles in a particular rural area, and strive to reduce the obstacles while capitalizing on and improving the existing incentives. What is important from the extension worker's perspective is that he or she should be able to identify the types and strengths of various obstacles and incentives.

REDUCING RESISTANCE TO CHANGE

One of the major goals of extension workers is to influence the way farmers and farm families make decisions. And it is important for extension workers to recognize that women in farm families influence men's decisions about agricultural activities (Mickelwait, Riegelman and Sweet, 1976). Whenever individuals are asked to change their decision-making habits, either by incorporating new information or excluding some traditionally used criteria, it is likely that resistance will occur. There is, of course, the fear of the unknown and the reluctance on the part of many farmers to learn new behaviour, especially when change involves risk. Resistance may also arise when anticipated new behaviour conflicts with existing traditions and practices, or is perceived as a threat to other village activities that are invested with a high value in the eyes of the people. For example, farmers have been known to be resistant to multiple-cropping or adoption of a different rotation pattern when the change infringes on village festivities at a particular time of year.

Resistance may also be a direct consequence of the technology being promoted, its costs, ease of use, labour requirements or perceived improvement over existing technology. Extension workers should not assume that resistance is necessarily indicative of "backwardness" or "conservatism." In many cases resistance is not irrational behaviour. Finally, a certain amount of resistance is built into the extension worker's role, especially where the worker is viewed as an outsider who may know very little about farmers' specific problems or of farming itself. From this discussion one important fact becomes clear, that overcoming resistance is one of the principal tasks of extension workers. They have to anticipate that some resistance will occur and determine how they will react to it beforehand.

Resistance to technological change can take different forms. In some cases it shows up as a lack of interest or apathy among farmers, and in other cases as outright opposition. Different types of resistance require different approaches on the part of the extension worker. Apathy may be a result of inadequate information or of a poorly designed technology, while overt opposition may be a result of perceived threats to a person's or group's status or security. In the former case, resistance may be overcome by providing more information, or through modification of the technology. In the case of outright opposition, neutralization of the opposition may be the best the extension worker can hope for. There are not many good suggestions as to how this can be done. Millikan and Hapgood (1967) suggest that one may ultimately have to mobilize peer pressure against the individual(s), exclude the person(s) from the change programme, or even resort to some form of coercion when the success of the whole effort is threatened.

There are several conditions under which resistance can be reduced (cf. Watson, 1969);

1.   when farmers or villagers are made to feel that an activity is at least partly their own, not one devised and operated solely by outsiders;

2.   if there is support from influential people and leaders in the village;

3.   if participants see the proposed changes are reducing rather than increasing their burdens, materially improving their lifestyle, and in general having some practical benefits;

4.   if participants have helped in diagnosing prevailing conditions, and the programme reflects their demonstrated needs, objections, and fears, and

66

5.   if there is group consensus on adoption of a new practice or on participation in a particular change-related activity.

Despite the many reports devoted to projects that have failed because farmers were opposed to them, it is probably true that people are more likely to accept change than resist it.   Resistance is not a constant element in most cultures, but arises as a result of inadequate understanding, misperceptions, unrealistic expections, and poor relations between extension workers and farmers, supposed threats to farmers' security, or from farmers being forced to change.

Encouraging Participation

A common complaint among extension workers is that it is difficult to get farmers to participate in meetings and projects, but most difficult getting the interest and involvement of those for whom the activities were specifically intended, the smaller, poorer farmers.   When there is participation by the poorer farmers, it is often irregular, or passive.

Because it is so difficult getting and expanding participation in extension efforts, participation should be given more attention.   It should be treated as a limited goal in itself, because it is obvious extension workers can have little influence on the lives of people without access to them.   Workers often, and erroneously, assume that what they are trying to promote is so attractive to farmers they will willingly get involved.   What the extension workers and what farmers perceive to be attractive are often quite different.

Because it is often difficult to elicit immediate enthusiasm from farmers, let alone their participation in the planning of activities, it is too easy for extension workers to infer that farmers are "backward" or "ignorant," do not want to improve their lifestyle, are resistant to change, not rational in their thinking.   From the point of view of the farmer there may be good reasons for not participating.   Participation can be costly, take time away from other activities or involve travel and other expenses; it may also expose farmers to criticism or ridicule by other members of the village, especially if the activity is new or controversial.

Many people do not become involved because they feel their participation is not wanted.   Women, for example, despite their dominant role in agricultural activities, have not been encouraged to participate in many projects until recently.   Other groups may also be excluded in particular settings either deliberately or unintentionally.   Finally, many people will not participate because they expect that others will do so, in particular those who have participated in the past.   Those who readily volunteer tend to be the ones whom the village serves best.   These are frequently the better off, the larger property owners, or the better educated.

The extension worker's task of getting farmer participation in change efforts is not likely to be an easy one, and sometimes, even after many efforts, only a small percentage of the target group will be involved.   Here are some general recommendations on how an extension worker might get broader participation in agricultural programmes.

1.   People are much more willing to participate in activities which meet their "felt needs."   A quick needs assessment can determine farmers' needs and priorities.   Again, the needs of all people should be taken into consideration, not just those who are accessible and co-operative.

2.   If farmers are encouraged to express their needs and provide some input into the structure of a programme, they should not then be ignored.   The price of widespread participation may be additional burdens on the extension worker, but if he or she wishes to sustain farmers' involvement, then the farmers' ideas should be taken into account.

3.   People are more likely to participate if actual benefits are directly tied to participation.

4.   Farmers, especially those with low incomes, are more likely to participate and remain involved in a development effort if the benefits are material, direct and immediate. People invest their participation in activities they believe will benefit them (Johnston and Clark, 1982).   There is evidence that in working with the uneducated benefits have to be obvious and tangible.   One of the best ways of getting farmers' interest is through the use of convincing and realistic demonstrations and trials.

5.   Extension workers should not expect that the participation of a small group of "progressive" farmers will ultimately lead to broader participation.   Progressive farmers do not have a lot of interest in providing other farmers with the necessary details for improving practices, do not often get called on for advice, and are viewed as being irrelevant to the small farmers' situation (Johnston and Clark, 1982).

STRATEGIES FOR INTRODUCING CHANGE

Developing agriculture by means of substituting new for existing technologies involves behavioural change on the part of the farmer. The amount of change involved will depend on the technologies and practices being promoted and the extent to which farmers' current behaviour is inconsistent with them. It is important that extension workers be able to identify as specifically as possible the behaviour they are attempting to change. Is it, for example, the proper use of a new input or practice, the adoption of a single or complex of inputs and practices, or the substitution of one practice for another? If one cannot identify the specific behaviour, it is not likely there will be much opportunity to assess the progress being made.

When it comes to devising a change strategy there are two basic questions confronting extension workers. First, what is the target of change? Is it the individual, the group, or the agricultural environment? Second, what methods of change are to be used? Are they education, demonstrations, use of power and coercion, attitude or environmental change? Extension workers seldom have a free choice with regard to these options, because the functions they are assigned to carry out may specify one target group or method more than another, and political or economic conditions may influence their choices.

Strategies for bringing about change have generally focused on altering the environment in which agriculture is carried out, or in the direct transformation of farmers themselves (Rogers, 1969). These two contrasting strategies are shown in Figure 4.3. Both emphasize the same result, a change in farmer behaviour. Individual-oriented strategies are based on the belief that the main obstacles to getting farmers to change are rooted in farmers themselves and their traditional culture. If change is to occur, it must be directed at individuals as the recipients of extension efforts and at those aspects of the culture that are viewed as obstacles. For example, attitudes and values may have to be changed so that farmers will perceive certain conditions as problems, and aspire to greater income for themselves and their families.

Figure 4.3   Contrasting approaches to behavioural change among farmers

STRATEGIES FOCUSING ON INDIVIDUALS AND GROUPS

1. Changing Attitudes, Beliefs Values
2. Education/Literacy Training
3. Demonstrations
4. Information Dissemination
5. Exhortation/Appeals/Persuasion
6. Forced Compliance/Coercion
7. Skills Training

BEHAVIOURAL CHANGE

Examples:

-using new practices
-adopting new technologies
-discontinuing established techniques
-employing different decision-making criteria

STRATEGIES FOCUSING ON FARMING ENVIRONMENT

1. Introduction of New Forms of Social Organization
2. Infra-structural Provisions:
   -irrigation and public water
   -markets
   -transport
   -storage facilities
   -processing facilities
   -extension research
   -credit, insurance
   -price incentives
3. Availability of New and/or Complementary Technologies
4. Availability of Additional Resources to Farmers
5. Legal Changes

For the most part, extension workers have concentrated on direct attempts to change people through education and training. Information is provided by word-of-mouth, demonstrations and the media, as well as in flyers, handouts and brochures. In some cases pressure, threats, coercion, and even exhortation are used. Occasionally ideological and charismatic appeals are used to get farmers to grow more so that the nation could be "self-reliant" or able to "feed the people." While these individual-oriented change strategies appear to be fairly widespread, there is a general feeling that they are not very effective for purposes of changing individuals. Where they have been effective, it was only for short periods of time, if they had not been coupled with other complementary changes. Demonstrations and other types of skill training are fine but do not go far enough. Unless they are accompanied by other changes such as credit or provision of new technologies, it is unlikely that farmers will change.

Group-Oriented Strategies

The question arises about whether one should focus change efforts on individuals alone or in groups (Cartwright, 1972). The reason for working with groups is based on the fact that individuals are also members of groups, and in many cases these groups can exert social pressure for or against change. Farmers are often cautious about co-operating individually with outsiders in change efforts out of fear of being laughed at for trying something different, or out of concern over how they will be viewed by their peers if they produce more, become wealthier or more progressive. When farmers take into account how family, village or friends will react to a decision they are, in effect, taking into account their group affiliations. In this way groups act as a constraint on individual decision-making, and in order to change individuals, the extension worker may have to change the groups in which they are involved. On the other hand, there are also opportunities to capitalize on group pressure to affect change. The following are some of the ways in which groups can be utilized (Cartwright, 1972).

1. Groups, as well a individuals, may be appropriate targets of change, but extension workers must know which significant groups exist in a particular cultural setting. They cannot ignore existing groups or assume they do not have much influence in people's lives.

2. Groups can be used as a "medium of change," that is, extension workers can work through groups to get change accepted. Decisions made as a result of a group process are much more binding, especially when group pressures enforce the decisions.

3. Where extension workers have created a shared group perception of a need for change, and the belief that change can improve their lives, it is more likely that change will occur.

Strategies Emphasising Change in the Farm Environment

A second broad change strategy emphasizes changes in the social environment or the agricultural system in which farmers work. The assumption implicit in this strategy is that if the material conditions or setting changes, then individuals will also change (Coleman, 1970). Arguments on behalf of this type of strategy have been heard many times in comments as the following.

1. "If farmers only had the opportunity and resources to purchase new technologies, they would."

2. "If prices were higher, markets available, there were access to credit, and farmers had some security against failure, they would willingly produce more."

3. "If farmers could depend on the timely availability of new inputs and services, they would adopt new practices."

The common element in these comments is the belief that "if only" the opportunities or incentives were available farmers would take advantage of them. In other words, they have not changed due to force of circumstance. This view has been reflected in a wide range of agricultural projects emphasizing removal of constraints in the general farming environment.

This strategy focuses on removing the barriers or obstacles to more modern behaviour on the part of farmers. In effect, farmers then begin to operate in a new, altered environment which is conducive to and supportive of change. In that respect, this strategy has an advantage over individual-based changes that are less effective when there are unfavourable incentives, and which require much more extensive individual contact. Whether or not farmers respond to such alterations in the agricultural environment depends on their ability to take advantage of such changes and, in many cases, their ability to learn about them.

The extension worker's role usually involves using both change strategies. There will be cases in which necessary environment changes will have been made, but many farmers will not be in a position to take advantage of them. For example, new technological inputs may be available to farmers, but for reasons of ignorance or lack of resources they may not be utilizing them. The extension worker's role in this case would be to first educate people to the availability and use of resources, and second, to put farmers into contact with input suppliers. It is not always that simple, and extension workers may very well have to effect behavioural change among agency personnel and input suppliers as well. The role of the extension worker is to bring new ideas and practices to farmers, explain their use, supervise their introduction, put farmers into contact with the institutions serving farmers, and seek answers to farmers' problems. This involves using a combination of strategies.

Impacts of Technological Change in Agriculture

The introduction of new technologies into agriculture is likely to have both <u>direct</u> and <u>indirect</u> effects. These latter effects are frequently referred to as "secondary" or "spread effects." Often the effects are discussed in terms of whether they were "intended" or "unanticipated."

The <u>Green Revolution</u> in Asian agriculture illustrates many of the consequences of introduction of new agricultural technologies. The term Green Revolution describes the introduction of new varieties of high-yielding wheat and rice in Asia. As a result of research at the International Maize and Wheat Improvement Center (CIMMYT) in Mexico, a high-yield dwarf variety of wheat was developed which proved to be highly responsive to the application of fertilizer, had reduced sensitivity to day length, and matured earlier than traditional varieties. Similar improvements were made in rice varieties at the International Rice Research Institute (IRRI) in the Philippines. Acreages planted in Asia with the new varieties of wheat and rice increased dramatically, as did production. With a shorter growing period for the new seeds, many farmers began to double-crop or plant year round. In many instances these improvements in technology increased the demand for farm labour, and farmers began investing in tube wells, drying equipment and ploughing services. Development in agriculture thus stimulated a demand for output from the domestic manufacturing sector.

Unquestionably, the new technologies have led to change, enabled farmers to alter practices, increased yields and created a desire and willingness to invest, take risks, and improve farming methods and standards of living (Ladejinksy, 1970). The Green Revolution has also had a variety of <u>unintended</u> consequences. It has highlighted the inequalities and differential access to resources among farmers and demonstrated that there are strong constraints on the expanded utilization of new technologies and continuation of high rates of adoption. While the costs of improved seeds are minimal, there are high costs associated with the complementary inputs, fertilizer and water. Farmers on small subsistence holdings, many of whom are tenants with few savings and no access to credit, have been unable to take advantage of the opportunities presented by new technologies (Frankel, 1969). As a result, sustaining the "Green Revolution" will require institutional changes to overcome costs of complementary inputs. There are the additional requirements of eliminating bottlenecks in the distribution systems so that farmers will get the seeds or chemicals at the exact time they are required and of training and supervision of farmers in new farming skills and expertise of a "higher order than was needed in traditional methods of cultivation" (Wharton, 1969, p. 466).

Various unexpected social problems have been attributed to the shift to more productive agriculture through technological change. Traditional landlord-tenant obligations and relationships have changed; women's traditional rights to land use have been altered; rents and crop shares have increased; some tenants have been evicted from the land; a class of "gentlemen farmers" has emerged; and tensions have arisen between landlords and tenants and between agricultural labourers and migrant labourers. These tensions have led to unrest, causing evicted tenants and displaced labourers to migrate to urban centres. Strains have also been placed on existing transportation, marketing and storage facilities. All of these are part of what has been referred to as the indirect and unanticipated consequences of social change.

This illustration demonstrates that technological change has impacts beyond what was intended, and that if technologies are to be distributed more equitably, they may have to "undergo some modification in order to be optimally useful or the institutions through which they must filter or be delivered may need to be fundamentally changed" (Theisehusen, 1974, p. 36). The lessons taught by the Green Revolution include the following, (a) that technological changes will have to be viewed from the perspective of the investments required to use the new technologies properly, (b) the agricultural conditions of the majority of the farmers will have to be taken into consideration, and (c) that there is a need for providing the resources with which farmers can adopt a necessary "technological package." The Green Revolution also demonstrates the need for modernizing distribution systems, upgrading government extension services, making fertilizer and pesticides widely available, and assuring more stable water supplies.

Unintended Consequences of Technological Change

Technological change "is a little like throwing a pebble into a pond; the effect of the throw on the water will extend in ever-widening circles until the impact loses force. In the same way, the impact of an introduced (technology) will extend ... until its effect is felt in areas of culture far removed from the point of contact" (Foster, 1973, p. 95). There are countless examples from change agents around the world of well-intended efforts aimed at introducing new technologies ending up displacing people, reinforcing already existing inequalities or forcing changes on other areas of the culture. As an example, Foster (1973) relates cases of attempts to introduce cooking methods which created less smoke in villagers' houses. The new methods worked, but in turn the homes became infested with mosquitoes and wood-boring white ants. The end result was a need for home repairs and the spread of malaria. The solution of one problem created greater problems. In a recent example from Nigeria, Williams (1982) points out how technological change in gari processing, which has traditionally been the occupation of women, has shifted the job of operating machinery from women to men and boys, depriving women of an important source of income.

The question arises as to why many impacts are unanticipated. There are two reasons. First, there is often a failure to take into account the fact that cultures are integrated systems, and that the changes which occur in one part are likely to have ramifications in others. Second, there is often very little, if any, forethought given to the question of what additional cultural changes would be required if the desired change was successful.

While it is true that one can never completely foresee all possible consequences, it is also true that many consequences can be anticipated with thought and planning. It is much easier for extension workers to focus on the tasks they are responsible for than to contemplate what might happen as a result of their efforts. Many view these types of impact as simply one of the "costs" of development. It is becoming common to see agricultural projects, which involve the introduction of new technologies, requiring detailed statements of potential impact based on a knowledge of agricultural and socio-cultural features of the project area. Detailed statements include identifying who will be the direct and indirect beneficiaries, the potential impact on persons and institutions not directly involved in the project, and impact on groups in adjoining areas. These impact statements are supposed to employ economic as well as socio-cultural criteria and provide insights into how negative impacts can be minimized.

Technology Transfer

Historically, the transfer of technology has occurred by means of cultural contact and migration. Currently the introduction of new technology is more deliberate, planned, and institutionalized, but remains largely unregulated. Many of the technologies introduced into Third-world agriculture have been expensive, labour displacing or unsuitable to the conditions of farmers. This has led to discussions about the appropriateness of technologies in light of employment problems in rural areas, the unavailability of resources for further adaptation and development of technologies, and the lack of capacity for disseminating technology to the intended recipients (Das, 1981).

It was stated at the beginning of the chapter that there is a gap between available knowledge and existing practices. This being the case, one would expect that the technology transfer process would be greatly accelerated by simply adopting technologies from either the developed nations or the international agricultural research centres. Applications of new technology in developed countries have not been made to tropical and subtropical environments, and many of the problems significant to the developing nations have not been sufficiently studied in the developed nations (Arnon, 1981). Research from the large international centres is also several steps removed from farmers in most countries. What is needed in either case is an intermediate testing state or preextension stage where

> "research recommendations on various crops and their respective production practices (crop mix, timing of operations, planting patterns such as intercropping, plant populations, crop protection, etc.) need to be incorporated into a complete farming system, or alternative systems, and then tested and evaluated under conditions similar to those encountered by the farmer, thereby making it possible to identify the technical, economic, social, and institutional constraints which prevent implementation by the farmer." (Arnon, 1981, p. 180)

Too often "adaptation" stops at the experimental plot of research stations, with little likelihood of easy transfer of the new technology. This transfer of technology from experimental plots has been hampered for several reasons; (a) the research does not (or cannot) take into account the complete farming system, (b) there is difficulty identifying the social and institutional constraints, (c) it is difficult to take into account the capabilities or small farmers and (d) it is not often conducted with the labour and equipment available to farmers. To overcome some of these conditions modifications have been introduced in the testing stage to more accurately approximate actual farming conditions. Among these testing and evaluation innovations have been the use of pilot farms, on-farm trials, verification trials on representative sites, decentralized demonstration plots or, as in the case of ICTA (Institute of Agricultural Science and Technology, Guatemala), research is conducted on plots loaned out by farmers.

The key elements in technology transfer are an adaptive research capability and an extension network with an institutionalized linkage between them and an effective delivery system. If research on new technologies is to be successfully transferred, it must be able to respond to results from verification trials in the testing and evaluation stages, as well as react to problems or emergency situations that may arise. On the extension side, since "much of the technology required by farmers must come from the research organization ... an essential function of extension is to select information derived from research or from other sources that can be beneficial to the farmers they serve" (Arnon, 1981, pp. 217-218). Ideally, extension workers should be linked to the transfer process. They need the technical skills in using information, understanding scientific recommendations, and interpreting and simplifying knowledge, in addition to skills in group dynamics, human relations and communication. If farmers are going to accept new technologies, it means that recommendations have to be translated into terms they can understand and that recommendations must be specific to their farming requirements. Unfortunately, the lack of training among extension workers in such skills is a serious barrier to a successful transfer of technology. Without training in interpretation of data, and without good interpreted data, there is nothing to deliver to farmers (Cusak, 1983).

ROLE OF THE EXTENSION WORKER

There have been suggestions concerning the role of the extension worker throughout the chapter. To avoid repetition, the final section will summarize some of the main points previously identified and provide additional comments that emerge from the above discussion.

Two broad preliminary comments are in order about the role extension workers play in introducing technological change. First, there is an unlimited number of roles extension workers might be expected to perform, including advocate, teacher, organizer, enforcer of regulations, planner, catalyst, co-ordinator, fee collector and communications specialists. One of the major difficulties with extension workers' roles is that they are often poorly defined. Roles may be vague or so broadly defined and with such high expectations that no one could hope to accomplish much (Axinn and Thorat, 1972).

A second comment concerns the need for extension workers to realize that there are many factors which lie outside their control and about which they can do very little. This is a recognition of several factors: that villages are not closed systems isolated from the changes affecting villages elsewhere, or society as a whole; that extension workers have limited power and authority; and that agricultural improvement is a process requiring the elimination of numerous constraints. The extension worker may have the ability to introduce and demonstrate the use of new farm technologies. But, if the constraints include limited labour during peak periods, difficulties with transportation, export policy or price incentives, there is probably very little the extension worker can do to alter these. The best one can hope for is a better understanding among extension workers of the change process within the limits of what is possible. In many cases this exceeds what extension workers are currently accomplishing.

A main point in this chapter is that extension workers should try to look at the total context of farmers' situations while avoiding as much as possible preconceived notions as to what exists. Before introducing new technologies into a rural community it is essential to have qualitative information. Because there is a large amount of such information that might be collected in a particular agricultural setting, a choice must be made as to what is relevant and important. One might focus on some of the more obvious major constraints farmers perceive and experience and the environmental factors affecting agricultural decision-making. The following questions represent ways of getting at some of this information.

1. What is the nature of the total cropping system; what crops are grown and in what sequence?

2. Is hired or family labour available and used? What labour contributions are made by men, women, and children? On which crops?

3. How are decisions made regarding time allocated to various crops and the ways in which they are grown?

4. What are the farm household's food demands and food consumption habits? Are staples adequate and available? Are there any festive requirements?

5. With what types of village and farm groups are farmers linked?

6. What patterns of influence and authority exist at the village/community level? Who holds high status? On what criteria is status based?

7. How is information, particularly that pertaining to agriculture, disseminated through the village community? What links do farmers have with external sources of information?

8. Who, or what types of individuals, serves as opinion leaders and models for farmers?

9. How are farmers' decisions affected by the decisions of others, by household needs, goals, and resources, by recently experienced events, perceived alternatives, constraints, and the various simplifying procedures farmers employ ("routines," "rules-of-thumb," rituals)?

10. Which agricultural organizations and agencies operate at the local level? Who do they serve? How are they perceived by farmers? How effective do they seem to be? What needs are not met?

11. What marketing, credit and extension services are available? How good are they in practice? How are they viewed by farmers?

12. What are some of the likely impacts of the introduction of new and improved technologies, in light of their attributes and requirements?

Answers to these types of questions will provide a context for understanding farmers' behaviour, particularly in making agricultural decisions.

There are other broad responsibilities of the extension worker which go beyond understanding the local culture and agricultural situation. They must, first of all, establish a change relationship with farmers. If workers are to have any influence on the farmers' innovation decisions, they must be viewed by farmers as competent, sympathetic and as having something useful to disseminate. Empathy can be developed by encouraging discussion of farmers' problems with farmers and understanding production constraints from their point of view. The competence of extension workers, and the perception farmers have of them will be influenced by the training that they have had in the social and economic spheres, their actual experiences with farming, and their interaction with the institutes and research agencies responsible for generating new or adapted technologies. Another major role of the extension worker is to assess farmers' needs, both with respect to the types of technologies that will fit into their farming scheme and the skill levels and information needed to promote successful transfer of the appropriate technology. In many cases the worker has to get farmers to develop a need for change (Rogers, 1969). A third major role of extension is that of a liaison between the farmer and the ministry, agency or research organization promoting change. This linking role has often been downplayed by extension agencies, yet it gets at the core of what knowledge transfer involves. It is essential, if workers are to translate ideas into action, for them to establish ties with the sources of knowledge. The problem of "having nothing to extend" is related in part to the insulation of workers from input suppliers as well as from research reports, technical information and results from trials.

Extension workers need to realize that such linkages are a part of their role, and extension administrators need to realize that workers require training in the use of scientific information. Finally, extension workers' roles involve stabilizing change by providing reinforcing messages to farmers, following through with new information, monitoring use of new technologies and assisting with the solution of problems that arise.

An extension worker's role in technological change calls for a unique blend of skills and talents. He or she must have an understanding of the culture into which technologies are introduced, the ability to understand and empathize with farmers' situations, the ability to diagnose problems and come up with solutions, a willingness to interact with other parts of the farming information and delivery system and the competence to understand, modify and apply scientific and technical information. The multidisciplinary role of an extension worker with these characteristics makes him or her a "change agent" or an agricultural and rural development agent.

REFERENCES CITED

Allen, F. R. (1971). Socio-cultural dynamics. New York: Macmillan.

Arnon, I. (1981). Modernization of agriculture in developing countries. New York: John Wiley.

Axinn, G. H. & Thorat, S. (1972). Modernizing world agriculture. New York: Praeger.

Barlett, P. E. (1982). Agricultural choice and change. New Brunswick, N.J.: Rutgers University Press.

Barry, I. L. (1971). Preliminary note on an agricultural innovation in Mando chiefdom, Sierra Leone. Human Organization, 30, 73-78.

Cartwright, D. (1972). Achieving change in people: Some applications of group dynamics theory. In G. Zaltman, P, Kotler & I. Kaufman (Eds.), Creating social change (pp. 74-80). New York: Holt, Rinehart, and Winston.

Coleman, J. S. (1970). Conflicting theories of social change. American Behavioral Scientist, 14, 633-650.

Cowan, L. G. (1970). The political and administrative setting for rural development. In F. S. Arkhurst (ed.), Africa in the seventies and eighties (pp. 87-129). New York: Praeger.

Coward, E. W., Jr. (1973). Authority innovation-decisions in rural society: Expanding the concept. Paper presented at the Rural Sociological Society Meetings, College Park, Maryland.

Cusack, D. F. (Ed.) (1983). Agroclimate information for development: Reviving the green revolution. Boulder, Colo.: Westview Press.

Das, R. (1981). Appropriate technology. New York: Vantage Press.

Feldman, D. (1970). An assessment of alternative policy strategies in the agricultural development of Tanzania and their application to tobacco farming in Iringa. East African Journal of Rural Development, 3(2), 1-29.

Foster, G. M. (1973). Traditional societies and technological change. 2nd ed. New York: Harper and Row.

Frankel, F. R. (1969). India's new strategy of agricultural development. Journal of Asian Studies, 28, 693-710.

Hirschman, A. O. (1965). Obstacles to development: A classification and quasi-vanishing act. Economic Development and Cultural Change, 13, 385-393.

International Women's Tribune Center. (1979). Women and food. Newsletter (International Women's Tribune Center), no. 10.

Johnston, B. F. & Clark, W. C. (1982). Redesigning rural development. Baltimore, Md.: Johns Hopkins University Press.

Kahl J. (1968). The measurement of modernism. Austin, Texas: University of Texas Press.

Ladejinsky, W. (1970). Ironies of India's green revolution. Foreign Affairs, 48, 158-168.

Leagans, J. P. (1971). Extension education and modernization. In J. P. Leagans & C. P. Loomis (Eds.), Behavioral change in agriculture (pp. 101-147). Ithaca, N.Y.: Cornell University Press.

Messerschmidt, D. A. (1981). Nogor and other traditional forms of cooperation in Nepal: Significance for development. Human Organization, 40, 40-47.

Mickelwait, D. R., Riegelman, M. A., & Sweet, C. F. (1976). Women in rural development. Boulder, Colo.: Westview Press.

Millikan, M. F. & Hapgood, D. (Eds.) (1967). No easy harvest: The dilemma of agriculture in underdeveloped countries. Boston, Mass.: Little, Brown.

Mosher, A. T. (1971). Agricultural development. In J. P. Leagans & C. P. Loomis (Eds.), Behavioral change in agriculture (pp. 12-26). Ithaca, N. Y.: Cornell University Press.

Mosher, A. T. (1976). Thinking about rural development. New York: Agricultural Development Council.

Niehoff, A. H. (1969). Planned change in agrarian countries. Alexandria, Va.: Human Resources Research Organization.

Patel, S. J. (1974). The technological dependence of developing countries. The Journal of Modern African Studies, 12, 1-18.

Rogers, E. M. (1969). Modernization among peasants. New York: Holt, Rinehart and Winston.

Sanders, I. T. (1977). Rural society. Englewood Cliffs, N.J.: Prentice-Hall.

Theisenhusen, W. C. (1974). What changing technology implies for agrarian reform. Land Economics, 40, 35-49.

Theodorson, G. A. & Theodorson, A. G. (1969). Modern dictionary of sociology. New York: Thomas Y. Crowell.

Uchendu, V. C. (1970). The impact of changing agricultural technology on African land tenure. Journal of Developing Areas, 4, 477-486.

Watson, G. (1969). Resistance to change. In G. Watson (Ed.), Concepts for social change (pp. 10-25). Washington, D.C.: National Training Laboratories, NEA.

Wharton, C. R., Jr. (1969). The green revolution: Cornucopia or pandora's box? Foreign Affairs, 47, 464-476.

Williams, C. E. (1982). The effect of technological innovation among rural women in Nigeria: A case study of 'GARI' processing in selected villages of Bendel State, Nigeria. Journal of Rural Development, 5, 247-257.

Chapter 5
Extension Communication and the Adoption Process

F. C. Fliegel

The primary purpose of this chapter is to sketch out the rudiments of what is known with some degree of certainty about the process by which modern agricultural technology spreads through a rural population and is ultimately adopted by some or all of the farmers who can put it to use. Within that broad purpose there will be special attention devoted to the role of extension workers at all stages in the process being described.

THE DECISION-MAKING ENVIRONMENT OF FARMERS

Any discussion of communication between extension workers and farmers must begin with some understanding of the context in which farmers live, operate their farms, and make day-to-day decisions. Such an understanding cannot be fully conveyed here because there is enormous diversity among people and their settings. Some of the major types of such diversity can be specified, however, and Figure 5.1 represents an attempt to do just that.

The major actor in the agricultural production drama is, of course, the individual farmer. Male or female, young or old, more or less well-educated, each farmer is ultimately an unique individual with a host of characteristics that may well affect how information is received, processed and either used or not used in the production process. If the farmer is central to the production process, and therefore centrally positioned in Figure 5.1, the second category of essential elements is that of physical resources, at lower left in the figure. Land resources are obviously variable in quality, though explicit information about such variability, crucial to decision-making, may or may not be available to decision-makers. The relationship of the farmer to the land, as owner, tenant or labourer, sets limits on the range of production options that are either possible or desirable. Variability in climatic conditions, the amount and temporal distribution of rainfall, the amount of sunlight, and so forth are also determinants of the options available to decision-makers.

Once the individual farmer and his or her resource base have been described, our level of knowledge about critical elments in the farmers' decision-making context becomes less certain. As a matter of fact, it can be argued that historically our approaches to agricultural development have tended to treat the farm-and-farmer complex as a closed system, relatively independent of any larger context. As agriculture has shifted from a resource-based to a science-based industry (Ruttan, 1983), however, it has become increasingly apparent that the farm-farmer system is necessarily involved in a host of linkages with a broad social and economic environment.

Some of the major institutions which constitute the necessary infrastructural environment for a science-based agriculture are listed to the right of Figure 5.1. One of the fundamental correlations of a science-based agriculture is that it involves an increase in the capital intensity of production. Chemical fertilizers, pesticides and high-yielding varieties of seeds, head the list of products that must be purchased from off-farm sources, and that means not only an array of linkages with input suppliers but also a much more prominent role for linkages with financial institutions, particularly as sources of credit.

If one thinks of agriculture as resource-based, with land and labour as the major elements, and if one further narrows one's conceptualization to a subsistence agriculture, it is not entirely unreasonable to view the farm-farmer complex as a closed system. Exchanges of surplus product and labour link that system to its environs, but the decision-making arena is essentially local. If, on the other hand, one takes a broad view of agricultural development as involving the emergence of a science-based industry, then one's view of the production process necessarily broadens. The farmer's decision-making context must be redefined to include linkages with input-supply organizations, output-marketing institutions, and, by no means least important, the basic and adult educational

Figure 5.1   The decision-making environment of the farmer

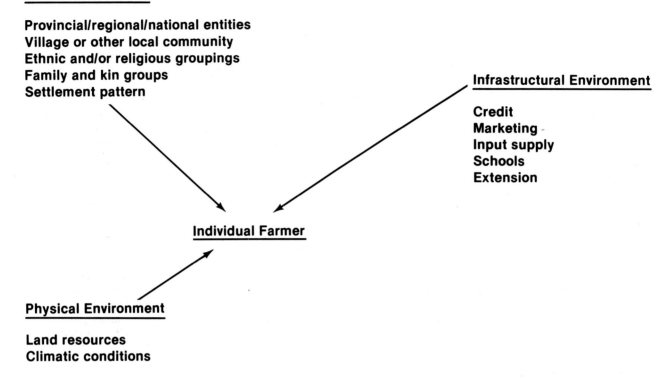

Social Environment

Provincial/regional/national entities
Village or other local community
Ethnic and/or religious groupings
Family and kin groups
Settlement pattern

Infrastructural Environment

Credit
Marketing
Input supply
Schools
Extension

Individual Farmer

Physical Environment

Land resources
Climatic conditions

institutions which provide information and information-processing capability for the central figure, the decision-maker. Extension plays a key role not only in terms of education and provision of information but also as a catalyst in establishing and strengthening the linkages between farmers and the assorted institutions that reinforce, undergird a science-based agricultural industry.

Least understood as part of the farmer's decision-making environment is the category labelled Social environment in the upper left-hand corner of Figure 5.1. The items listed are familiar enough--family, kin group, religious groupings and the local community, but perhaps by virtue of familiarity they tend to be taken for granted, even ignored. To the extent that agriculture is being transformed into a science-based industry, it is being geared to bring about increases in human and physical resource productivity. That increased productivity, in turn, is intended to enhance human welfare. It is surely easy to avoid definitions of what we mean by welfare or well-being, and we often do, because we have reasons to believe that it is highly variable. The point is that changes in agriculture are a means to a range of ends. Unless those ends are specifically taken into account the farmer's decisions which implement and change can hardly be understood.

The social fabric in which decision-makers are elements is a critical base for information about the ends to be served by production decisions. Effective extension communication not only requires substantial knowledge of the complex means-ends linkages that affect decisions, but can also capitalize on the information-transfer capabilities of that social fabric and minimize the negative consequences of factors that impede information flow. Social ties form networks. Networks can be utilized for information transmission, and the privileges and obligations of network membership can directly affect decisions. These are all familiar themes, but two points deserve stress in discussions of communication and decision-making. First, firm knowledge of the social environments of farmers in Third World settings is scant. And second, explicit consideration of family, kin, community and other social ties as part of the open system which comprises science-based agriculture is still in its infancy. In absence of firm knowledge there is little choice but to proceed with caution in bringing information to farm audiences. Reasons for acting or failing to do so are not often directly observable, but they can be discerned if the observer is open minded and willing to learn.

BASIC COMMUNICATION PROCESSES

It is useful to conceptualize communication processes in terms of the S-M-C-R model depicted in Figure 5.2. The letters, in order, stand for <u>Sender</u> or source, <u>Message</u>, <u>Channel</u>, and <u>Receiver</u>. The imagery of the electronic mass media which is conveyed by those terms is not inappropriate, but communication among human beings is by no means as simple as the mass media imagery might imply.

For purposes of explication one can use the extension worker as a prime example of a sender, the source of some communications. The extension worker as a teacher must know his or her audience, as was indicated in the preceding section. An extension worker must, of course, rely on others for information to initiate communication with a farm audience. One could trace such a chain of origins for information back almost endlessly, but for practical purposes it is useful to view extension personnel not only as one of many initiators of communication with farmers but also, in a tactical sense, as key initiators (senders) in the development process.

Figure 5.2   Elements of the communication process

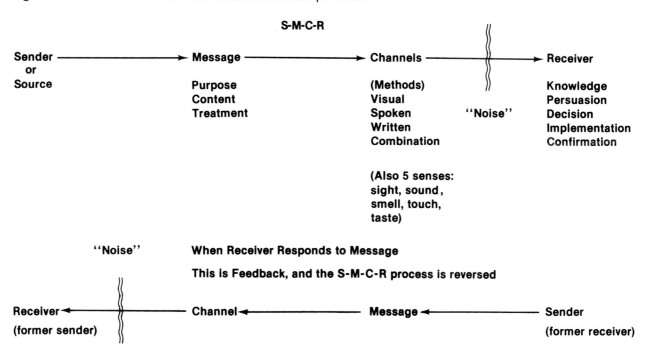

Note: From <u>Agricultural extension: A reference manual</u> by A. H. Maunder, 1973, Rome: Food and Agriculture Organization.

Referring again to Figure 5.2, the message prepared by an extension worker must be clear as to its purpose. Objectives must be specified, the content of the message must be relevant to the audience and directly linked to the intent or purpose of the communication. In addition, the treatment of the message must be such as to be intelligible to the intended audience. Complex ideas are not easily encoded in such a way that an intended audience can, in turn, decode and derive the information contained in a message. The training of extension workers is a critical factor here. Preparation of a message which can be understood by an audience requires a considerable depth of understanding of the content of the message. Such depth of understanding ideally includes practical experience with the implementation of ideas involved in the message, and also assumes considerable knowledge of how particular message elements fit into the aggregate agricultural production process of farmer clients.

Channels of communication are the various methods available to any communicator in reaching an audience with a message. Written communication has obvious limitations in those Third World settings where literacy levels are low, but cannot be rejected out-of-hand in view of the considerable evidence that print messages are read to non-literates in areas of low literacy (see, e.g., Deutschman, 1963). Direct, face-to-face interaction via the spoken word is preferable in that it allows for questions to be raised and, in general, two-way communication to be easily and successfully accomplished (see Feedback in the lower portion of the figure). Face-to-face interaction is expensive, however, in that extension workers are commonly expected to serve rather large farmer audiences. It is for that reason that mass media methods, radio, and more recently television, have come into increasingly wider use to reach audience with the spoken word. Visual means of communication include slides, films and television, plus the many variants of field demonstrations which are probably the most effective method of communication available to extension personnel. To be effective, result demonstrations require the use of both visual and spoken communication, and can easily benefit from the use of written material as well; a combination of methods, in other words, is the ideal.

One of least appreciated contrasts between farmers in industrialized settings and less developed settings is that there is an enormous amount of redundancy in both messages and channel usage in industrial societies. Farmers are exposed to similar information from a variety of senders in both the public and private sectors. Those senders use a range of channels to reach audiences and consciously use message repetition to make an impact on their audiences. In Third World settings it is not uncommon for extension workers to be almost literally the only sources of information about modern production technology. That places an awesome burden on the individuals involved, and emphasizes the need to use combinations of methods, and variations in messages, to reach farm audiences effectively.

The receiver of main interest here is the farmer. Several items are listed below the receiver heading in Figure 5.2, which essentially describe the desired impact of a message on a farmer. The terms listed are intended to specify the farmer's mental and physical responses, evoked by effective communication. They can be thought of as stages in the process of adoption of improved agricultural technology (Rogers, 1983), which are the preferred outcomes of the communication process. In the first instance it is necessary to convey information and consolidate that into knowledge. The new knowledge is intended to persuade, and to result in a decision to at least try out a new idea or practice. The communicator's task is not over, however, for farmers' information needs are most acute at the point of implementation or adoption of a new idea. Finally, farmers who adopt a new idea continue to seek information about the merits of their adoption decision, to assess whether the intended improvement performs as expected.

From the communicator's point of view it is clear that confirmation of the farmer's decision is the desired goal, but it should be equally clear that farmers may well not proceed from step to step as outlined in theory in Figure 5.2. They may opt against trials, for example, or reverse a decision if expectations are not confirmed. Farmers also may or may not proceed through the stages of decision-making in the precise manner outlined; for example, any one stage might be left out. Part of the utility of conceptualizing the decision-making process in terms of stages is that the stages imply different kinds of information needs and different messages. This, therefore implies a continuing process of communication if success is to be achieved. Effective communication is not a once-only matter but a goal-oriented and continuing process.

The bottom portion of Figure 5.2 illustrates another element of effective communication, the process of information feedback which, ideally at least, makes the communications' process two-way rather than one-way only. If the farmer is viewed as the receiver as in the preceding discussion, then he or she must also be given the opportunity to function as sender, with the extension worker, in this case, as receiver. In the absence of any reactions from the farmer (feedback), it is virtually impossible to gauge the appropriateness of the message content, or channel selection, for example, in the implementation of an information campaign.

All too often the feedback process is simply allowed to happen and is treated as a relatively passive aspect of the communication process, with major attention being given to the extension worker as initiator of the important messages in the process. Figure 5.3, below, can serve to illustrate the active pursuit of information feedback. In the discussion which follows, researchers, extension personnel, and farmers are each, in turn, senders and receivers of messages. In other words, the knowledge-transfer process is viewed from a broader perspective.

AN EXAMPLE OF TECHNOLOGY TRANSFER AND FEEDBACK

Figure 5.3 illustrates a small part of the results of an aggressive and systematic technology introduction programme in the Chiang Mai valley of northern Thailand (Kellogg, 1977). Part of the purpose of the programme was to develop technologies for better two-season crop combinations and, if possible, for three-season combinations. Research and extension personnel worked together, in close interaction with a select group of small-holder farmers. Technology packages were developed for the major crops grown in the area, rice being the most important. Co-operating farmers were asked to try out the packages, under the direct supervision of research and extension workers from Chiang Mai University. Detailed records were kept on costs of purchased inputs, labour requirements, yields, and the prices obtained for the product. Identical information was obtained from a more-or-less matched group of nonco-operating farmers who used conventional technology in producing the same crops. In short, a systematic effort was made to obtain comparative information on the performance of conventional and improved production technology as experienced by the farmers themselves. Feedback was sought with two goals in mind (a) research and extension workers needed information on any problems encountered by farmers in using the improved technology, and (b) extension workers needed information on the performance of the relatively successful technologies to formulate educational campaigns for eventual wider diffusion of those technologies.

Figure 5.3 presents what might be called the "bottom line" comparison of the major crop in the study area, which was rice in the rainy season. When all costs were considered, including those for land and labour, how well did the new technology package perform? Net returns in baht (the local currency) per rai (the local land measure) clearly favoured the new technology, with the experimental group averaging 710 baht/rai as against 430 baht/rai for the farmers using conventional technology. The higher average net-returns attributable to the new technology package (in this case significantly higher in a statistical sense) are an obvious element in an extension message directed toward more widespread adoption. A second and more subtle message can also be derived from the figure. It illustrates the value of a deliberate quest for performance information (feedback). Note, in the upper portion of the figure, that a minority of farmers using conventional technology experienced negative returns; they lost money in an accounting sense. That is, they may well have recovered cash outlays for seed and so on, but their returns on land and labour inputs were low. None of the farmers trying out the new technology had net earnings of less than 200 baht per rai. The second message is that the new technology package is less risky, and that's an important piece of information in the context of smallholder farming.

Still a third message can be deduced from the information given in Figure 5.3. It again illustrates the importance of actively seeking feedback information. If one were to superimpose the distribution of net-return results of the control group of farmers on the distribution of results for those using the new technology package, it would be apparent that some of the farmers using conventional technology did as well as or better than the average for those using the new technology. Conversely, some in the experimental group of farmers had net returns lower than the average for the control group. The relevant question here is: what can farmers directly observe among their neighbours? The answer is that the comparative performance message will be mixed; the "demonstration effect" will not unequivocally favour the new technology because there is a decided overlap in the distribution of results obtained from the two technologies. Some farmers will note that they have done as well with their conventional techniques as their neighbours did with the new technology; others will note superior performance for the new technology. What should the farmer as interested observer conclude?

Erasmus (1961) made a useful distinction between casual observation and technical observation. What the farmer can directly (casually) observe, with reference to the above example, is a mixed bag of results. Technically, the average across a range of performances clearly favours the new package of technology. But averages cannot be observed directly. Systematic information gathering, not necessarily on a large scale, can fill in the broader picture and provide the basis for a fairly unequivocal message about the superior performance of the new technology. The role of the extension worker in formulating and transmitting such an unequivocal message is extremely important.

Before leaving the topic of feedback, it is important to stress that information feedback from farmers to extension workers is only part of the process. The designers of the specific technology, agricultural scientists, must also be involved in the flow of information, both as senders and receivers. Some technologies are more effective than others; some must be redesigned to enhance their effectiveness. While the importance of informa-

Figure 5.3   Frequency distribution of farmers' net returns from rainy season paddy, Chiang Mai Valley, Thailand, 1976

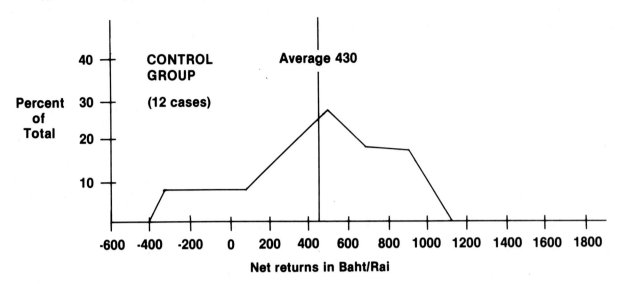

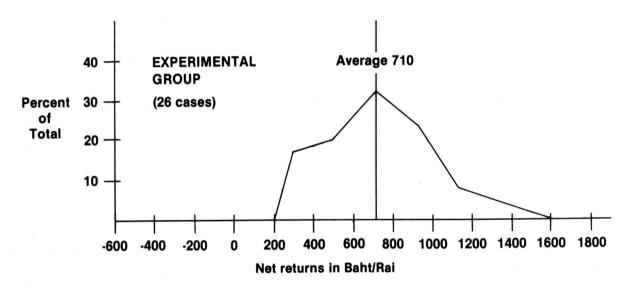

Note: From   Farm level testing of cropping systems: An economic analysis of multiple cropping project experience (Agricultural economics research report no. 2) by E. D. Kellogg, 1977, Chiang Mai, Thailand: Faculty of Agriculture, Chiang Mai University.

tion feedback from extension field workers to researchers is often noted, it is not clear that the feedback process works as it should.   In a recent study in the United States, Busch and colleagues asked over 1400 agricultural scientists, represented all of the agricultural sciences, to rank the criteria they used in selecting problems for research. "Enjoy doing this kind of work" was given the highest rank; "Importance to society" came in second; "Funding" was ranked ninth; and "Feedback from extension personnel" was ranked twentieth out of twenty-one possible criteria.   The study concluded that "Scientists' perceptions of the importance of research to society are based on the scientists' commonsense assessments of societal needs...[they] are not formulated through systematic observation or formal feedback through extension; rather they are often projected onto the larger society" (Busch, Lacy, and Sachs, 1983, p. 199).   Such conclusions are unsettling.   They indicate that a passive view of feedback, that will happen more or less automatically, is not enough.

THE ADOPTION PROCESS

The farmer's decision for or against adoption of science-based production technology was described as a mental process, consisting of several stages, in the discussion involving Figure 5.2. The objective of extension communication, as outlined in that figure, is to provide firm knowledge on which action can be based, to persuade the farmer to make a decision to try the new technology, to provide the information necessary for actual implementation, and to provide the information needed by the farmer to assess the results of that decision, and hopefully to confirm the decision.

It is a fact that people, farmers in the context of this discussion, do not all accept a new idea at the same time. However desirable it might be to have an entire population decide on some new course of action immediately upon pronouncement of the new idea, that does not occur. As a matter of fact, most new ideas come and go without causing much of a stir. Some new ideas are perceived as having sufficient merit that at least some people accept them and try them out. Those few trials may lead to more widespread acceptance if others perceive the new idea to have merit. This fact of gradual acceptance has led to an inference that people differ in their willingness to accept new ideas. In other words, it has been inferred that a psychological trait exists which might be called innovativeness. Whether the trait is described as innovativeness, venturesomeness or a willingness to take risks is immaterial.

The categorization illustrated in Figure 5.4 follows the conventions of statistics, dividing the total curve into segments based on distance from the central point in the distribution. Dividing lines might well have been drawn at other points on the curve, but it is useful to accept some standard classification in order to begin to give meaning to such category labels as are used here.

Figure 5.4 represents a widely used characterization of farmers on the basis of differences in an underlying (not directly observable) psychological trait which can be called innovativeness. Based on observations of farmers' behaviour (earliness or lateness of adoption), it is possible to classify farmers as possessing more or less of that trait. Those few who are first to try out a new idea are called Innovators at the left of the figure. If the new idea survives for an appreciable length of time and is accepted by more than the first few, one can identify a second category of farmers, here called Early Adopters. Then, if the new idea continues to spead, the bulk of farmers who ultimately accept the new idea can be classified as Early and Late Majority, depending on the time (relatively early or late) at which they make the decision to adopt. Finally, some minority of farmers accept the idea very late, and are conventionally called Laggards.

Several points can be made with respect to the types of adopters represented in Figure 5.4. Firstly, given the fact that some farmers are very early in adopting a new idea, and some are very late, with the majority in between, it is possible to use that factual information to describe adoption behaviour in terms of the familiar bell-shaped, or normal curve. The important point here is that, if adoption behaviour is observed over time, it can be seen to follow a certain pattern, and is predictable. Secondly, given the predictability of adoption behaviour, it is useful to characterize individuals who make the decision to adopt at different points in time. Classification with respect to some dimension (taxonomy) is an early step toward gaining scientific understanding. Thirdly, the particular category labels chosen, and the dividing lines between categories, have gained acceptance as being useful, but are not to be interpreted as representing sharp differences between types of people.

The central question with respect to Figure 5.4 relates to the possibility of identifying distinguishing characteristics of farmers who can be classified as early or late in adoption; do they differ in other ways besides time of adoption? And the answer is "Yes," although one must always remember that the differences are not absolute. There is a great deal of evidence (Rogers, 1983) to show that innovators tend to be relatively young, better educated and "better off." They tend to have more land and other physical resources at their disposal. They tend to have more contact with farm-related organizations such as co-operatives, and to have more contact with a range of information sources. Conversely, laggards tend to rank at the opposite extreme on each of the characteristics just listed, and the other adopter categories rank between the two extremes.

Innovators, early adopters, laggards, and so on differ systematically in a variety of ways. The practical significance of these systematic differences is that extension workers can approach a population of farmers and begin to differentiate that clientele with respect

Figure 5.4   Adopters categorized with respect to earliness or lateness in adoption

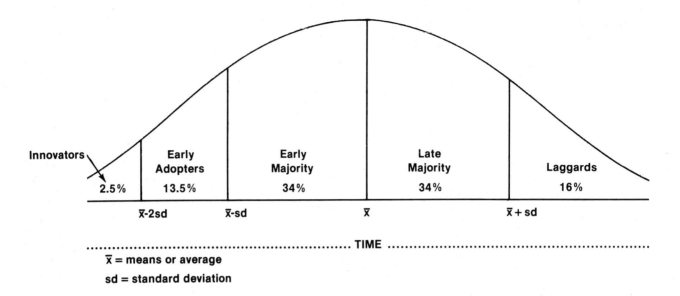

Note: From <u>Diffusion of innovations</u> by E. M. Rogers, 1983, New York: The Free Press.

to their likely receptiveness to information about new technology. A strategy for an information campaign can be worked out with respect to such questions as; Who should be approached first? What type of client may require longer term effort?, or, Which channels of communication should be most useful early in the campaign, or later in the campaign? With reference to women farmers, the problem may well be one of differential access to co-operatives where they exist. Many other questions could be listed, but the immediate point is to indicate a basis for differentiating among audience members and planning an information programme to serve the various segments of that audience.

One might stress that an obvious characteristic on which audiences might be differentiated is gender. The research on farmers as early or late adopters has rarely considered gender, however. This research gap has been tackled only in recent years, as, for example, in Fortmann's (1977) study of Tanzanian Ujamaa villages in which it was shown that women farmers were using improved farming techniques.

The socio-economic factors related to adoption of new technologies include both the personal characteristics of farmers mentioned above and the characteristics of the context in which farmers act. Efforts to explain adoption behaviour have, in the past, tended to focus on innovativeness as an individual trait, and on other personal characteristics that tend to coincide with differences in innovativeness. In recent years there has been a realization that the focus on individual characteristics may be too limiting. Galjart (1971), from his studies of small-holder farmers in Brazil, was among the earliest to suggest that in addition to unwillingness to adopt innovations, one should raise questions about inability to adopt. The fact that a farmer is a tenant, for example, may mean that he or she is not free to make certain decisions. Information about improved technology may be more appropriately directed to the landlord in such cases. Similarly, an individual may not be a member of a supply or credit co-operative simply because no such institution exists in the area. Again, messages to persuade the farmer may be of little use because the support mechanisms which facilitate implementation are not in place. Development strategy, in such instances, ought to include efforts to develop service institutions.

The point at issue here is that an information campaign directed towards persuading farmers to take certain action may fail, at least partially, because of farmers' inability to implement decisions to adopt the recommended technology. A good illustration of this type of situation is described by Havens and Flinn (1975) on the introduction of an improved coffee variety in Colombia. Those smallholder producers who could not obtain credit that would have permitted them to plant the new variety, and to wait three years for the new plantings to come into production, were left behind. Innovativeness in this case hinged on farmers' differential ability to forego income on at least part of their land while new plantings matured and/or to obtain credit to tide them over. The laggards were not necessarily unwilling to adopt the new variety, but were unable to implement their adoption decision. The example given points to a need for a broad view of the context in which farmers are expected to make decisions. External constraints on adoption, such as a lack of credit or a lack of labour at crucial points in the production process, may be the limiting factors, and argue for a development strategy that goes behond an assumed unwillingness to take action on the part of the farmer.

THE DIFFUSION PROCESS

Diffusion of innovations refers to the spread of those innovations through a population, and is simply the result of a host of individual adoption decisions. If individual adoption decisions are, to an extent, predictable, then the larger diffusion process is also predictable. It follows a pattern, and that element of predictability has substantial implications for action programmes and for extension educational campaigns. Figure 5.5 presents a graphic view of the diffusion process. It relies on the same basic information as that employed in Figure 5.4, but presents it in terms of the cumulative percentages of farmers making adoption decisions over time, resulting in the characteristics S curve shown in Figure 5.5. With the passage of time, the few innovators, at lower left in the figure, are joined by others until only a few remain, the laggards.

The specific information on which Figure 5.5 is based is taken from a relatively old study of diffusion in the United States (Fliegel and Kivlin, 1962). In this case, dairy farmers in Pennsylvania were the target population. The particular innovations shown in the figure (use of the mechanical hay baler and keeping farm records) are typical of the technologies concerned at that time in Pennsylvania. Mechanization has had a key role in the transformation of agriculture in many of the industrialized nations, particularly in North America; there, labour, rather than land availability, has been a serious constraint in improving agricultural productivity (Berardi, 1981; Council for Agricultural Science and Technology, 1983). In that context, the investment in labour-saving machinery has been an important element in a general increase in the capital intensity of agriculture, and that carries with it a need for formal farm accounts. The latter element, keeping accurate farm accounts, is a universal element in the agricultural transformation. Labour-displacing machinery, on the other hand, is a more location-specific type of technology, depending ultimately on the price of labour relative to capital. In less developed nations the most common measure of productivity is based on land, the scarce resource, rather than labour.

The generic patterns illustrated in Figure 5.5 start slowly, with a few innovators who try a new technology for a season or two. The speed of the process increases as others are able to observe results, and as interaction between innovators and others take place. It is important to note that the "break" in the diffusion curve, that is, the point at which diffusion tends to progress at a more rapid rate after a slow start, results from a social process of interaction among farmers. Existing communication networks and ties between people are central to this. Extension workers can influence the point at which upward inflection of the curve occurs, and can influence the steepness of the upward slope, by providing the knowledge for decisions to be based on. Extension communication is powerfully reinforced by the informal communication which takes place among farmers on a day-to-day basis, however. It is also the case that informal communication about performance would have to be consistent with persuasive messages from extension if the diffusion rate is to increase. That is, unless casual observation of results supports the persuasive message, rapid diffusion is unlikely to occur.

The two diffusion curves shown in Figure 5.5 differ substantially in the steepness of the upward slope in the curves. In either case the time span involved is a matter of several decades. However, it can be seen that the hay baler became more widely diffused in a shorter time than did the use of a farm account book. The innovations shown in the figure were chosen in part to illustrate the considerable variability in rates of diffusion of different innovations, even within the same area during the same period. From a communications perspective, the challenge is to provide the knowledge which will speed up the rate of adoption, assuming that performance in the field warrants a continuation of the persuasion process.

Another characteristic of both curves shown in Figure 5.5 is that they tend to level off after some years. As more farmers adopt, the potential for further adoption

Figure 5.5   Cumulative percentages of farmers adopting two improved practices over a 17-year period, Susquehanna County, Pennsylvania

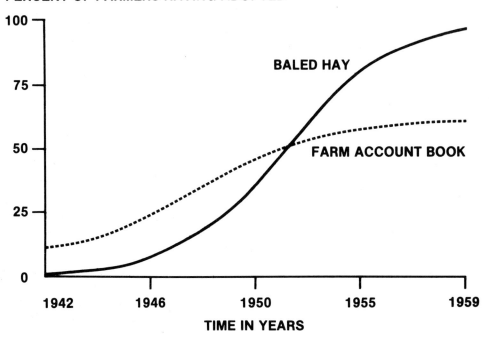

Note: From   Differences among improved farm practices as related to rates of adoption (Bulletin 691) by F. C. Fliegel and J. E. Kivlin, 1962, University Park, Penn.:   Agricultural Experiment Station, Pennsylvania State University.

decreases; thus, the rate of diffusion slows down over time.   Such a slowing down may occur well short of total audience penetration, however, and therein lies a challenge.   As a matter of fact, slow-moving diffusion processes are the major challenge to extension.   It has been argued, and not entirely without merit, that extension programmes are not needed when radically better technology is introduced to farmers, who will seek it out and put it to use (the input supply then becomes the major impediment to rapid diffusion).   Two sets of facts argue against that point of view, however.   Firstly, radically improved technologies are not always available.   The general case is one of incremental improvement over time, with limits depending on the state of scientific knowledge in a given area.   Secondly, agricultural innovations can be described as separate, discrete entities, but their potential typically lies in their appropriate utilization in combination with a range of other items of technology, conventional and/or newly introduced.   Innovations achieve their potential in packages, in other words, and that places a heavy burden on management.   When one speaks of implementation in the adoption process, one is speaking of a complex learning process, of learning to manage a range of resources in an optimally productive way.   It is at that point that the need for extension efforts is greatest and the extension communication task most complex.   It is at that point, also, that the need for feedback from farmers is most critical, if mistakes are to be avoided.

SUMMARY: SOME PERSPECTIVES ON THE ADOPTION/DIFFUSION PROCESS

Adoption was viewed as a decision-making process in the above discussion, with the timing of the decision to adopt any changes influenced by an underlying psychological trait called innovativeness.   Some of the characteristics of innovative people are that they tend to be younger, better-educated, more involved in organizations of various kinds, and have more productive resources under their control. There is substantial research to document that perspective.   A communication strategy which follows from that perspective is that potential innovators can be sought out early in an information campaign and persuaded to try new technologies.   The broader diffusion process is then enhanced by the early action of innovative people; others can observe what has been achieved, and obtain information from the innovators as well as from extension personnel and other formal information sources.

The most widely documented fact about innovators and early adopters is that they tend, on average, to be higher in socio-economic status than those who wait longer before adopting innovations. For that reason, the communication strategy which attempts to capitalize on innovativeness is sometimes called a trickle down strategy. Innovators are sometimes leaders in a formal sense, leaders of organizations, for example. This is not always the case, however. Whether they are leaders in a formal or informal sense, their actions can be and are observed, and others are influenced by those observations. Extension workers can capitalize on those early adoption experiences and speed up the diffusion process.

One challenge to the adoption/diffusion process as described above was also noted in the discussion. It was noted that some farmers may be inclined to be innovative, that is psychologically predisposed to accept innovations quickly, but unable to act because of situational constraints. Insecure land tenure, lack of access to credit, and lack of access to markets are some of the factors which can effectively make it impossible for a farmer to adopt certain technologies. An extreme but obvious example would be the farmer with insecure rights to land faced with a decision to invest in a tube-well. A communications and broader development strategy which follows from the above puts stress on removal of key external constraints, in conjunction with educational programmes beamed to the farmer.

Farmers' inability to adopt innovations can stem from a variety of causes, but lack of resources, inability to qualify for credit, and, in general, poverty, loom large in developing countries. Still another view of the development process as a whole argues that the resource-poor farmer is not only unable to capitalize on the benefits of improved technology but is also placed at a further competitive disadvantage by those who can. The possibility that the "rich get richer and the poor get poorer" as a result of technological change in agriculture is an extraordinarily broad question, and cannot be adequately addressed here. There is evidence that capital intensive technologies tend to benefit primarily those farmers with most resources, thus increasing economic disparities. The case of credit for bringing a new variety of coffee trees into production in Colombia was cited earlier (Havens and Flinn, 1975). Other examples involving irrigation wells, tractors, and so on, could be cited. There is also evidence to suggest that use of new seed varieties and other less capital intensive technologies do not increase existing income disparities among farmers (see Swenson, 1976; Shingi, Fliegel & Kivlin, 1981). The broad issue remains controversial and requires further work.

An alternative view of development strategy places central emphasis on lack of resources as an inhibitor of farmers' adoption decisions, and also emphasizes the possibility of increasing economic disparities as a result of technological change. Some adherents of that view argue that many technologies designed for agriculture are basically inappropriate for a large segment of the clientele, and that the solution lies in the design of appropriate technology (see Morrison, et al., 1984, for a general view). The communications strategy which follows from that view of development is basically one "from the bottom up." That is, if technology which is appropriate for the resource poor farmer can be designed, adoption and diffusion can proceed from that type of farmer to the aggregate rather than the other way around (Rogers, 1983).

Whether one of the approaches sketched out above is better than another is not at issue here and in any case cannot be proven given our current state of knowledge. Different communication strategies follow from the different approaches, but the principles of communication remain the same. Audiences can be differentiated in various ways, but it is always advisable to know the audiences and to select channels and design message content in terms of particular audiences. Similarly, it makes sense to take into account the farmer's context for action, that is, whether that farmer is high or low in socio-economic status. And finally, the active pursuit of feedback information, from farmer to extension worker and from both to the designer of technology, not only enhances technology transfer in the short run but also lays the groundwork for redesign of technologies which may prove to be inappropriate.

REFERENCES CITED

Berardi, G. M. (1981). Socio-economic consequences of agricultural mechanization in the United States. Rural Sociology, 46, 483-504.

Busch, L., Lacy, W. B. & Sachs, C. (1983). Perceived criteria for research problem choice in the agricultural sciences. Social Forces, 62, 190-200.

Council for Agricultural Science and Technology (CAST) (1983). Agricultural mechanization: Physical and societal effects and implications for policy development (Report no. 96). Ames, Iowa: CAST.

Deutschman, P. J. (1963). The mass media in an underdeveloped village. Journalism Quarterly, 40, 27-35.

Erasmus, C. J. (1961). Man takes control: Cultural development and American aid. Minneapolis, Minn.: University of Minnesota Press.

Fliegel, F. C. & Kivlin, J. E. (1962). Differences among improved farm practices as related to rates of adoption (Bulletin 691). University Park, Penn.: Agricultural Experiment Station, Pennsylvania State University.

Fortmann, L. (1977). Women and Tanzanian agricultural development (Economic Research Bureau paper 77.4). Dar es Salaam, Tanzania: Economic Research Bureau.

Galjart, B. (1971). Rural development and sociological concepts: A critique. Rural Sociology, 36, 31-41.

Havens, A. E. & Flinn, W. L. (1975). Green revolution technology and community development: The limits of action programs. Economic Development and Cultural Change, 23, 469-481.

Kellogg, E. D. (1977). Farm level testing of cropping systems: An economic analysis of Multiple Cropping Project experience (Agricultural economics research report no. 2). Chiang Mai, Thailand: Faculty of Agriculture, Chiang Mai University.

Maunder, A. H. (1973). Agricultural extension: A reference manual. Abridged edition. Rome: Food and Agriculture Organization.

Morrison, D. E., Lodwick, D. G., Harris, C. G., & Stommel, M. (in press). Appropriate technology as a conceptual framework for social impact assessment. In H. Schwarzweller (Ed.), Research on rural sociology and rural development. Vol. I. Greenwich, Conn.: JAI Press.

Rogers, E. M. (1983). Diffusion of innovations. 3rd ed. New York: The Free Press.

Ruttan, V. W. (1983). The global agricultural support system. Science, 222 (4619), 11.

Shingi, P. M., Fliegel, F. C. & Kivlin, J. E. (1981). Agricultural technology and the issue of unequal distribution of rewards: An Indian case study. Rural Sociology, 46, 430-445.

Swenson, C. G. (1976). The distribution of benefits from increased rice production in Thanjavur District, South India. Indian Journal of Agricultural Economics, 31, 1-12.

Chapter 6
Extension Strategies for Technology Utilization

B. E. Swanson N. Röling, and J. Jiggins

The purpose of this chapter is to formulate a strategy for designing and implementing an extension programme aimed at technology transfer and utilization. In doing so, the objective will be to integrate some of the main themes from earlier chapters to develop a realistic extension strategy that will result in broad-based technology utilization. The main themes to be covered are (a) specifying extension objectives, (b) identifying client categories, (c) formulating extension strategies, and (d) selecting appropriate extension methods. The focus of this chapter is at the operational level of a provincial or district extension programme, but the concepts should also apply to the level of the individual extension worker.

Before discussing these four main themes, it is useful to review extension's role in a functioning technology development, transfer and utilization system. Most people would agree that extension should be involved in a two-way process of transmiting problem solving information to farmers and information on farmer problems back to agricultural research (see Figure 6.1). However, it proves difficult to translate this theoretical conviction into actual practice. Therefore, the technology transfer function is frequently stressed, with little or no concern with extension's role in farmer feedback.

Figure 6.1 A simple conception of a technology development, transfer and utilization system

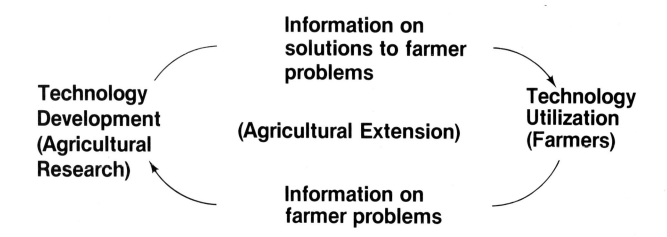

Note: Adapted from <u>Planning for innovation through dissemination and utilization of knowledge</u> by R. G. Havelock, 1976, Ann Arbor, Mich.: University of Michigan.

An extension service that is to function as part of an interdependent technology development, transfer and utilization system must achieve a two-way flow of information. Therefore, strengthening extension is not just a process of training and deploying more extension workers, providing better technical information or improving the teaching skills of extension workers; rather, it is a process of strengthening the whole system. For example, in cases where field extension workers are poorly trained, it may be overly optimistic to expect them to be able to clearly identify and then articulate farmer problems back to researchers. An alternative approach, depicted in Figure 6.2, might be to have agricultural researchers become directly involved in identifying farmer problems and then working to solve them through a farming systems research approach (see Chapter 3). Under these circumstances, potential solutions to farmer problems (which result from farming systems research) could be considered by a technical committee involving farmers, researchers and extension specialists (as well as representatives from agri-service firms or agencies and agricultural banks) to formulate technical recommendations that would be subsequently disseminated by extension and utilized by farmers.

Figure 6.2   An alternative conception of a technology development, transfer and utilization system

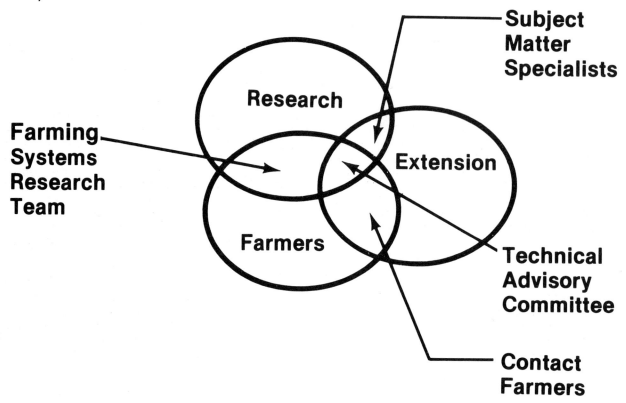

Thus, improving the flow of information about farmer problems either directly to research, or indirectly through extension, and assisting farmers to improve their organizational and leadership skills so they can effectively articulate their problems and needs are essential features of an effective technology development, transfer and utilization system. Regardless of how these functions are organized in a country they must be present if a national system is to increase farm production and income. It is within the context of this overall conceptual framework that the four main themes of this chapter are discussed.

SPECIFYING EXTENSION OBJECTIVES TO ACHIEVE
AGRICULTURAL DEVELOPMENT GOALS

It is important to clarify and align extension objectives at the policy level so they reflect the overall agricultural development goals of the country concerned. Then, it is possible to develop an overall extension strategy and select extension programmes and activities that are consistent with these policy objectives. Failure to do so may result in policy conflicts between different agricultural agencies, inefficient use or waste of resources, and undesired or unintended consequences. The following section examines different agricultural policy and development goals, as well as different types of technology. The objective of this discussion is to provide a context for, and to sharpen the focus on different extension strategies.

Agricultural Policy and Development Goals

As pointed out in Chapter 2, government policy-makers who establish agricultural and food policy for a country can have a major impact on the agricultural development

90

process. If a nation wants to develop a productive agricultural sector, it is very important that agricultural policy be supportive of, and consistent with, agricultural development goals. Furthermore, there should be congruency and continuity between these agricultural development goals and agricultural extension objectives. Finally, these policies, goals, and objectives should reflect the resources and realities of the agricultural sector in general, and the majority of agricultural producers in particular. If agricultural policy is not consistent with agricultural development goals and vice versa, then it will be very difficult for agricultural extension to operate effectively.

For example, many governments pursue a policy of providing cheap food for urban consumers (Todaro, and Stilkind, 1981). This policy reflects the relative political power of urban consumers, as contrasted with a dispersed rural population. In this case, farm prices are frequently held at artificially low levels. Under these conditions it will be difficult, if not impossible, to increase agricultural production and to achieve broad-based agricultural development. In some cases, the agricultural sector stagnates, and production actually decreases. Therefore, no matter how effective extension may be, it will be very difficult to disseminate new agricultural technology, because there is no incentive for farmers to adopt more productive practices.

Another policy alternative is where a government wants to increase national agricultural production rapidly, but has little concern with equity issues, such as broad-based agricultural development. In this case, the government may formulate policies and direct new resources to large production units, such as government-operated farms or to larger farmers to increase national food production. Under these conditions, extension workers may also focus their activities toward the medium- and large-scale farmers who can have a more rapid impact on total national production. This alternative, however, would soon result in a dual-structured agricultural economy, where a relatively small number of large-scale commercial (or government) farms would produce most of the food for the urban population, while the large number of small farmers would remain at the subsistence level and not be brought into the development process.

Another alternative is where a government is committed to increasing agricultural production as well as broad-based agricultural development. In this case, the goal would be to improve the livelihoods for the bulk of the rural population by increasing farm productivity and incomes. This policy implies that there should be increased investment in the agricultural sector, both in those institutions that serve farmers and obtaining new resources (especially credit and inputs) for the majority of farmers who wish to increase their producitivity and incomes. In this case, extension would need sufficient numbers of trained personnel to work with the large number of small farmers found in most developing countries. This alternative would also imply the need for new technology that is well-suited or appropriate to small farmer conditions.

Alternative Types of Agricultural Technology

The types of agricultural technology that are developed for and adopted by farmers in a country can also have a direct impact on the agricultural development process, and affect the structure of the agricultural economy. For example, if a country develops and promotes labour-saving technology, such as larger scale mechanization, this approach would generally favour large-scale production units. This approach might be appropriate in a sparsely populated area where labour is the important limiting factor (or scarce resource). However, in areas where small-scale farming is the predominant unit of production, labour-saving technology can be socially disruptive and can contribute to an increase in rural-urban migration.

In the United States, labour-saving technology has continued to play an important role in shaping the structure of the agricultural sector. The larger, more progressive farmers have captured the benefits of this new technology and have become progressively larger and more efficient. However, in the United States, there were urban jobs in the industrial sector to absorb the large number of small farmers who couldn't compete and were forced out of farming. While the transition may have been difficult, there were employment alternatives for those people who left production agriculture. But the situation in densely populated developing countries is quite different, and large-scale agricultural mechanization would be very inappropriate to promote broad-based agricultural development.

On the other hand land-saving technology, such as more intensive farming systems, which may involve multiple-cropping techniques, will tend to favour smaller producers.

These systems, because they absorb the excess labour of farm families or households in a productive manner and because they generally involve many manual operations, tend to be very appropriate for small farm agriculture. At the same time, they are generally less suitable for larger farmers, unless there is an ample supply of inexpensive farm labour available.

New varieties of crops that are grown in mono-culture, such as wheat and rice, are generally considered to be scale-neutral technology. Both large and small farmers can adopt these new varieties equally well and reap similar benefits. The rapid spread of the high-yielding wheat and rice varieties clearly demonstrated that neither farm size nor land tenure conditions were important constraining factors with respect to adoption or productivity (Ruttan and Binswanger, 1978 and Luning, 1981). Therefore, small farmers will adopt appropriate technology, if it is economically and technically superior for their farming systems, just as rapidly as the larger, so-called "progressive" farmers.

Reconsidering the Basic Extension Strategy

The prevailing extension strategy employed in most countries is built upon diffusion theory. In operational terms, this means that extension workers concentrate their attention on the early adopter category of farmers, and new technology is expected to trickle down through the early majority, late majority and, finally, to the laggards (see Chapter 5). This strategy might be acceptable if government policies and institutions did not favour larger farmers and if the technology was scale-neutral. However, these conditions are seldom present.

In reflecting back on the adoption behaviour of small farmers, it appears that different agricultural policies and different types of agricultural technology have had a major, if not a determining impact on the adoption behaviour of different categories of farmers. For example, in explaining the different categories of adopters (such as innovators and early adopters) rural sociologists largely focused on the social and psychological characteristics of farmers. The characteristics of specific types of technology were included in these analyses as a variable, but the technology itself was a constant; in other words, alternative types of technology were generally not considered in the analysis as a variable in terms of the different adopter categories.

In addition, the resource endowment of different groups of farmers continues to be an important factor affecting adoption behaviour. In interpreting studies of the adoption behaviour of small, subsistence farmers, it now appears that many of them were more concerned with minimizing risk rather than maximizing profit. However, given their limited resource base, their personal and/or family goals, and what they considered to be an acceptable level of risk, they generally behave in a rational way. However, because their traditional technology is not competitive with certain types of new technology, many continue to be forced out of farming. However, based on the Ruttan and Binswanger findings cited above, it appears quite probable that if these small, subsistence farmers are given technology that is appropriate for their circumstances, they will, in fact, adopt this new technology and increase their productivity and incomes.

What this premise suggests is that given appropriate types of improved technology and supportive agricultural policies, agricultural extension can simultaneously pursue the goals of increasing national agricultural production as well as broad-based agricultural development. However, to do so, both research and extension must be strongly committed to pursuing a balanced approach that focuses on both increased production and equity considerations.

If, on the other hand, research and extension focus on increasing agricultural production, particularly by increasing the output of large-scale farmers or units, they are likely to contribute to more acute rural poverty, hunger and/or forced rural-urban migration of small farmers. This situation can have very serious and undesirable consequences for the farm families and households affected. In fact, rural areas may have to act as "holding grounds" for millions of small-farm households for decades to come. A policy oriented only to overall national production and efficiency, which favours labour-saving technology, might make this latter alternative impossible.

The need for broad-based agricultural development was clearly reflected in one of the primary recommendations of the World Conference on Agrarian Reform and Rural Development (WCARRD) in 1979 that called for "growth with equity." Therefore, to achieve broad-based agricultural development, extension must focus on the technological needs of all categories of farmers, not just a single category of large, progressive farmers, who are already high access farmers (differential access will be discussed in the next section). The goal of extension should be to serve the needs of all farmers, but especially the small low access farmers. To do so will result in more broad-based technology utilization that will be a much more efficient and effective use of the vast human resources found in the rural areas of most developing countries.

IDENTIFYING UTILIZER CATEGORIES

The need for broad-based agricultural development has shifted attention to a more careful consideration of extension strategies and methods that are "appropriate" for small famers, including women, young people, and other disadvantaged or ethnic minority categories in a community. This awareness explicitly recognizes that there are different socio-economic categories in rural communities, and that their circumstances are sufficiently different to mean that appropriate technology must be developed and adapted by agricultural research and then targeted, through appropriate extension programmes, to reach these major categories. Therefore, instead of working with selected progressive farmers in a community, and assuming that technology will trickle down, it is important that each extension worker understand both the agricultural conditions in his or her area of assignment, as well as the major socio-economic categories of farmers. Only by taking both of these factors into consideration can new technology be more effectively developed and targeted to these different categories and, thereby, reduce some of the more negative and disruptive social consequences of technological change.

Criteria for Specifying Target Categories and/or Zones

Agro-ecological Zones

Classifying a target area, such as a district or province, into major agro-ecological zones is the first criterion that should be used to differentiate a target area, as most crop technology is, to a certain degree, "location specific." Thus, certain varieties, crops, and/or farming systems will be well-suited for one agro-ecological zone, but, because of a variety of factors, such as rainfall, soil type and slope, they may not be suited to a nearby zone. Therefore, the target area must be mapped to show the major agro-ecological zones, so that new technology can be adapted to and then verified for each major zone (see Chapter 3). This classification of target areas should be carried down to the village level, where each extension worker would know if there are different agro-ecological zones in his or her area of assignment.

Differential Access to Resources

As pointed out above, it is now clearly recognized that there are a variety of different socio-economic factors, in addition to socio-psychological factors, that affect the adoption behaviour of farmers. Therefore, rather than grouping farmers as being progressive or traditional (on the basis of socio-psychological criteria), Röling (1982, p. 95-96) suggests that a more accurate categorization would be high access and low access farmers. These terms explicitly recognize that farmers have differential access to land, water, labour, inputs, markets, capital and information.

"It has been a persistent error of extension workers across the globe to explain differences caused by differential access as if they were caused by psychological differences. This error allows extension workers to use diffusion strategies in situations where they are totally unwarranted. In rural communities which are very differentiated in terms of access, as indeed most rural communities are nowadays, it is nonsensical to expect innovations to diffuse from high access farmers to low access ones. In general, innovations will only diffuse within groups of people who are homogeneous in access" (Röling, 1982, pp. 95-96).

It is obvious that while a large number of factors will affect the diffusion of new technology, to be effective, extension workers must work with rather homogeneous categories of farmers. Some of the broad socio-economic factors that should be considered are farmer access to

1. land - size of holding: small, medium, large.
    - type of tenure: owner-operator, renter/share-cropper;

2.   water - irrigated, non-irrigated;

3.   labour - family, hired (cost and availability), communal or customary;

4.   inputs -availability of improved seeds, agricultural chemicals, fertilizers, and so forth; cost/unit;

5.   markets - location, availability of storage and transport;

6.   capital - sources and cost of credit, type of collateral needed, and ease of obtaining credit;

7.   information - availability of extension service (worker/farmer ratio); appropriateness of the technology, and

8.   influence - ability to affect technology development, transfer to be appropriate to user needs (i.e., user control, claim making capacity).

Differentiation by Gender

It is now widely recognized that women play an important role in both agricultural production and marketing, as well as in food processing, storage and preparation. For example, see Table 6.1 for recent estimates of women's participation in the agricultural labour force. The role of women in food production varies considerably between countries, ethnic groups, and even agro-ecological zones.

Generally, women tend to concentrate on food crops rather than cash crops, and/or they may specialize in certain farm practices, such as planting, weeding, and harvesting, rather than land clearing, preparation and bunding. However, there are numerous examples where women play key or predominant roles in cash crop or commodity production (for example, cotton in Malawi, coffee in Colombia, milk and silk in India). Also, they are often responsible for the post-harvest processing and storage of agricultural

Table 6.1
Sex Composition of the Agricultural Labour Force in 32 Countries, Organized by Geographic Regions

	No. of Countries	Females as percent of agricultural labour force
Sub-Saharan Africa	11	47.2
North Africa, Middle East	6	25.2
South, South-east Asia	5	40.2
Central, South America	8	19.0
Caribbean	2	54.0
Total	32	35.6

Note: From   Follow-up to WCARRD: The role of women in agricultural production   by FAO, Committee on Agriculture, 1982, Rome: Food and Agricultural Organization of the United Nations.

products, as well as feeding and caring for livestock. They generally use hand tools, and seldom have access to mechanized equipment, such as small tillers or tractors. It is unlikely that they will have direct access to inputs and credit, either because most farm income, especially from cash crops, is generally controlled by the male head of household, or because certain institutional factors deny women access to inputs and credit. Furthermore, women have little or no access to extension services because most countries have mainly male extension workers, and women's contribution to agricultural production is not fully appreciated. In some countries, social customs limit or prohibit the contact of non-family males with women.

Inasmuch as women tend to grow specific crops (especially food/subsistence crops) and/or have specific cropping patterns or farming systems, and they are frequently limited in their access to resources, (including extension), it is essential that the technology and related needs of this important group of food producers be addressed by those institutions serving the agricultural sector (especially research and extension), so women farmers will not be further disenfranchised from the development process. Women farmers should be considered as an equally important target for extension services as other small farmers, young farmers and other significant categories of farmers (such as different ethnic groups).

Differentiation by Age

As indicated in Chapter 1, young people (15-24 years of age) make up about 20 percent of the population in rural areas. Those who are fortunate enough to complete primary school, and especially those who complete secondary school, are those young people most likely to migrate to urban areas in search of better jobs. The World Fertility Survey found that primary education for females is one of the most powerful factors associated with higher family welfare (across both cultures and situations). On the other hand, educating boys in low income agricultural areas contributes to outmigration and the weakening of customary family obligations. The rural youth which remain behind, particularly those from small-farm families or households, are generally illiterate and have very low access to production resources. While single, these young people may work at home as unpaid family labour or be hired out as farm labourers.

This large and growing number of poorly educated young men and women are the future agricultural producers in developing nations. They are very poorly equipped to continue the transition from traditional to more science-based agricultural production. In addition, they are in a precarious position to be the mothers and fathers of the next generation, because many lack basic resources, including education, to provide adequate nutrition, health care and the other basic needs of infants and young children. There is evidence that the health and nutritional status of female infants and girls is usually worse than that of male infants and boys. Poor nutrition and health leads to lower academic performance in school. Moreover, parents are usually more ready to take girls from school when money is scarce or when extra labour is needed.

The urgency to increase overall agricultural production has frequently resulted in rural young people being another neglected clientele group of agricultural extension. In nations where rural youth programmes have been organized, they are generally small in scope and tend to focus on young people from high access families, who have received some education already.

Thus, the rural youth most in need of extension programmes are seldom served. The short-run impact of extension programmes for rural young people on national food production may be minimal, but the long-range impact could be substantial. These young people are the future, and if they learn more about improved agricultural technology and gain increased confidence and pride, they can become a major force for long-term agricultural development, including increased agricultural production and improved living conditions in rural areas.

Categorizing the Major (Homogeneous) Target Categories and/or Zones

Given the four major types of factors listed in the preceeding section--agro-ecological zones, access to resources, and differentiation based on sex and age in agricultural production--the task of extension, at each administrative level, is to focus on the major categories of farmers and agro-ecological zones. By doing so, extension workers can be more efficient and effective in promoting the utilization of technology that is appropriate to each area and category.

At present much progress is being made by international agricultural research centres (IARCs), such as the International Maize and Wheat Improvement Center (CIMMYT), the International Rice Research Institute (IRRI), as well as by other researchers, to develop feasible, low-cost, low-expertise methods of segmenting small rural administrative units into homogeneous categories or "recommendation domains" for which appropriate packages can be developed. Segmentation is done on the basis of cropping patterns, the availability of animal traction, the nature of off-farm employment, on the one hand, with, on the other, categories of farmers (such as "commercial," "emergent" and "traditional") (CIMMYT, 1979). Such approaches are called farming systems research and seek to introduce a user-oriented approach to technology development (see Chapter 3).

If agricultural personnel differentiate rural communites, areas and regions into both specific agro-ecological zones and specific categories of farmers, the focus of extension will increasingly shift to identifying and solving farmer problems through the application of appropriate technology. This shift to focus more on farmer problems will be a major improvement over the more top-down, technology-dissemination approach that has been pursued in many developing countries. Also by differentiating between major categories of farmers, extension should be more responsive to all major types of farmers in each community. As extension personnel begin to make this shift, the essential feedback system will be created. This system can articulate more effectively the needs of different categories of farmers in each agro-ecological zone to research personnel and other agricultural development agencies (such as credit, input supply and service, and marketing, as well as policy makers) to find a solution.

Identifying the Problems and Technological Needs of Farmers in the Target Area

Once the different categories of intended users or "recommendation domains" have been identified, it is necessary to carry out further analyses of these different user categories. One alternative of doing this analysis is to collect extensive production data from a representative sample of farmers in each category (see Chapter 3). Since farmers in each category are assumed to be similar, it may not be necessary to carry out the costly sample surveys (especially when expanding into new, but similar agro-ecological zones), but to use instead a case study method to assess the problems, needs, ideas and opinions of representative farmers in each target category. This activity should be carried out in close co-operation with, or under the supervision of, the research division. This division should also be involved in processing and interpreting these data.

Preparing a Map of The Service Area

One of the important activities that extension workers should undertake as soon as they are assigned to a new post is to obtain and/or prepare a map of their service area. This map should show the major agro-ecological zones, based on the predominant cropping or farming systems in each zone, as well as other factors which might affect these cropping patterns (for example, irrigated areas, or major topographical features). This exercise is an excellent tool, enabling extension personnel to better understand the agro-ecological features of their service area in a minimum amount of time.

Broad-Based Participation in Setting Extension Objectives

After the survey or case study data have been analysed and the results made available to each extension worker, it is very important that the major problems and needs that were identified be discussed with farmers in each village. The objective of these discussions is to validate the survey findings from the farmers' point of view, and then to move the discussion towards establishing programme objectives and priorities for future research and extension programmes.

It is very important that all major categories of farmers in a village participate in these discussions. In some cases, it might be necessary to meet small groups of farmers, so that specific categories, such as small farmers, women and young farmers, can all express their particular problems and priorities. By doing so, farmers can add additional information and insight about those factors that are limiting agricultural production, and thus help set extension priorities. The results of these local level meetings should be communicated to extension supervisors and subject matter specialists (SMSs), so that they can be aggregated for the major agro-ecological zones and categories of farmers in the region. Extension workers can use similar participatory methods to identify contact farmers (in a Training and Visit System), as can other paraprofessionals working in extension to develop different categories of farmers into effective constituencies within an overall technology utilization system.

Developing Appropriate Technical Recommendations

Technical recommendations for promotion in each domain must be based on the results of on-farm research. Generally, SMSs, working locally with extension workers, would be expected to co-operate closely on these on-farm research activities. The extension personnel should help locate representative farmers (including women and young farmers) for these trials, and then they should follow the progress of these research activities through the growing season.

Based on the on-farm findings, a technical committee should be formed of research personnel, representative farmers, SMSs and senior-level extension personnel, to determine what technical recommendations are to be promoted during the next growing season (see Figure 6.2). Representatives of supply agencies and/or input dealers, as well as representatives of the agricultural development bank and/or credit societies should also be invited to attend these meetings.

Recommendations that result from meetings of the technical committee should be made available to extension workers, so that they can plan a series of on-farm demonstrations in their respective service areas. The subject matter specialist should work with individual extension workers to modify general recommendations to the specific conditions existing in each target area. Extension workers in turn should locate representative farmers throughout the target area on whose farms these different demonstration plots can be placed. Agricultural extension officers who supervise extension workers must be vigilant to ensure that demonstrations are carried out on farms operated by women and other important categories of farmers. In the case of the T & V System, these result demonstrations would generally be carried out on the fields of contact farmers, including women.

Once each extension worker has planned a series of demonstrations for his or her service area, the inputs (seeds, fertilizer, pesticides, etc.) needed for these demonstrations should be made available, in advance of the growing season, by their extension supervisors, working in close cooperation with the SMS. The SMS, in turn, should provide the necessary training and technical backstopping to ensure that workers know how to lay out each type of demonstration. In the case of the T & V System, this training would be done incrementally throughout the growing season through the fortnightly training sessions conducted by the SMSs and during the alternative weeks by supervising Agricultural Extension Officers (AEOs).

At the conclusion of the growing season, these result demonstrations should be used for farmer field days, as well as to collect data for the next year's educational campaign. These findings should also be collected and aggregated by the SMS, and serve as additional input into the meetings of the Technical Committee when planning trials and demonstrations for next year's cropping season.

In addition to working with contact farmers to put out several demonstrations in each village, extension workers would also be expected to carry out other extension activities, including the dissemination of currently recommended practices. These extension activities would generally be planned as part of an overall extension campaign (introduced in the next section and discussed more fully in the next three chapters) that would use an appropriate combination of mass, group and individual extension methods. The combination of extension methods would depend in part on the type of technical recommendation(s) and/or package of practices being promoted, the types of farmers being served, and where farmers are positioned in the adoption process.

FORMULATING EXTENSION STRATEGIES FOR PROMOTING TECHNOLOGY UTILIZATION

Extension Programmes for Small Farm Households

Mosher (1966) defined ten factors of agricultural development that were classified as either essentials or accelerators. In addition to improved new technology, the other essentials to agricultural development included adequate markets, supplies and inputs, transport, and incentives. When these essentials are in place, extension (as well as four other factors) can accelerate the agricultural development process. In Chapter 1, six different models of extension were described briefly. Huizinga (1983) and Haverkort and

Röling (1984) have also classified extension into similar models, taking into account the role and type of extension included in the organizational mix.

Based on Mosher's factors of agricultural development and these different models of agricultural extension, it is possible to begin formulating an extension strategy (or strategies) that will result in the successful utilization of improved technology by small farmers. Essentially what we know or assume is that farmers will utilize new agricultural technology only if they want to, know how to, and have the capacity to do so. However, motivation and knowledge are seldom sufficient conditions, especially when working with small farmers in developing countries. In fact, as pointed out earlier and in other chapters, they generally farm in an optimal way, given their conditions and goals. In other words, tangible new opportunities are essential before changes in farming practices will occur.

Kulp (1977) elaborated on this idea when he formulated a model based on what he viewed as the six stages of agricultural development (see Table 6.2). Within this model, he differentiates farmers into three basic groups: subsistence, mixed and specialized. Although Kulp's model tends to be deterministic and top-down, it does suggest the approximate percentage of farmers who were likely to be using the various types of core services, as well as the typical type(s) of innovation that should be introduced at different levels of agricultural development.

In terms of the extension strategy reflected in this model, the first innovations to be introduced to small, subsistence farmers should deal with the staple food crop in the target area. Innovations to be introduced during the experimental stage would generally include a new, more productive variety and/or improved cultural practices (such as changes in the plant population, date of planting, improved seed bed preparation or better weeding). Once the production of the staple food crop has increased and the surplus sold, the farmer may be ready to move to the next stage and further increase his or her production through the use of purchased inputs.

The major crop improvement stage would involve purchased inputs, particularly fertilizer. As fertilizer use is initiated and/or expanded by small farmers, the resulting larger surplus should further increase farm income. At the same time, however, improved growing conditions and more intensive production frequently result in more problems, particularly weeds, and possibly new or more serious insect and/or disease problems. These problems will require the use of additional purchased inputs and/or more labour. This latter alternative could be a particular problem for women, because they are frequently expected to do the (additional) weeding.

As the production of the food staple increases in the target area, the price will generally decline. This situation should set the stage for the farmer to move into the comprehensive improvement stage, which may involve reducing the land area allocated to the food staple (since the productivity has increased substantially) and thereby increase the production of other crops. Again the improvement process with these other crops would follow essentially the same process, starting with improved varieties and cultural practices, and then further expanding production through the use of purchased inputs, including fertilizer and pest control measures.

Farmers in the comprehensive improvement stage have substantially increased the intensity of their farming systems, possibly including double- and/or multiple-cropping techniques. These changes frequently exceed the availability of family labour, particularly during peak periods. Some small equipment and/or tractor services may be required for harvesting and ploughing. In some cases, mechanical cultivation (or herbicides) may be required so the excess burden of weeding does not fall on women.

As farmers increase the intensity of their farming systems through the use of new varieties, improved cultural practices, purchased inputs, and small labour saving equipment, their management skills will also need to improve. They may then further intensify their farming systems by moving into the high value diversification stage, through the production of high-value crops and livestock for sale. The last stage of Kulp's model (capital intensification) is expected to occur as farm size increases; therefore, it's not relevant to this discussion.

At this point a serious warning should be given regarding a possible wrong interpretation of Kulp's model. He has developed this model on the basis of both empirical studies in developing countries and by analyzing the agricultural development experience in Europe and North America. The stages should be seen as illustrations of what might happen at different levels in the development process, not as predetermined stages of

Table 6.2
The Stages of Agricultural Development

Stages	Pre-development PSC	Experimental Exp	Major Crop Improvement MCI	Comprehensive Improvement CFI	High Value Diversification HVD	Capital Intensification CIn
Coverage of Core Services*						
Extension	0%	1-20%	40- 80%	100%	100%	100%
Seed Distribution	0%	10-80%	70-100%	100%	100%	100%
Co-op Supply--Agricultural Chemicals	0%	1- 5%	30- 80%	70-100%	100%	100%
Co-op Marketing--Single Crop	0%	1- 5%	20- 80%	100%	100%	100%
Other Crops	0%	0%	1- 10%	40- 80%	100%	100%
Credit--Short Term	0%	1- 5%	10- 60%	100%	100%	100%
Credit--Medium Term	0%	0%	1- 5%	30- 80%	100%	100%
Typical Catalytic Innovations		Improved Seed---------	Small Equipment---------		Horticulture---------	Large Equipment
		Fertilizer---------			Livestock---------	
Orientation of Farmers						
Subsistence	100%	20-90%	1- 5%	0%	0%	0%
Subsistence-Mixed	0%	5-80%	50- 80%	10- 30%	1- 5%	0%
Mixed	0%	1- 5%	5- 15%	60- 80%	80- 90%	10- 40%
Specialized	0%	0%	0%	0%	5- 10%	60- 90%
Cash Income of Farms Per Annum	$10-$30	$20-$80	$50-$150	$200-$400	$300-$1500	$1200 plus

*Percentage of farmers reached by the service:
Note: From Designing and managing basic agricultural programs by E. M. Kulp, 1977, Bloomington, IND: International Development Institute.

99

agricultural development through which all nations will pass. It is possible that developing countries will have to choose different paths of development.

For example, Kulp's model assumes a development process that will result in high rural-urban migration at the later stages. In many developing countries there is already a very high rate of migration to urban areas, as well as high levels of unemployment and under-employment in the cities. It may be essential for governments to improve the livelihoods of the masses of rural people and to keep them in the rural areas. The example illustrated by the People's Republic of China clearly demonstrates that each country must pursue a path that is appropriate to its respective resources and conditions.

Although Kulp's model can be criticized, it is still a useful tool for agricultural policy makers to use in identifying levels of agricultural development in their own countries, and to determine which types of technology (innovations), and which mixes of agricultural development essentials and facilitators would be most appropriate to introduce at each stage. For instance, if a target category, or even individual farmers, are in the experimental stage of agricultural development, it would not be advisable to introduce technology that is associated with more advanced stages and vice versa.

Benor and Harrison (1977) imply a similar conception in suggesting the types of technical recommendations to be introduced into an area where traditional agricultural practices predominate. Initially they suggest starting with low-cost, improved management practices, such as better seed-bed preparation, use of good seed (including improved varieties), improved weeding, increased plant populations and other essentially low cost inputs.

These practices should increase the incomes of farmers. As farmers gain confidence in these new practices and the extension worker, they will be more willing and financially able to try other types of technology that require purchased inputs, such as fertilizer and pesticides. Farmers will not be able to take full advantage of purchased inputs, of course, unless they are using improved management practices.

As small, subsistence farmers make the transition from using improved varieties and better cultural and management practices to the use of purchased inputs, it may be necessary to organize farmers into functional groups to increase their access to inputs, credit, and possibly marketing and other services. These self-managed farmer organizations might later take the form of co-operative and/or credit societies, depending on the key limiting factors to the provision of these different inputs and services in the local community.

One reason why farmer organizations are necessary is that the costs of dealing with large numbers of small farmers is very high (particularly increased number of staff, vehicles, and other facilities and equipment). By developing self-managed farm organizations, farmers can buy inputs in much larger quantities jointly, at lower per unit prices and acquire larger group loans from banks with lower service costs per farmer.

Farmer organizations are also necessary to increase the participation of rural people in the political processes of the country, especially in influencing agricultural policy and in articulating the needs and problems of the farm community to the agri-service agencies. In reviewing the development of agricultural extension in North American and Europe (see Chapter 1), it is difficult to imagine that an effective technology development, transfer and utilization system can emerge without farmers becoming better organized.

While extension should not become involved in managing or operating these farm organizations, it can provide the necessary leadership and management training and advice to enhance the success of these groups. These leadership and training extension activities may be beyond the capability of most field extension workers, but they can identify the groups that need assistance and then coordinate the provision of these services by extension specialists. While agricultural extension can only play a limited role in helping farmers to become organized, these farmer organizations can play a major future role in gaining additional resources for agricultural research and extension services (if these institutions are responsive to farmer needs and problems).

In summary, extension programmes must be farmer-oriented and must reflect the introduction of specific practices and new technology that will increase farm income. Furthermore, these recommendations must enable the farmer to move toward his or her own goals (and the goals of the farm family or household) and to be consistent with a farmer's position in terms of his or her level of agricultural development.

Extension Programmes for Women Farmers

Most women farmers are also small farmers. Therefore, part, if not most, of the preceeding section should also apply to women who are engaged in agricultural production. In some countries women are largely responsible for food crop production, while men primarily grow cash crops. In other cases, joint farming is carried out, and there is a particular division of labour between men and women that has been worked out over time; often men do land clearing, ploughing and bunding, while women do planting, weeding and harvesting.

Agricultural Extension Has Neglected Women Farmers

Although women make a major contribution to world food production, they seldom benefit from agricultural extension services. This is a particularly serious problem in certain countries and regions where women produce most of the food crops and/or where women are frequently the heads of household in charge of all agricultural production (due to temporary or permanent absence of the men in search of other work, or other factors). Furthermore, where women have little or no contact with extension services, then the production problems they face are seldom known to the extension feedback system; thus research cannot develop technology suited to their needs.

There are numerous reasons to explain why women have not benefited from agricultural extension. These reasons range from extension personnel scheduling meetings and demonstrations at times and/or places which are inconvenient or inaccessible to women, to gender bias in extension staffing, where women are not hired for agricultural positions, or are assigned to home economic assignments, even though their training is in agriculture. In some countries, where colonial governments concentrated research efforts in the past on export crops, there is still too little research being carried out to improve food crops. Therefore, agricultural extension has little or no improved technology to extend to women farmers who grow the traditional food crops. In other cases, technology is available, but women are not able to utilize it because extension inappropriately directs its efforts to men, and/or women are unable to obtain the credit and purchased inputs needed to utilize the new technology. These are just some of the many reasons why women farmers have been seriously neglected by agricultural extension in the past. However, these obstacles must be overcome if nations are going to expand food output.

Improving Extension Services for Women Farmers

The general extension strategies and methods included in this manual should work equally well for women farmers as they do for male farmers; therefore, new or different strategies or methods for women farmers are not needed. What is recommended is more awareness and sensitivity on the part of extension administrators, supervisors, technical specialists and field workers of the role and importance of women to food and agricultural production in each country, so that they are no longer neglected by agricultural extension workers. Some of the actions agricultural extension could take to resolve these problems are listed below.

1. Clearly identify the role of women in food and agricultural production when the major socio-economic groups of farmers are categorized in each target area. In the process, their production problems and technology needs should be clearly identified, and they should be given full opportunity to participate in setting extension objectives for the target area.

2. The extension service should be gender-sensitive when organizing extension activities so that women farmers have full and appropriate access to meetings, demonstrations, field days and other activities that can increase their farm production and incomes. In some cases, this may require separate meetings; in other cases, opportunities for joint participation of both men and women in common extension activities may be culturally acceptable, and then women should be strongly encouraged to attend.

3. Informal communication networks and channels are frequently gender-specific; therefore, extension personnel need to learn how to use these traditional networks more effectively to disseminate improved technology to women farmers. This may require exten-

sion to give more careful consideration to the message, the channel and the sender (see Chapter 5).

4. In the case of the T & V System (see Chapter 1), women farmers must be fully represented as contact farmers in each village. These women contact farmers, in turn, should be strongly encouraged to discuss the fortnightly "impact points" (technical recommendations) with other women farmers in their respective communities. These fortnightly training visits should be organized in such a way that women can attend meetings, and these visits should not conflict with their other roles in the household.

5. In some cases, it may be necessary to organize women into functional groups to increase their access to credit, inputs, and even marketing services. In these cases, they will need leadership and management training to operate these self-managed groups effectively. In addition, extension may need to work closely with regional or provincial agri-service agencies to insure that these self-managed groups or societies of women have access to sources of inputs and credit.

6. Extension may need to develop new extension training materials that are gender-sensitive and appropriate for women farmers. For example, the literacy rate among the female population in rural areas is generally much less than that for males. In addition, the placement of extension announcements, posters, and so forth should be in places that are easily accessible to both rural men and women.

7. Finally, much more attention should be given to the training and employment of women in all aspects of agricultural extension work. These efforts should not seek to create separate extension systems for men and women. Rather, the objective should be to develop an integrated system that is both gender- and cultural-sensitive, and will thus effectively serve the needs of all farmers.

Extension Programmes for Rural Youth

Rural youth, as mentioned in Chapter 1, are defined as those young people from 15 to 24 years of age who live in rural areas. During this ten-year period, these young people change quickly from boys and girls to young men and women. In the rural areas of many developing nations, most of these young people are out of school and are involved in either unpaid labour around the house and/or farm or are periodically employed in low-paying menial jobs, many in agricultural production. The employment opportunities for them are limited, and poor. They have few employable skills, and most are destined to a life of poverty.

In some countries, rural youth programmes have been organized. These programmes include "4-H" type rural youth clubs, frequently for the younger clientele, young farmer programmes that are more occupationally oriented, and a variety of full-time youth programmes, such as national youth service organizations and youth settlement schemes. This latter group of programmes generally concentrate on a particular age group (16-18) for a limited period of time; the programmes are more highly structured, and are relatively costly in comparison with voluntary rural youth programmes organized within agricultural extension.

One difficulty encountered in organizing rural youth clubs is their limited impact on the majority of rural youth. Frequently, they concentrate on the youth of high access farmers who already have an educational base to build on. Another major problem is that national extension organizations and most donor agencies tend to give low priority to rural youth programmes, because of their limited short-term impact on agricultural production. Therefore, few agricultural extension organizations commit adequate staff resources to these programmes.

The future of national agricultural development rests with today's rural youth and young farmers. The development strategy of shifting from a traditional to science-based technology of agricultural production assumes the concurrent development of the technical and managerial skills of farmers. How can this occur if little or no investment is made in this generation of rural youth?

In reviewing the historical development of agricultural extension in Europe and North America (see Chapter 1), it appears clear that the application of science to agricultural production was closely paralleled with the development of formal and/or non-formal agricultural education opportunities for rural young people. It appears very short-sighted to expect today's youth in the Third World to continue progress towards developing a strong agricultural sector and to develop positive attitudes towards agriculture and improving life in rural areas if they have little or no access to educational programmes that will increase their agricultural skills and knowledge.

Developing Effective Extension Programmes for Rural Youth

The type of rural youth or young farmer programmes that are needed and appropriate for one country may be different from the programmes needed in another country, due to socio-cultural and other factors. The objectives of these programmes, however, will tend to be similar. These objectives can be categorized under four major headings (a) leadership development, (b) citizenship, (c) personal development, and (d) career or occupational development. The relative emphasis given to each of these objectives will again vary from country to country.

Leadership development. Rural youth programmes, such as young farmer clubs, are generally nurtured by continuing, voluntary adult leadership in the community. Extension personnel play a catalytic role in getting these programmes organized and then in providing a broader organizational structure, at sub-district, district, provincial and/or national levels for programmatic input into local programmes. In addition, they provide the mechanism for communication and the sharing of ideas and successes between programmes.

Because these programmes are largely self-sustaining, leadership development becomes both a means and an end. Younger members learn leadership skills from the older, more experienced members, first by serving on committees and later by taking on leadership roles in the organization themselves. In the process, the participating rural youth learn how to organize and operate a self-managed organization. These skills will be particularly valuable when they are older and will need to organize and/or operate co-operative groups, credit societies and so forth.

Citizenship. This objective can vary from promoting national identity and consciousness to community orientation and development. Generally, most rural youth programmes will seek to encourage young people to become involved in improving their local communities. By fostering this attitude among the youth, it is expected that this community orientation and pride will continue into adulthood. This dimension of rural youth programmes generally focuses on community improvement projects that, in turn, are recognized and praised by adult leaders in the community, (the community elders, the chief or the mayor, for example). In the process of carrying out these projects, young people will increasingly identify with the local community and the need to improve community life.

Personal Development. The social interaction that occurs in a rural youth club or young farmer organization helps develop both social skills, such as talking in front of groups or working with others on a committee, to developing self-esteem and self-confidence. This dimension of rural youth organizations is extremely important, given the low status of agriculture, and rural life itself. By developing strong rural youth organizations, group solidarity will develop and members will begin to take pride in being associated with a positive force in the community, and in the nation.

The use of visible symbols, such as a common colour and type of shirt, blouse, or jacket with the name of the member and/or club written on the article of clothing, and the use of pins, medals and ribbons to recognize distinguished service or achievement, may all contribute to strengthening group solidarity and personal growth.

Career Development. This area is particularly important for an organization that serves rural youth in the 15-24 age group. During this period, these young men and women should be acquiring technical and management skills in agricultural production. By the time they leave the organization, they should be on the road to becoming established in farming.

The common method of developing these agricultural and management skills is through project work. Members are expected to carry out some agricultural activity, generally on the family farm. This might involve a small garden plot for younger members, or operating a small farm for older members. This project work should be developmental in character and should seek to help the member become established in some aspect of agricultural production or related activity.

Each meeting of the club or organization should include an educational programme dealing with some aspect of agricultural production. Members are expected to apply this new knowledge in their respective projects. Furthermore, record keeping and analysis will teach members management skills, while project competition at local fairs, which focuses on high quality (for example superior animals) and quantity (highest yield), promotes the use of improved technology. The overall effect of these projects should be to develop a positive attitude towards change, increasing agricultural production, and farm income.

To summarize, rural youth organizations appear to be essential to the long-term development of the agricultural sector and farming communities. Continuing to increase agricultural production through the use of improved technology will require an increasing level of education on the part of rural people. Until agricultural education is more widely available in rural areas, the logical alternative is that agricultural extension organizations should organize and support rural youth clubs and/or young farmer programmes. These activities must be organized in a socially acceptable manner, and should be open to both rural young men and women (either through separate or joint organizations), and to all socio-economic groups in the community. Unless this latter point is adhered to, these extension programmes will not contribute to the overall goal of "growth with equity."

In conclusion, it is the target group(s) and the unit of analysis used in measuring progress that become important in determining how agricultural extension is organized and what objectives and programmes are pursued. If the objective is to increase total food production and to provide relatively cheap food for an urban population, then extension's focus will probably continue to emphasize the larger, more progressive farmers where more rapid, short-term progress can be made. On the other hand, if the goal is to pursue broad-based agricultural development by increasing the income-generating opportunities for the mass of small farmers, including women and young people, then they must have access to new technology, inputs, credit, markets and any other factors that are appropriate to their needs for increasing their productivity and incomes.

SELECTING APPROPRIATE EXTENSION METHODS

Appropriate Methods in the Adoption Process

It is generally accepted that most farmers go through a logical, problem-solving process when considering new technology. This process can be divided into five clearly defined stages or steps known as the "adoption process" (see Figure 6.3). Those five steps are awareness, interest, evaluation, trial and adoption (or rejection). Extension personnel must understand this process regardless of whether they are planning an overall extension programme or working with an individual contact farmer.

Figure 6.3   Steps in the adoption process

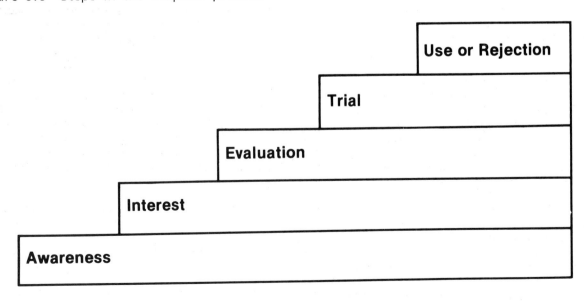

Note: From Agricultural Extension: A reference manual by A. J. Maunder, 1973, Abridged ed.   Rome: Food and Agricultural Organization of the United States.

Awareness

Awareness of a new idea or practice is the first stage in the adoption process. A farmer might hear about a new variety of wheat or maize over the radio or from a neighbour. While farmers may hear about a new type of technology through a random or unorganized diffusion process, it is extension's responsibility to communicate with farmers through several channels to create awareness of some new idea or practice. To insure wide and rapid coverage of the farm population in a district or province, mass media is the logical method to be used.

Interest

If the new practice, idea, or input sounds applicable to the farmer's situation, he or she may become interested in getting more information about it. At this stage, mass media can supply some of the needed information, but the farmer needs a source of information that can respond to his or her questioning. To be efficient and effective at the interest stage, extension workers should organize group meetings at the village level, or hold meetings on market days, or at the same time as some other on-going activity where a substantial number of farmers can learn more about the new technology and ask questions. In the case of the T & V System, the fortnightly visits of extension workers to contact farmers in each village provide an ideal setting for interested farmers to get more information about the approved practices being recommended.

Evaluation

The evaluation stage for each farmer is a process of weighing the likely advantages and disadvantages of the technology vis-à-vis his or her own situation. What does the new practice cost? Will it mean more work? How much extra yield seems likely? Will it require other changes in the farming system? At the evaluation stage, opinions of neighbours, friends, and family will probably be very important. Most farmers want tangible evidence or data about the new practice. It is at this stage that opportunities for direct observation become important. In particular, result demonstrations and farmer field days are effective extension techniques. Furthermore, it is important that these demonstrations and trials be carried out on a variety of different farms that are representative of all the major groups or categories of farmers in a local community. Individual farmers will trust or put higher value on the results and opinions of farmers in his or her own reference group.

Trial

Assuming that the risks appear manageable and the potential benefits outweigh the estimated costs, the farmer is at the trial stage of the adoption process. At this point, the farmer may need individual help or advice about how to utilize the new technology. This assistance may be provided directly by the extension worker, or possibly by a contact farmer in the case of the T & V System. Generally farmers should be encouraged to try the new technology on a small portion of their land both to gain experience using the technology and to easily observe results. Depending on the results of this trial stage, the farmer may either adopt or reject the new technology.

In considering the adoption process, the extension methods used generally move from mass media approaches at the awareness and information stages to group methods at the information and evaluation stages. The decision to try a new system or crop is an individual household or family decision and may require one-to-one assistance. In some cases, this individual advice may be given by the local extension worker, but in other cases it may be given by a neighbour, friend, or relative. In the case of the T & V System, contact farmers are expected to provide much of the individual assistance to other farmers in their community.

Since the adoption process occurs as a logical sequence of steps, it is important for the extension worker to remember that different farmers are frequently at different steps in the process. Therefore, if one farmer is at the trial stage, he or she can be very useful to other farmers who are at the evaluation stage and who want more information about the actual field experience of others.

In most countries small subsistence farmers are illiterate and are at the subsistence level of production; therefore, individualized and small group methods, particularly using on-farm demonstrations, may be the most important methods to use in disseminating new technology. Most farmers want to see the results of a new variety or input utilized under conditions similar to their own situation. This is the reason why extension workers must always attempt to work with a representative group of contact farmers in each village, so that on-farm demonstrations produce believable results.

REFERENCES

Benor, D. & Harrison, J. Q. (1977). Agricultural extension: The training and visit system. Washington, D.C.: The World Bank.

CIMMYT (1979). Demonstration of an interdisciplinary approach to planning adaptive agricultural research programmes (CIMMYT Eastern Africa economics programme report 4). Nairobi: International Maize and Wheat Improvement Center (CIMMYT).

FAO (1980a). Key principles for operational guidelines in the implementation of WCARRD Programme of Action. Rome: Food and Agriculture Organization of the United Nations.

FAO (1980b). WCARRD: A turning point for rural women. Rome: Food and Agriculture Organization of the United Nations.

FAO, Committee on Agriculture (1982). Follow-up to WCARRD: The role of women in agricultural production. Rome: Food and Agriculture Organization of the United Nations.

Havelock, R. G. (1976). Planning for innovation through dissemination and utilization of knowledge. Ann Arbor, Mich.: Institute for Social Research/ Center for Research on Utilization of Scientific Knowledge, University of Michigan.

Haverkort, A. & Röling, N. (1984, January). Six approaches to rural extension. Paper presented at the Seminar on Strategies for Agricultural Extension in the Third World, International Agricultural Centre, Wageningen, The Netherlands.

Howell, J. (1984, January). Issues, non-issues and lessons of the T & V extension system. Paper presented at the Seminar on Strategies for Agricultural Extension in the Third World, International Agricultural Centre, Wageningen, The Netherlands.

Huzinga, B. (1983). Vernaderende percepties in de voorlichting: Implicaties voor voorlichtingskundig onderzoek in de derde wereld. In H. Lamers (Ed.), In de ban van de voorlichtingskunde. Wageningen: Landbouwhogeschool.

Jiggins, J. (1983, December). Agricultural extension and training for rural women. Technical report prepared for the Expert Consultation on Women and Food Production, Food and Agriculture Organization of the United Nations, Rome.

Kulp, E. M. (1977). Designing and managing basic agricultural programs. Bloomington, Ind.: International Development Institute.

Luning, H. A. (1982). The impact of technological change on income distribution in low-income agriculture. In G. E. Jones & M. J. Rolls (Eds.), Progress in rural extension and community development. Vol. 1, Extension and relative advantage in rural development (pp. 21-42). Chichester, UK: John Wiley.

Maunder, A. H. (1973). Agricultural extension: A reference manual. Abridged ed. Rome: Food and Agriculture Organization of the United Nations.

Moris, J. R. (1983). Reforming agricultural extension and research services in Africa (Discussion paper no. 11). London: Agricultural Administration Network, Agricultural Administration Unit, Overseas Development Institute.

Mosher, A. T. (1966). Getting agriculture moving. New York: Praeger.

Röling, N. (1982). Alternative approaches in extension. In G. E. Jones & M. J. Rolls (Eds.), Progress in rural extension and community development. Vol. 1, Extension and relative advantage in rural development (pp. 87-115). Chichester, UK: John Wiley.

Ruttan, V. & Binswanger, H. P. (1978). Induced innovation and the Green Revolution. In H. P. Binswanger & V. W. Ruttan (Eds.), Induced innovation: Technology, institutions and development (pp. 358-408). Baltimore: Johns Hopkins University Press.

Todaro, M. P. & Stilkind, J. (1981). City bias and rural neglect: The dilemma of urban development. New York: The Population Council.

Chapter 7
Extension Programme Development
J. L. Compton

The planning and development of extension programmes is a continuous and interrelated series of processes. The purpose of extension is to facilitate learning and action among the members of farm families and communities, extension educators and administrators, and personnel of other service agencies and groups in order to promote agricultural production and improvements in the general quality of rural life. Extension programme planning is conducted through a systematic sequence of actions to promote the diffusion and utilization of science-based and indigenous knowledge throughout a particular geographic area for which that knowledge is relevant.

The extension educational sequence includes (a) the assessment of existing problems, needs and interests of farming communities, (b) the identification of available resources, (c) the planning of a programme, (d) the preparation of work and teaching-learning plans, (e) action to implement the plans, (f) steps to monitor and evaluate the action, learning processes and the resulting achievements (or failures) and (g) follow-up meetings to adjust on-going programme plans in light of new awareness gained through the previous processes. A major emphasis in extension is given to active participation by clientele in decisions about both the immediate and the long-range planning processes.

SOME ASSUMPTIONS IN EXTENSION PROGRAMMING

1.   Development is an endless process.

2.   Ways to solve most problems and improve the quality of life of humankind, no matter where, can be found.

3.   It is possible to select, organize and administer certain resources of knowledge, technology, personnel, the physical environment and teaching-learning methods to help people achieve a more desirable quality of life.

4.   The knowledge and skills of professionals can be meshed with the knowledge and skills of the people to find optimum solutions to development problems and issues.

5.   Change is sometimes desirable and necessary, but change for the sake of change is not always desirable. People can be helped to make wise choices in adopting new behaviour or in preserving old and cherished ones.

6.   The bases of decisions for change should not be taken lightly, but rather should be considered carefully.

7.   People will usually accept new modes of thinking and doing in favour of present ones if the new ones are perceived as offering certain advantages and having sufficient aesthetic appeal.

8.   Learning sometimes occurs best as a result of choice and deliberate effort to pursue that choice. Sometimes it occurs as the result of interaction within an environment or social climate conducive to incidental learning.

9.   It is often possible to create opportunities and supportive socio-emotional climates to enhance the learning of new attitudes and behaviour.

10. Education can be a means of empowering people to take greater control over the course of their lives.

PURPOSE OF PROGRAMME PLANNING

Agricultural extension works to promote desired changes in a geographical area over time. It usually assumes a systematic nature to planning by objectives, specifying targeted client groups, delineating an interrelated set of roles and professional specializations, utilizing a variety of methods to promote learning, identifying and mobilizing available resources, and a continuous upgrading of the competencies of extension personnel.

Objectives

Continuous involvement with the population of an area, combined with a careful monitoring of on-going extension operations, will enable an accurate identification of the interests and needs of the people and of the programme itself. Such an identification of interests, needs and available resources will make it possible for extension personnel and the people to agree upon programme objectives.

It is necessary to bear in mind that extension programmes cannot and should not try to respond to all of the interests and needs of rural people. To attempt to do so would reduce the effectiveness of an extension service in helping the people to meet those needs to which an extension education service is best equipped to respond. Extension should not establish programme objectives which fall into such improper categories as (see Figure 7.1) (a) those for which there is no knowledge base, (b) those not in the public interest, or which are of low priority, (c) those from other agencies which are more readily available to citizens, and (d) those which are services, not education.

Figure 7.1   Programme categories extension should not address

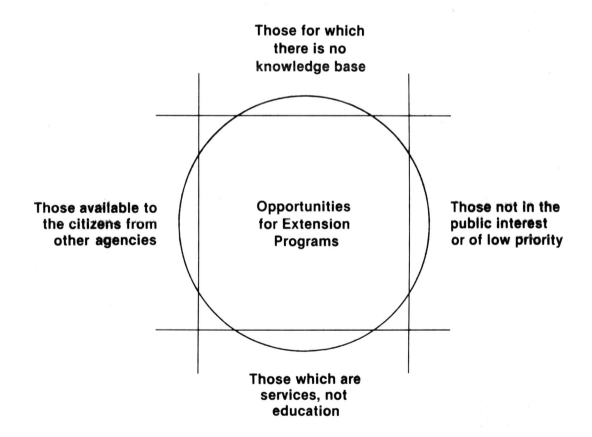

The identification of needs as one of the steps leading to specifying objectives has been characterized as the process of differentiating between "what is" and "what ought to be" (see Figure 7.2). Often the needs identified are such that considerable time and

effort would be needed to satisfy them, such as the development and application of new technology suitable for revitalizing depleted soils. Such a long-range objective usually provides the basis for guiding research and the development of new practices over a long period of time, with results observed in stages. In such cases it is more rational to envisage a sequence of objectives, the accomplishment of each providing a foundation for the pursuit of the next. In this way, research and extension activities become functional parts of the process.

Figure 7.2 Needs identification and establishing programme goals

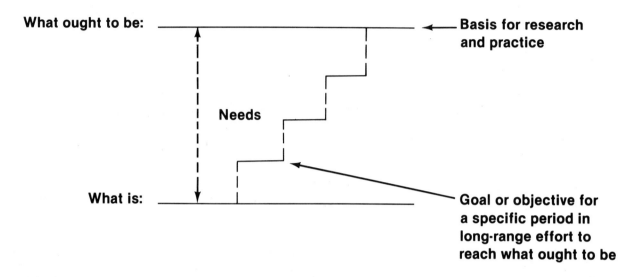

Programme planning by objectives should also be seen within the context of the mutual influences and interaction between the processes of rural research, programme evaluation, local participation, field staff management, actual programming and implementation, and plan formulation. It is wrong to think of these processes as working in a single direction and sequence from research to programme evaluation, by way of plan formulation and other processes. Programme planning is part of a dynamic system of interactive processes, as illustrated in Figure 7.3.

Targeted Client Groups

Extension programmes serve a variety of client groups, each having its own characteristics and needs. Both men and women farmers as well as rural young people, collectively and individually, constitute clientele for extension services. For example, women may be farm managers and/or workers within the context of the family farm and thus represent specific target groups of extension. These farmers may have sufficient resources and capabilities to be characterized as production agriculturalists, or they may be small farmers whose amount of land is comparatively small and whose market interaction is of a subsistence nature (the buying and selling of only a small portion of inputs and products). The landless poor, men and women, who often serve as labourers or tenant-farmers on other people's lands are also the clients of extension and usually require special services tailored to fit their circumstances.

Rural women frequently have several roles, such as farm manager, agricultural labourer, and/or homemaker. Therefore, in addition to specific information about agricultural production, such subjects as food preservation, storage and preparation for home consumption, and market exchange, home management and family health are of concern for many of them. Rural youth, as future farmers and homemakers, also represent a special client group for extension.

Extension personnel work collectively with the client groups in a three-way transfer of knowledge and information; from the base for scientific knowledge provided by research institutions to farm communities; from farm communities to research, training and administrative centres; and laterally among members of farm communities themselves.

Central to extension programming are the needs of the people. It is an extension responsibility to translate new technology or indigenous experience into information that

Figure 7.3 A dynamic systems perspective in the programming process

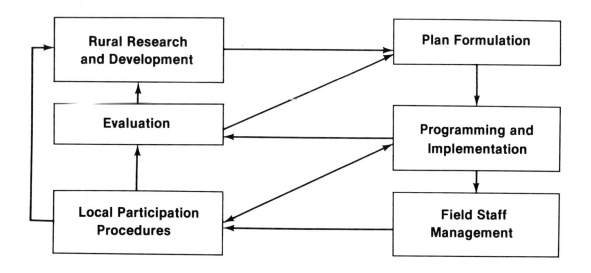

Note. From Managing rural development: Ideas and experiences for East Africa by R. Chambers, 1973, Uppsala: Scandinavian Institute of African Studies.

can be understood and applied by a large number of clients. To accomplish this, it is necessary for extension personnel to receive professional in-service education at frequent intervals, followed by enthusiastic and effective programme delivery, in order to have a positive impact on the people's needs.

STEPS IN THE PLANNING PROCESS

Extension programme planning involves the following steps which will be discussed in detail (a) an examination of the situation of the farm family and the services available to them, (b) the identification of their needs, wants and interests, (c) the determination of problems, resources and priorities, (d) the co-ordination of activities and plans with other agencies serving the farmer, (e) the planning of learning experiences, (f) the implementation of the programme, and (g) the monitoring and evaluation of the programme.

Situational Analysis

There are several situations which need to be assessed carefully by extension staff in collaboration with the farm families. The division of labour within the farming unit, based on age and sex, requires close attention. Such agro-ecological factors as crop-soil-animal relationships, and limitations of the existing farming system, need to be understood. Terrain, climate and water availability, among other factors, will often determine the agricultural potential within a given area (see Chapter 3 for more detail). It is extremely important that extension staff be concerned about the fit of recommended technology to these local conditions. Large, specialized monocrop systems with high energy and capital requirements call for quite different technological inputs from small, mixed crop-animal integrated systems with low energy and capital requirements.

The political-economic context within which extension is operating is also an important matter. Land tenure, for example, is usually a determining factor in efforts to promote improved agricultural practices and increased production. Often tenant farmers cannot be expected to plan in long-range terms. The nature of landowner-tenant relationships will also affect tenant farmers' willingness and ability to try out new recommendations. Other political-economic factors such as market forces will influence what types of agricultural inputs (seeds, fertilizers, pesticides) are available to the farmer and at what cost. Along these same lines, farmers may have limited access to credit, or access under terms which are prohibitive. It is not logical for extension to recommend practices to farmers that do not take into account the political-economic context of the farmers' situation.

The local socio-cultural context of farm families and households must also be both understood and appreciated by extension staff. Ignorance of local values and belief systems, the nature of family relationships, and the organizational structure and social interactions among various factions within a community all influence farmer behaviour. Extension staff must understand the socio-cultural system of any farming community if they are to be successful in establishing the kind of rapport necessary to have effective dialogue with local farmers' groups.

The context of the national development programme and its goals also should influence to a certain degree the nature of extension staff's interactions with farm families. At the same time it is important and necessary that extension staff maintain a sensitivity to local needs and interests; these can then be fed back for consideration in the establishment of programme goals.

The availability of other agencies providing services (health, education, technical) that are needed by the farm family should also influence the nature of extension work. The frequency of contact that extension has with farm families makes it imperative that the extension worker be aware of non-extension needs of the farm families so that information about these needs can be relayed to the proper service agency. Sometimes farm families are prevented from giving serious attention to an extension recommendation because of the distraction caused by some other problem which extension itself cannot solve. An effective extension staff will be constantly aware of all of the above contexts.

Identification of Needs, Wants, and Interests

As previously mentioned, needs assessment is an important step in the planning process. It is also important for extension staff to make clear distinctions among needs, wants and interests. Keeping in mind that it is important to maintain good rapport with farm clientele, extension staff will give as much attention to the wants and interests of farm families as they will to expressed or real needs. The subtle distinctions of meaning among these three words are critical. It is often necessary to deal with people's interests (or aspirations) before they will willingly state their wants, which must then be analyzed to provide an identification of real needs. Obviously, as previously stated, extension cannot relate to all of the interests and wants of a rural clientele nor should it attempt to do so. However, an outright rejection of those interests and wants may destroy the kind of personal relationships underlying much effective extension work. In short, extension field personnel must be willing and able to deal with the totality of farm family concerns.

Determining Problems, Resources, and Priorities

An assessment of interests, wants and needs should result in an identification of a farmers' related problems. People's inability to satisfy some lack or need, want or interest may be seen as a problem. This inability to achieve satisfaction (or solve their problem) may result from the absence of certain human or physical resources, or from lack of understanding of complex social or economic conditions, or from lack of previous experience. But the problems of any human group are usually numerous, and it is impossible to concentrate on the solution of all of them at one time. Problem-solving itself is usually a very complex process which first entails the determination of a list of priorities, to discover which problem is to be solved first and why. The solution of one problem is sometimes basic to the solution of another. In the extension education programme planning process, then, we move from situational analysis to the identification of real needs and learning needs, and the learning of knowledge and skills necessary to the solution of specific problems. These learning needs are discussed in greater detail in a later section.

Of great importance to the process of problem-solving is the initial stage of gaining a clear understanding of the nature of the problem itself. It has been noted that most people have a peculiar tendency to draw premature conclusions about the nature of a problem, and its possible solution. There is a general tendency to draw such conclusions quickly, without first gathering all, or sufficient, available information related to the problem and assessing that information carefully. A general procedural principle is to identify and list all available information, and to resist evaluating any of that information until it has all been listed. Once all available information has been so listed, decisions can be made as to the value or validity of the different items of information. The most relevant information can then be used in making decisions about how to best solve the specific problem.

112

A related technique is that of a force-field analysis of a problem situation, when an effort is made to identify those factors or forces which work against the solution of a problem. This is followed by identification of ways to decrease the restraining forces and ways to increase the driving (for solution) forces. Action steps can be identified for accomplishing this. This approach to problem solving, if carried out co-operatively by extension personnel and farmer clientele, can result in a clearer understanding by everyone of the nature of the problem and a mutual agreement as to steps to be taken to solve problems.

Co-ordinating With Other Agencies

Ignorance, poverty, disease and civic inertia are often intertwined. Programmes of community development, non-formal education, literacy, self-government, vocational training or agricultural production are needed to help rural people solve this complex of problems. But the simple multiplication and provision of a wide range of government services may cause as many problems as it solves. Failure to co-ordinate the delivery of such services to rural clientele results in waste and confusion, and a lowering of service effectiveness. Extension personnel need to co-ordinate their activities closely with those of other service agencies, and to share information about mutual clientele. Regularly scheduled information-sharing sessions need to be arranged, in addition to the pursuit of information sharing on a more informal basis. In agricultural matters alone, banks which provide credit to farmers, input suppliers, plant protection specialists, and soil, water and forest conservation services collectively need to analyse and discuss clientele problems, needs, resources and plans. Such sharing of information among the providers of such services can result in higher staff motivation levels, improved timing in the delivery of services, anticipation of bottlenecks and subsequent improvement in project performance.

Planning Learning Experiences and Action Programmes

The identification of learning needs and related learning objectives requires inputs from all who are going to be involved in, or responsible for, the implementation of the educational effort to achieve those objectives or satisfy those needs. Extension personnel have sometimes been guilty of specifying what a farmer is supposed to learn without having conferred with the farmer, particularly in the case of women farmers. Not surprisingly, the result is usually one of poor participation by the farmer in the proposed learning experience, if not outright rejection of it.

Once learning needs have been determined and objectives specified, it is then necessary to reach an agreement on the types of learning strategies and methods which will be employed. Because people have different learning styles, it is important to match them with the appropriate teaching strategy. Sometimes the learning objective may simply call for an exchange of information; at other times repeated physical manipulation of some device may be required. The desired learning may occur best within a formal context, or in an informal setting. Regardless of the teaching strategy decided upon, it becomes necessary to specify and obtain the resources needed to help carry out the learning. Instructional models, devices, expert testimony, demonstrations by experienced farmers, filmstrips and many other instructional resources may be selected based upon their appropriateness and availability.

The acquisition of knowledge and skills is usually an on-going process that calls for continuous evaluation of the achievement, and the relative effectiveness and efficiency of the facilitative instructional effort. In the planning of learning experiences, an effort should be made by extension staff to specify how the results will be evaluated and the criteria which will be used to determine the evaluative decision. A specification of indicators (criteria) should be made part of an overall educational plan.

Implementing the Programme

The extension educational process utilizes a wide variety of educational and communication methods, as described above. The centre of much extension work is the field demonstration, wherein an effort is made to demonstrate both the methods for improving production and the results obtained after recommended practices have been followed. Ideally farmers are encouraged to form themselves into dialogue groups organized around the trials, to observe and monitor the method and result demonstrations. It sometimes becomes necessary for extension to provide group leadership training for contact farmers chosen by their peers. Extension may also provide supportive communication services which provide timely and useful information for the dialogue groups. In this local educational process, it is important that extension staff encourage the emergence of both a scientific attitude and a respect for indigenous knowledge.

Monitoring and Evaluating the Programme

As planned programme activities are implemented, activities are monitored carefully by extension field staff and their supervisors. Men and women farmers are encouraged, individually and collectively, to observe programme operations and provide feedback for the extension staff. In any planned programme there are things which may go wrong, or projected activities which may be found to have been based upon false assumptions or inaccurate information. Programme adjustments then become necessary. But there has to be an attitude on the part of all participating parties that positive efforts will be made to improve the programme constantly, in the light of the emergence of new knowledge and insights.

Both the provider and the receiver of feedback must be tolerant and open-minded. Observers of field operations who provide feedback must understand that it may take some time to bring about desired programme adjustments. The receivers of such feedback must accept such information as beng offered in a positive spirit. It may be necessary for extension personnel and members of the farming community to reach a clear understanding of the desire and need for the monitoring of programme activities and the provision of feedback. They should agree that such information should have functional value; the provision of a massive amount of information alone can often be disfunctional. A guiding principle is one of seeking optimal simplicity, that is, to provide only as much information as is necessary to clarify a situation. Simplicity is sophistication.

Programme evaluation in extension usually takes two forms, formative and summative (see Chapter 13). We have discussed the formative dimension. Summative evaluation should occur at key periods during the year. Usually it involves meetings where representatives from different parts of the research, extension and farming communities come together to assess the data on the functioning and the results of the extension programme. Such summative evaluation provides information which is used to justify the continued existence of a programme, to suggest new or additional inputs, or to adjust the thrust and scope of a particular set of activities.

THE CONTENT AND LEVELS OF PROGRAMME PLANNING

The content of extension programme planning is ever-changing, related to such matters as the development or discovery of new knowledge or technology, the changing international, national, or local economic or political climate, government shifts in policy or programme support, agro-ecological changes such as pest or disease infestations or natural disaster, changes in the level of farmers' interests or awareness, or the professional readiness of extension personnel to pursue or promote a given objective. It is useful to discuss the content of programme planning from the perspective of the different levels and contexts of planning, how such planning is (or can be) conducted at each level, and how plannng at the different levels can be linked together.

Levels of Programme Planning and Different Contexts

Production agriculture as a context of extension programming differs considerably from programming to meet the needs of subsistence farm families. It is usually easier to reach and interact effectively with larger, wealthier, and often more highly motivated farmers than it is with small landholding, limited resource, low-income farmers. Previous experience or familiarity with guided experiments, perceived risk and uncertainty, knowledge, and availability of (or access to) equipment, supplies and credit, are all factors which may be different in the two groups.

In most countries, both groups of farmers and all those in between are important to the welfare of the nation as a whole. Large-scale production is sometimes pursued to feed large urban populations, or to provide goods for export (thus generating foreign exchange). But small farmers also contribute more to the national welfare than is usually acknowledged. In fact, in most countries they still provide the vast majority of foodstuffs to feed their nation. Women farmers in particular play a significant role in food crop production. Additionally, by staying on the farm rather than migrating to join the growing mass of urban unemployed, the small farm family's self-reliance does not place additional strain on national coffers already providing welfare and services for the urban poor.

Another context for programming is the distinction between agro-technical objectives and the provision of family improvement education, such as farm- and home-management training. Finally, there is a context of the degree to which extension activities should be focused solely on crop and animal production, rather than on such matters as nutrition, and post-harvest technology for processing, preservation, storage, transporting and marketing of food crops.

These issues aside, we can focus on the various levels of programme planning.

1.   The farm and farm family, in which decisions are made about labour allocation and responsibility, about goals, and the plans to achieve them.

2.   Informal learning-interest groups, within which participants share feelings and insights about new opportunities, or the value of existing practices.

3.   Men and women farm groups, and youth associations, in which problems are analyzed, plans are established to solve problems, and decisions are taken to carry out the plan.

4.   Community-based organizations, such as co-operatives, which facilitate a pooling and sharing of resources to make possible, and to achieve optimal benefits for all members; and local community government that works to integrate and secure all the interests of all members of the community.

5.   Sub-district councils, which co-ordinate the resources of several communities in order to deal with problems or opportunities requiring resources which are beyond the capability of one or a few communities to provide. VEWs and AEOs usually work with such councils.

6.   District inter-agency councils in which decisions are made about how best to correlate the services of the various agencies in order to reduce waste and confusion and increase the efficiency and impact of service delivery. District extension officers play an important role in representing the agricultural sector.

7.   Regional and/or provincial (or state) technical advisory groups, that strive to find solutions to technical agricultural problems confronting farm communities and extension personnel in their particular area. Regional and provincial extension officers, subject matter specialists, training officers, and selected farmers usually play a critical role in problem-solving and planning at this level. Their inputs are especially helpful in linking knowledge of indigenous practices, underlying knowledge and prevalent problems with the scientific knowledge and technical expertise of advisory group members from the research institutes, universities, colleges and experimental stations within the region. It is critical to recognize that women (as farmers, workers, mothers, wives and community members) have a vested interest in programme planning that attempts to affect their lives. They have valuable contributions to make at each of the different levels of the programme planning process. Their participation should be solicited, supported and expected.

How Planning is or Can Be Conducted at Each Level

1.   Farm families can be encouraged to keep records relating to home and farm management. Such records, properly kept and periodically analyzed, can yield valuable insights into worthwhile practices, and changes or improvements which are needed. Daily accounts do not usually mean very much until accumulated over a long enough period of time to make an identification of trends or patterns possible . Such information can help farm families make wiser decisions regarding their investments of time, money and energy.

2.   Learning-interest groups (LIGs), once they are organized around discernable topics (such as field crops, small animal husbandry or fruits), can differentiate among group members and identify those most interested or skilled in relation to a particular aspect of their topic (for example, for field crops there may be one member for mung

beans, one for cassava, etc.). Such identification/selection can be linked to special training for a particular member. After this training, the member would be expected to share what he or she has learned with other members of the group. They, in turn, would be expected to share any other information with the group following their own special training.

3.   Community-level farmers' associations can undertake tasks which cannot be carried out effectively by an individual or a small group. There are many examples of those tasks, including the establishment of production, marketing or consumer co-operatives, the arrangement of co-operative work schedules to accomplish some agricultural project of community-wide importance such as constructing a road, negotiating credit with a local bank (by putting up collective collateral, or providing the association's guarantee of repayment), or the representation of collective grievances, requests or approval to government agencies or authorities. Such actions usually necessitate considerable thought, discussion and planning by association members. Participation in such matters is often a valuable educational experience for association members, an experience which can increase their skill and value as citizens.

4.   Within any community there is some degree of splintering into factions or sub-groups according to the special interests of different people. This fact combines with that of the diversity of functional problems or needs (e.g., health, literacy, housing, etc.), to make it imperative that the community's chosen leaders strive to achieve some degree of functional interface between problems and programmes. Farmers in poor health themselves are poor producers, and those who cannot read or write cannot avail themselves of valuable information contained in printed documents that might lead to a cure for their illness, or a solution to an agricultural production problem. The achievement of social solidarity and integrated development often go together.

5.   Sub-district councils, comprising the popular representative leadership of the various communities in the sub-district, can often enhance or strengthen the chance of farmer associations and individual community councils' achieving of their goals. Credit, the supply of agricultural inputs, and the provision of other government services can be made more efficient, and make responsive to local initiative. Irrigation facilities and roads have sometimes been extended to farm communities if the government has been made aware of an articulated demand. This is also the level at which the greatest prospects exist for governments to realize the benefits of decentralization as government resources (block grants, services, and technical advice) are integrated with local resources (labour and materials).

6.   District level inter-agency correlation of services requires a clear understanding on the part of all agency leaders of the mandate, goals, resources, methods and limitations of every other agency, and an open and mutual respect for each agency's right to pursue its own objectives. Often this understanding and respect are missing, and it becomes necessary to arrange seminars to achieve them. Planning sessions at this level should occur at least monthly, in order to share information and resources, and evolve more realistic, and functionally integrated, plans.

7.   The technical advisory committees at the provincial and regional level should meet at least quarterly to deal with technical problem-solving, determine messages and recommendations, and plan future research and extension programme needs.

PARTICIPATION IN THE PLANNING PROCESS

The participation of rural clientele in the development, implementation and evaluation of extension programmes has been a long-standing, fundamental, guiding principle. It is a principle which represents the very essence of the nonformal, popular nature of effective extension work. And yet, there has not been a wide degree of adherence to this principle throughout the world.

People in the Planning Process

It is imperative that the clientele of rural farming communities be involved in the planning process, to assure a proper balance of programme inputs, proper timing or sequence of those inputs, and the bridging of cognitive, social, and geographic distance between more formally educated extension personnel and less formally educated clientele. All too frequently, programme inputs are determined by one agency or another without

116

the consultation of the people themselves, and this leads to an imbalance of inputs or activities. One agency may emphasize a particular programme objective which actually may be less important to the client than some other service. Often an agency will provide a service at a time ill-suited to the agricultural cycle of the rural population. Extension personnel and the members of rural farming communities are often separated by different thinking and behaviour, as well as by miles or kilometres of rough terrain. Without the proper kind and amount of popular participation by youth and adult men and women in the planning process, it is impossible to obtain an integrated programme of services, to achieve delivery of inputs synchronized with the people's readiness for them. Only through the people's participation can cognitive, social and geographic gaps be closed.

The resources needed to develop diverse and populous areas of the world are so massive that a major proportion of these resources must be mobilized from within these areas themselves. Additionally, the reception and use of delivered services is largely dependent upon demand, which is based upon the receiver's perceptions of the fit of the product or service to his or her needs. We have also found that highly centralized strategies of development decision-making have not been able to accommodate local variations, indigenous experience and knowledge, and the absolute need of people to have a chance to practice and improve their planning and management skills. People have a right to, and the need of, self-determination. It is clear that there is a relationship between participation and the development of an ensuing positive self-concept, a sense of control, and a sense of commitment and responsibility to others, which together provide a motivational basis for change and improvement.

The composite knowledge and resources of local areas throughout a country represent an untapped or under-utilized national resource. Such local knowledge and resources are likely to continue to be ignored by extension personnel whose training has led them to believe that scientific or valid knowledge is the scope solely of institutionally based scientists, and those who would be their messengers. This flaw in extension programming continues to prevail in spite of the growing number of well-documented analyses of the value and validity of indigenous knowledge. Ignoring such indigenous knowledge, whether intentional or not, further restricts the ability of extension personnel to communicate effectively with farming populations. A lack of awareness of farmer-derived knowledge-classification schemes for biological systems severely limits the ability of extension personnel to communicate effectively with local farmers.

The Role of Extension Workers

Within the context of the extension education system, it is useful to assess the complementary roles played by subject matter specialists, village extension workers, extension trainers, or district-level agricultural extension officers. One increasingly prevalent extension strategy is the Training and Visit System wherein VEWs are provided regular fortnightly training by subject matter specialists and training officers and then proceed to visit a given number of contact farmers within the farming communities on a regular basis during the following two-week period. This T & V System might be contrasted with an exchange and follow system wherein subject matter specialists and training officers provide training at residential training centres for farmers selected by their peers. The trained farmers then share what they have learned with their learning/ interest groups back in their community setting, followed up by facilitative and communicative work on the part of the VEWs. Regardless of the system or strategy, or the variety of roles played by extension personnel, it is important that all understand and appreciate the value and necessity of farmer participation in planning at all levels.

The Role of Local-Level Client Groups

The role of dialogue (learning-interest) groups must never be underestimated. Careful research on the effect of reference-group relationships on individual farmer behaviour is documented. People need time, and the confidence provided by the presence of friends and accepted peers, in order to analyse and think through some prospective change in existing behaviour patterns, whether the change be of a social or a technical nature. In any community there are a variety of age- and sex-related categories of people who usually come together in a very natural way to form groups wherein a great deal of sharing of ideas and beliefs take place. Village elders may identify and associate with each other as members of a common reference group, as may youth, women or middle-aged farmers. The compatibility of interest serves to draw these persons together, and these interests need to be recognized and related to by extension person-nel.

Often within such groups, members will differentiate among themselves those who have special skills relating to some particular phenomenon. It is usually possible for farmers to acknowledge which one of them possess special knowledge or skills (a green thumb) in relation to a particular crop or animal. Usually these are individuals who have proven their capabilities to the satisfaction of their peers. Their advice is sought and adhered to. These indigenous specialists should be recognized by extension staff, as they play an important role in learning occurring within the local community. In short, men and women farmers, as well as scientists, may be seen as valid sources of knowl-edge.

Another important role is played by farmer associations that can work to prepare and articulate collective demands to government authorities so that better decisions can be made about relevant and timely programme inputs. Extension personnel should work actively to encourage the formation and operation of such farmer associations.

Participatory Forms

In the past few decades, much extension work has relied upon the so-called pro-gressive farmer, the assumption being that helping these progressive farmers would make it easier and more efficient to demonstrate the value of certain recommended practices to the mass of farmers. It has now been shown, however, that this strategy has resulted in an increase in the gap between those progressive farmers and the others. It is now generally understood that recommendations suitable for such progressive farmers were not necessarily relevant to the situation of the remainder. Gradually we have come to realize the importance of the collective involvement of farmers in programme planning, imple-mentation and evaluation. We now realize the necessity to move away from a strategy based on dealing with the best farmers to a more appropriate strategy emphasizing the needs of the majority.

Local-interest learning groups of farmers are an important medium for trying out new ideas, seeking and obtaining clarification, or gaining the encouragement or confidence basic to taking necessary risks. Community councils are now understood as being impor-tant mechanisms for providing a more complete perspective on the complex of changes being considered by a given community at any one time. Such councils often serve as an umbrella organization for functional committees dealing with such topics as health, liveli-hood (agriculture) or education. Such functional committees guarantee the focused atten-tion of individuals interested in the particular topic area on deriving solutions to related problems. But sometimes the problems of concern to a given group exceed the boundaries of that group's community, and are common to the interests and problems faced by similar groups in other nearby communities.

Sub-district councils, comprising a network of popularly elected village or group leaders, are often needed to conduct the planning, implementation and evaluation of development or extension projects to service needs common to all the villages in an area, or which exceed the capabilities of one or several of the village communities to accomplish by themselves. In addition to these types of local participatory forms, members of the farming community should also be invited to serve on district services advisory groups and provincial, or state technical advisory groups. It is always useful to have the voice of the people represented in the deliberations normally undertaken by such groups.

Problems in Maintaining Effective Participation

It is important to understand why people drop out of groups or projects supposedly designed to service their needs. Frequently projects evolve in such a way that there is a failure to move as fast as emerging states of awareness, understanding, appreciation, readiness, and knowledge and skills of the participants, simply because someone started to make decisions for the group rather than with the group. The pressure that is some-times placed on government agencies to plan and budget ahead for one, if not two years, often defeats the intention of utilizing farmer inputs into the planning process.

There is a general failure to establish plans and budgets using broad categories and allowing local input into determining the specifics of those plans and budgets at the micro level. As a result it has often become very difficult to fit the extension planning process into the ebb and flow of the seasonal cycles of farmers' lives. But there have been many successes in extension programme planning as awareness and understanding of the elements essential to the process have increased. The trend is sure to continue.

REFERENCES

Boyle, P. G. (1981). Planning better programs. New York: McGraw-Hill.

Chambers, R. (1981) Understanding professionals: Small farmers and scientists. (Occasional paper). New York: International Agricultural Development Service.

Chambers, R. (1973). Managing rural development: Ideas and experiences for East Africa. Uppsala: Scandinavian Institute of African Studies.

Cole, J. M. & Cole, M. G. (1983). Advisory councils: A theoretical and practical guide for program planners. Englewood Cliffs, N.J.: Prentice-Hall.

Freire, P. (1973). Extension or communication. In P. Freire (Ed.), Education for critical consciousness (pp. 85-164). New York: Continuum.

Mbithi, P. M. Agricultural extension as an intervention strategy: An analysis of extension approaches. In D. K. Leonard (Ed.), Rural administration in Kenya, (pp. 76-96). Nairobi, Kenya: East African Literature Brueau.

119

Chapter 8
Planning Extension Campaigns

J. F. Evans

In this chapter attention is focused on planning a comprehensive extension campaign, in an integrated manner, that utilizes different educational and communications methods to achieve an extension objective. For example, assume that you have used the extension programme development process to identify an important problem or opportunity facing your clientele. You may have decided to place high priority during the coming year on helping growers reduce grain losses caused by insects, disease and spoilage. You may have also decided to help families in your area improve their diets for better health. Now you want to plan a communication programme to achieve the stated goals. This is the situation where you can use the campaign planning approach to advantage.

WHAT IS A CAMPAIGN?

Co-ordinated communication and educational efforts often are called campaigns. Campaigns have been defined in various ways, but basically a campaign involves co-ordinated use of different methods of communication and education aimed at focusing attention on a particular problem, and its solution, over a period of time.

A campaign may take many forms (such as the charity campaign, sales campaign, political campaign or image-type campaign), but the kind used most widely in extension is the self-help campaign. It is intended to provide information and education which people can use to improve their lives.

ADVANTAGES OF THE CAMPAIGN APPROACH

Several features make the campaign approach valuable in extension.

1. A campaign approach is the only way to handle large, complex programmes of public information and education. Haphazard communications cannot be effective when working with large and varied audiences, varied messages, and varied communication methods.

2. The campaign approach permits the use of resources (time, funds, personnel), more effectively because it helps you co-ordinate them.

3. The campaign approach is unique in the way it permits the use of combinations of methods, all directed toward the same programming objective. It can add unity to educational efforts.

4. The campaign approach produces a planned schedule of co-ordinated activities, so it helps to adjust the efforts of personnel over a period of time.

5. It can help to reach more members of the intended audience, by using a combination of communication methods.

6. It can provide more impact than a haphazard educational approach because it helps to reach audience members through multiple channels, and in a repetitive pattern that increases learning.

7. Because of its visibility as a campaign, it can help build enthusiasm and co-operation within an extension group and among organizations that might help to carry out the campaign.

WHEN TO USE A CAMPAIGN

The campaign approach is most useful (indeed vital) when the topic under consideration is important to the audience and sponsoring organization, when the education effort is complete and perhaps (not necessarily) large-scale, when a variety of communication methods will be needed, when the education programme extends over a period of time, and when resources are available to carry it out successfully. Actually, the principles of communication planning that will be discussed here can be useful in any extension programme, whether it is called a campaign or not.

AN APPROPRIATE FRAME OF MIND

The typical approach to communication planning is what we could call topic-centred. A sponsoring organization identifies some subject matter that it wants to communicate, then selects media and methods, prepares the information, and distributes it. For example, a planner's thinking might go something like this. "Our researchers have developed a new tillage practice that reduces soil erosion. Let's arrange some meetings to promote the practice among growers."

This frame of mind often leads to disappointment and failure, for the basic reason that it fails to grasp the essence of the communication process. Communication is not the same as information. Communication has an interactive, sharing quality (reflected in the root word, commune). So, in the example cited, effective efforts to communicate a new tillage practice would focus as directly on growers as on the practice itself.

In fact, one might think in terms of the growers and the extension organization as problem-solvers who interact in their searches for solutions. Both are actively searching. Your job, as communication planner, is to understand and relate to the problem-solving process, as does your clientele, topic by topic. The more closely you can help the extension organization and your clientele reflect each other's needs and problem-solving processes, the more effectively they can communicate to mutual advantage.

Note how this approach differs greatly from the one that says, "We have developed a new practice. Let's call meetings to promote it." The planning steps that follow will reflect a philosophy of the extension organization and its clientele as joint problem-solvers.

CREATING THE PLAN

The planning process for an educational campaign might be approached in the three stage of analysis, identification of objectives, and formulation of the plan (see Table 8.1).

Stage 1: Analysis

Careful analysis, of topic, situation, intended audience, and the local extension organization, can help keep a campaign simple and on target. Analysis requires some time, but can save much more time when the decision-making stage is reached.

The following are some questions that might be asked about each of the four areas for analysis.

Topic

    1. How familiar is the topic to the intended audience?

    2. How visible is it? How easy is it to see and describe?

    3. How readily can it be demonstrated?

    4. How strong or weak is the scientific base for it?

    5. To what extent does it agree or conflict with the current values and experiences of the audience?

    6. How much advantage does it offer to the members of the audience?

7. How simple or complex is it to understand and use?

8. How much expense is involved?

9. How divisible is it? Can it be acted upon in many steps, or does it require all-in-one action?

Situation

1. How severe is the problem, or great the opportunity?

2. What has created the problem or opportunity?

3. What previous efforts have the extension organization, or others, made to address the problem or opportunity?

4. What were the results of those efforts?

Audience

1. How many persons make up the intended audience?

2. Where are they located, geographically?

3. What are their major characteristics (such as age, sex, level of education, financial resources, facilities, occupation)?

4. How much do they know about the topic?

5. How interested are they in the topic?

6. How important do they consider the topic?

7. What are their feelings and opinions about the topic?

8. What are their current practices related to the topic?

9. What skills and capabilities do they have (or lack) concerning the topic?

10. What are their goals related to the topic?

11. How may families and friends influence them concerning the topic?

12. How may habits and customs influence them concerning this topic?

13. To what sources do, or would, they normally go for information about the topic?

14. What groups or organizations are important to them?

15. What mass media sources do they use (radio stations, newspapers, others)?

16. What are their opinions about the extension organization? (For example, how well-informed, prestigious, reliable, impartial, and easy to reach do they think extension is?)

17. When do they normally make decisions and take action related to the topic?

Sponsor

1. Why is the extension organization concerned with the topic?

2. How urgent is this matter, from an extension point of view?

3. How much priority will the matter receive within extension?

4. What resources (funds, time, personnel) are available to work on the programme?

Answers to such questions will help you make decisions about whether to conduct the campaign, to whom it should be directed, what should be said, what communication channels and methods should be used, when the campaign should begin and end, and so on. Such analyses can be extremely, and sometimes surprising, valuable to the communication planner.

For example, a study among farmers, village-level workers, and block-level personnel in India revealed a surprising gap in priorities. Among 12 topics, the one rated as most important by farmers (improved seeds) ranked 12th in perceived importance among village-level workers and 9th among block-level personnel. Another topic (water test) ranked 1st by village-level workers and 2nd by block-level personnel was ranked by farmers themselves as 12th in importance.

How is information gathered for an analysis? Personal or group interviews, surveys, study of organizational files, observation and other techniques have been used successfully.

Stage 2:   Identification of Objectives

Whereas analysis helps to determine "what is," communication objectives help to express change toward "what should be." A useful statement of objectives must do three things;

1.   specify the kind and amount of change desired,

2.   pinpoint the intended audience, and

3.   state the period of time involved.

Using these criteria, we can see that a statement of objectives such as the following is not adequate: "To help farmers control spiny bollworms in cotton." Here is a statement that would be more useful and acceptable: "By October 1, increase by 25 percent the average ability of all cotton growers in Nakhon Sawan Province to recognize spiny bollworms and prescribe a recommended control measure."

This statement involves change in knowledge and skills, but in specific campaigns a choice might be made from among at least six kinds of change.

1.   What people know (awareness, understanding).

2.   What people can do (skills and techniques).

3   What people feel (attitudes, interests, values).

4.   What people think they should do.

5.   What people intend to do (intentions).

6.   What people actually do (behaviour, action).

Notice how broad this view is. An extension campaign may aim for new patterns of action or increased levels of knowledge. But even a campaign that reaffirms one's current beliefs seeks change, in the sense of increasing the strength with which the person holds those beliefs.

Stage 3:   Formulation of the Plan

This stage involves planning how to reach your communication objectives. It has several dimensions.

Methods to Use

At this point the planner chooses from available communication channels, identifies messages to be communicated, decides on the amount and format of material to be used, selects a schedule and chooses ways to arrange feedback from audience members during the campaign.

Tasks to Perform

Each decision about method creates one or more tasks to be completed. So at this stage the planner decides exactly what needs to be done throughout the campaign. For example, meetings may need to be arranged, demonstrations set up, and radio material produced.

Organization

Now the planner needs to relate tasks to people. This stage identifies centres for action and responsibility, and states who will do what, and when.

Control

A control section of the plan shows how the activities will be monitored and adjusted as necessary throughout the campaign. It compares what actually happens with what was supposed to happen at various times.

Table 8.1
Outline for a Campaign Plan

I.   Analysis

      A.   Topic

      B.   Situation

      C.   Audience

      D.   Sponsoring organization

II.   Statement of Objectives

III. The Plan

      A.   Message content

      B.   Media and methods

      C.   Timing

      D.   Schedule of activities and responsibilities

      E.   Evaluation strategies (during and after the campaign)

      F.   Budget

SOME TECHNIQUES FOR CAMPAIGNS

Space does not permit a full discussion about techniques for planning extension campaigns, but some deserve special attention.

Timing the Campaign

In deciding when to begin and end a campaign, be guided by the patterns in which your audience members make decisions and carry out actions concerning the campaign topic. For instance, a campaign that involves how much fertilizer to apply for rice production might logically be timed to match the periods when growers make their fertilizing decisions and order fertilizers. Sometimes a campaign may be timed to fit the intensity of public interest in a given topic, such as a disease outbreak, while in other cases a campaign may be timed to coincide with designated public events, such as Rural Safety Week. Length of a campaign may be influenced by your educational goal. For example, a campaign to influence behaviour and actions may need to run longer than a campaign to create awareness.

124

Using Slogans and Symbols

Slogans and symbols can help add impact and unity to a campaign and are used often in extension. Samples of campaign slogans are:

Philippines: "16 Cavans More Unmilled Rice Per Hectare Through Good Management"

Honduras: "The Mother Who Breastfeeds is the Mother Indeed"

Tanzania: "Food is Life"

A slogan uses words to express all or part of a campaign theme. Graphic and audio symbols also are slogans in the sense that they, too, are unchanging, repetitive expressions of a campaign theme.

Effective slogans help attract attention to the topic and message, help audiences learn and remember information more easily, add unity, and sustain interest in the campaign among audience members and within the sponsoring organization. They should emphasize a single idea tied to the campaign objective. They should be memorable, easy to understand, versatile, and geared to the interests and needs of the intended audience.

Pre-testing Messages

It pays to pre-test possible slogans and other major campaign materials among audience members before you prepare and use them in a campaign. A pre-test need not be fancy or sophisticated to be useful in helping you choose your creative approach. At the simplest level, you might select a few finalist versions, show them to a sample of persons in your intended audience, and ask questions such as: Among the slogans that you see here, which one interests you most? Why? How would you rate it in terms of its appeal to you (excellent, good, fair, poor)? What point or points is it trying to make? You can use the same process, of course, to pre-test graphic symbols (such as campaign logos).

Selecting Media and Methods

Many arguments among extension workers have centred on the selection of media and methods. Some workers argue that personal contact is the only way to produce change. So they turn to meetings, demonstrations, personal visits, and other face-to-face methods. They consider mass media efforts as something extra, something to carry out only if time permits. They grant that mass media reach large numbers of people, but believe those media cannot help much in extension's basic educational programmes.

Other extension workers argue that an educator cannot possibly succeed in a public education programme based only on personal contacts. Face-to-face communicating can be effective, they concede, but how can you afford the time and cost for it when audiences are large, scattered, and difficult to reach? They claim that the use of mass media permits extension to multiply its efforts far beyond what a face-to-face approach can permit. The effective communication planner refuses to get ensnarled in this kind of debate and is slow to generalize about the impact and effectiveness of certain media. Instead, the planner uses certain approaches and guidelines to select specific media for specific campaigns.

1.   As an extension worker, you are basically in the teaching-learning business and have no loyalty to any particular method for information and education. A method is merely a tool which you use in a given situation.

2.   All educational media are best in certain cases, and all may be worst in other cases. A cook does not feel compelled to use a single "best" spice for preparing all foods, but instead uses the spice or spices best suited to an occasion. Similarly, as a communication planner, you want to choose the mixture of media best suited to a given situation.

3.   Each medium has inherent characteristics which define its strengths, limitations and capabilities. These should be taken into account when studying the results of analyses of the topic, situation, audience and sponsor. For example,

a. Is demonstration important? If so, audio methods (such as radio) and static audio-visuals (such as charts) may be less effective than live performances or other audio-visual approaches that involve motion (such as films).

b. Are the concepts detailed and complex? If so, audio methods may be limited, if used alone.

c. Will members of the audience need information in a form they can refer back to? If so, some kinds of printed materials may be superior to approaches that rely upon memory.

d. Are there low levels of literacy among audience members? If so, oral media such as meetings and radio have advantages over printed materials.

e. How important are immediacy and timing in this situation? If an outbreak of disease or other emergency is the subject, it may be better to use some of the speedier media available. There may be no time to print a leaflet.

f. How vital is give-and-take among teacher and learner, or among learners? Face-to-face methods offer advantages in this regard.

g. How important is hands-on practice? Some media can provide it easily, others cannot.

h. Will the purpose of the campaign be mainly to inform and create awareness (early stages of the adoption process), or to help audience members evaluate and arrive at decisions? This may influence the media to use.

i. How important is public visibility of the programme? Radio, newspapers, posters and other mass media offer more public visibility than the more private methods such as home visits.

4. The best guide is what your analysis shows about how your audience members gather information concerning the topic. Results of this analysis can help add a new and creative dimension to the use of communication media.

There may be ways in which to channel the campaign information through other organizations, such as co-operatives, schools or religious organizations. Dozens of methods have been used successfully in campaigns throughout the world, including audio cassettes, printed cloth posters in tea shops, comic books, folk theatre, blackboard newspapers, postcards, card games, mobile information vans, puppet shows, para-professional assistants, manuals, festivals, contests, buttons, inserts, tours, kits and super-sized flannelgraphs. Often a new method can be created, which is uniquely matched to a specific campaign.

5. Generally, more than one communication channel should be used in a campaign. Combinations of media offer several advantages. They help overcome the tendency for people to use certain information channels more than others. Messages may have more impact when they come from a variety of media. A multiple-media campaign may also be preferred when different messages must be delivered to different sub-groups in your audiences.

Combining Mass and Personal Methods

A proven technique among campaign planners involves what one could call funnelling from impersonal types of contact toward more personal types of contact. For example, you might use a radio message to refer listeners to extension personnel, or to a group meeting, where more personal contact is possible. This funnelling approach is consistent with the idea that mass media seem more important in the early stages of decision making (becoming aware and interested) and personal media more important in later stages (evaluating and deciding).

Providing Channels for Information-Seeking

Campaign planners often think only in terms of information-giving media, or media through which the sponsor transmits information to audience members. However, a problem-solver's approach to communicating requires the planner to think also about methods by which interested audience members can seek information.

For example, radio-listening groups or television-viewing groups might be provided with recording equipment and audio cassettes, which they can use to record their questions, concerns, and ideas for return to the radio station or sponsoring organization. Print and broadcast messages can give detailed instructions about how readers, listeners and viewers can get further information. Question boxes and information centres can be established and publicized.

Involving People

In planning, list all of the different people and organizations that should be informed of and involved in, the campaign; including community leaders, public officials and media representatives. People concerned with the campaign must be involved at all stages, in the planning, in the activities of the campaign, in evaluating results, and in publicizing the results.

Deciding How Much is Enough

Maximum learning or other changes usually require at least two exposures for each audience member, regardless of communication objectives. So media and messages must be used in ways that permit the campaign to reach most or all of the intended audience members, and to reach each audience member several times during the campaign period.

Pacing the Campaign

Should the effort involved in a campaign be distributed evenly over the period involved? Should the campaign begin strongly, then ease off? Should it begin slowly, then build to a climax? The options are endless and guide-lines are not fixed. However, here are several guide-lines suggested by results of research.

1.   Decide on the continuity of a campaign mainly on the basis of seasonality, as it relates to the topic and the audience members. The timing of messages should be matched with the prevailing seasonal pattern concerning the campaign topic.

2.   Messages could be conveyed in clusters during the campaign period rather than in a continuous, even flow. For example, messages might be clustered during Months 1, 2, 4, and 6 of a 6-month campaign. Within a cluster of activity, messages should be scheduled intensively to achieve maximum impact.

3.   The learning-forgetting process is dynamic, so campaign pacing must also be considered in terms of peaks, lows and averages, rather than of stable levels of audience awareness and learning.

Using Calendars and Work Charts

It is a good idea to prepare a campaign media calendar that shows when each communication method will be used during the campaign period. It can graphically demonstrate how audience members will be reached at various times and can help avoid any gaps in coverage during the campaign.

A work chart identifies each activity that must be carried out before, during and after the campaign period. If a chronological work chart is prepared, it will tell what needs to be done on a day-to-day or week-to-week basis, such as planting a demonstration plot, recording a radio announcement, posting copies of a wall newspaper, preparing visuals for a meeting, pre-testing a poster, or calling on co-operating organizations. The work chart not only shows activities and deadlines but also identifies those who are responsible for each activity.

Evaluating the Campaign

At the same time as any campaign is planned, its subsequent evaluation, in terms of the objectives, must also be planned. Evaluation during the campaign may deal with aspects such as the adequacy of resources, the degree to which deadlines are being met, co-operation with partner organizations is functioning, the extent to which media organizations are using the materials which are submitted to them, and the amounts and kinds of feedback from audience members.

Evaluation after the campaign should focus on the communication objectives, as stated during the planning process. "Did the campaign achieve the kind and amount of change desired?" is the critical question to be assessed.

REFERENCES

Buie, C. (1979). The communications process and effectiveness of health messages (PASITAM design notes no. 17). Bloomington, Ind.: Program of Advanced Studies in Institution Building and Technical Assistance Methodology, Indiana University.

Colle, R. D. (1976, August). Communicating with villagers. Paper presented at the Planning Seminar of Agriculture for Nutritional Improvement, East-West Food Institute, Honolulu, Hawaii.

Crawford, R. H. & Ward, W. B. (Eds.) (1974). Communication strategies for rural development. Proceedings of the Cornell-CIAT International Symposium, Cali, Colombia, March 17-22, 1974. Ithaca, N.Y.: Program in International Agriculture, Cornell University.

Entine, L. & Ziffern, A. (1980). Getting the word out: A handbook for planning a public information campaign. Madison, Wisc.: University of Wisconsin - Extension.

Evans, J. F. (1984). Education campaign planning. Manuscript submitted for publication.

Havelock, R. G. (1969). Planning for innovation. Ann Arbor, Mich.: Center for Research on Utilization of Scientific Knowledge.

Hyman, H. H. & Sheatsley, P. B. (1947). Some reasons why information campaigns fail. Public Opinion Quarterly, 11, 412-423.

Jamias, J. R. (Ed.) (1975). Readings in development communication. Los Banos, Philippines: Department of Development Communication, University of the Philippines at Los Banos.

Leagans, J. P. (1963). The communication process in rural development (Cornell international agricultural development bulletin 1). Ithaca, N.Y.: Cornell University.

Leslie, J. (1978). The use of mass media in health education campaigns. Educational Broadcasting International, 11, 136-142.

Lionberger, H. F. & Gwin, P. H. (1982). Communication strategies: A guide for agricultural change agents. Danville, Ill.: Interstate Printers and Publishers.

Mendelsohn, H. (1973). Some reasons why information campaigns can succeed. Public Opinion Quarterly, 37, 50-61.

Reeder, W. W., LeRay, N. L., Jr., & Mackenzie, S. T. (1974). Problem diagnosis: Applying sound theory to problem solving. Journal of Extension, 12(1), 20-35.

Reeder, W. W., LeRay, N. L., Jr., & Mackenzie, S. T. (1974). Planning powerful extension programs. Journal of Extension, 12(2), 36-47.

Rice, R. E. & Paisley, W. J. (Eds.) (1981). Public communication campaigns. Beverly Hills, Calif.: Sage Publications.

Rogers, E. M. (Ed.) (1976). Communication and development: Critical perspectives. Beverly Hills, Calif.: Sage Publications.

Salcedo, R. N. (1974). Blood and gore on the information campaign trail. _Journal of Extension_, 12(2), 9-19.

Salcedo, R. N., Read, H., Evans, J. F., & Kong, A. C. (1973). A test of some principles in information campaign planning. _American Association of Agricultural College Editors Quarterly_, 56(3), 15-27.

Spaven, J. W. (n.d.). _Campaigns in agricultural extension programs_. East Lansing, Mich.: National Project in Agricultural Communications.

Chapter 9
Individual and Group Extension Teaching Methods

J. T. Kang and H. K. Song

The primary responsibility of extension workers is education. A substantial number of proven educational methods or techniques exist from which the extension worker may choose to set up learning situations and to maximize the transfer of information and skills to young and adult learners. Once the needs of an area or community have been identified, it is the task of extension workers to choose the teaching methods that will be most effective in achieving their educational objectives. It is the purpose of this chapter to identify and explain various teaching methods commonly used in individual and group situations. Mass educational techniques will be dealt with in Chapter 10.

SELECTING A TEACHING METHOD

People learn through their own activity, through what they do. No one can learn for the clientele. Before choosing a teaching method, the following points should be considered.

1. No single teaching method is better than another. The extension worker should choose those techniques best suited to the situation. No one technique is considered superior to another.

2. Use a number of teaching methods to carry out the programme. Experi-ence in extension work has shown that the more ways new information is presented the faster an individual learns.

3. Methods will overlap. It is anticipated that the methods will overlap. If a demonstration stimulates group discussion, two methods are utilized which will reinforce the information contained in the demonstration.

4. Use visual aids and written material when possible. Teaching can be reinforced and supported by the use of visual aids and written material (Laird, 1972).

INDIVIDUAL TECHNIQUES

The extension worker, interacting on a one-to-one basis with the people, is utilizing an individual method of education. Although this approach is time-consuming, its importance cannot be stressed enough, because it is through working individually with the clientele that the extension worker learns about the people of the area, how they think, what their needs are, and how they carry on their work. Equally important is the opportunity individual contact provides for the local citizen to get to know the extension worker so that the personal bond between the extension worker and the community can be established. It is through the use of this method that the extension worker's credibility and integrity can be nurtured. These methods are widely used and have been found to be highly effective when dealing with illiterate farmers working small holdings who are not normally exposed to other educational techniques.

Method 1: Farm and Home Visit

The farm and home visit method involves meeting individually with the farmer or farm worker at the farm or home. A farm and home visit serves a number of purposes (a) to establish contact with men and women farmers and with others within the farm household, (b) to learn what practices and problems exist on the farm and in the farm household, (c) to provide information and assistance. This technique is costly in terms of time spent and the number of clients contacted, which will necessarily be few. However, the benefits are numerous enough to make this a highly recommended technique. The extension worker should visit many different farms and homes, and care should be exercised to visit with both men and women farm managers as well as with other members of the farm family.

An illustration of the effectiveness of the farm and home visit technique occurred in Venezuela (Ruddle & Chesterfield, 1974). A female extension worker worked with the women of an island village. She walked through the village stopping and talking with the women about everyday matters. As a result of the visits, the extension worker became aware of the daily routine of village life and of the role the women played in the traditional food supply system. She suggested the expansion of the gardens traditionally kept by the women for the purpose of selling any additional produce as a commercial venture. When the idea was discussed in informal meetings, the village women asked the extension worker for new seeds and starts from which commercial produce could be started. The potential for improvement in the village economy was made possible through the home visits and informal meetings. The groundwork was also laid for additional extension work, such as teaching better methods of cultivation.

As with any educational method, careful planning and preparation should be carried out. The following procedure is suggested.

Before the Visit

1. <u>Obtain or prepare a community map</u>. A map showing where each family lives or works will assist the worker in planning area visits. In densely populated areas where visits to all clientele are not possible, the location of individuals or families already visited should be marked on the map. The map will enable the extension worker to plan the visits efficiently in terms of time and mileage.

2. <u>Preparation and review of the visitation record</u>. A visitation record should be maintained; it should contain the date each client was visited, and notes describing the purpose and activities of the visit. When planning a follow-up visit to a client, the visitation record should be reviewed to refresh the memory. If a visit to a new client is planned, the visitation record and map should be consulted to learn if prior visits have been made in the vicinity. The review will help define the exact objective to be accomplished by the visit.

3. <u>Maintain an activity calendar</u>. The extension worker should always plan in advance. Maintaining a list of activities and objectives on a calendar which can be easily carried is highly recommended.

Conducting the Visit

1. <u>Greet the farmer and members of the farm family</u> upon entering the farm or home and accept hospitality offered, according to local custom. Spend some time talking with the clients. This approach gives each party an opportunity to become better acquainted.

2. <u>During the visit observe</u> the conditions and activities of the farm operation.

3. <u>Discuss observations</u> with the farmer. Be tactful if it is necessary to criticize something or to suggest changes. When offering solutions, discuss what should be done and how the changes could be carried out. If additional information is needed, suggest a return visit.

Follow-up

5. <u>Make notes</u> in the visitation record and record location on map.

6. <u>Prepare any additional information promised to the client</u>. It is essential to fulfil such promises if the extension worker is to build and maintain trust.

7. <u>Respect the client's privacy</u>, and do not discuss the family's business with others.

Method 2 :   Office Calls and Inquiries

This method is concerned with personal visits made by the clientele to the extension office, to seek information and assistance.   To encourage office visits, extension workers should consider the following.

1.   Place the extension office in a convenient location.

2.   Keep regular office hours so clients will know when the extension worker will be available.   Provide a visitors' record book, so clients may register their visit and inquiry should the extension worker be out.   The extension worker may then contact the visitors at a later time.

3.   Keep the office neat, orderly, and attractive.

4.   Maintain an up-to-date bulletin board and have information materials readily available.

5.   Make a special effort to put the visitor at ease, especially if the individual appears to be shy in the unfamiliar environment.   It may be necessary for the extension worker to ask questions in order to determine the visitor's concerns or questions.

A visit to the extension office is a statement of confidence in the extension worker and his or her advice, and should be handled carefully.

Method 3:   Informal Contacts

Informal contacts are unstructured and/or planned meetings with clientele in an informal setting.   Such meetings provide the extension worker with an opportunity to meet clientele in an informal situation, which facilitates the establishment of a personal bond, discussion of problems, and the recommendation of solutions.   Informal contacts can take place on the street, in the market place or at local celebrations.   These meetings often take place by chance and are casual in nature.   An effective extension worker is skilful in utilizing such informal teaching situations.

Indian extension offices actively promote informal contact by organizing social days. The purpose of these social days is to encourage the extension worker and administrative officer to meet the villagers.   The informal environment promotes discussion of village programmes, and gives the extension worker an opportunity to involve villagers in community projects (Krishan, 1965).

Method 4:   The Model Farmer

The model farmer method involves the identification of a farmer whose farming methods and personal attitudes are so superior that his or her operation can serve as a model for others to follow.   The purpose of selecting a model farmer is to demonstrate good farming practices by emphasizing an outstanding local example, to persuade the clientele to adopt better farming practices, and to create a learning situation.   The model farmer technique will be most effective of course, if the individual involved is well-liked and respected, and can follow his or her voluntarily.

Korea has enthusiastically adopted the model farmer technique.   The model farmer is chosen by the government to promote local farm initiative.   Once selected, the model farmer receives intensive instruction aimed at fostering a pioneer spirit dedicated to benefit the community.   The model farmer receives full governmental co-operation and support.   In effect, model farmers facilitate extension work by acting as volunteer civil extension workers.

Method 5:   The Field Flag

A final recommendation concerns an individual communication technique to use when farmers and farm families are not at home or in their fields when the extension worker visits.   This method, which was developed by Korean extension workers, functions as follows.   An extension worker visits a farm to help the farmer identify plant diseases and harmful insects in a rice paddy.   The results of the crop examination and recommendations for treatment, if any, are written down and placed in the pocket of a red vinyl flag that is attached to a thin pole or a stiff wire. The red flag is placed in the field where the farmer can easily find it. After reading the message, the farmer rolls up the flag and replaces in the same location. The extension worker recovers the flag on the next trip by the field.

Figure 9.1 The field flag

The obvious advantage in using this method is that the time and energy expended by the extension worker in travelling to a farm is not wasted if the farmer or farm family is not there. The extension worker may go ahead with the work, and then proceed with the next task scheduled for that day. As staff time is ordinarily limited, this technique can help the extension worker make more effective use of his or her time.

GROUP TECHNIQUES

Group teaching methods are more frequently used in extension work than individual teaching techniques. This is not surprising because, by utilizing group techniques, an extension worker can reach more people than is possible by following individual methods alone. This is an important factor when time and limited staff are limited. Group methods are especially effective in persuading extension's clientele to try a new idea or practice. A group decision to try a new practice, for example, is likely to carry more weight in an area than a similar decision made by an individual.

One category of group techniques discussed in this section includes the traditional, demonstration-based methods which have been, in many ways, the corner-stones of extension work. The methods fall into a grouping which could be informally called "seeing is believing", because they include the physical demonstration of practices the extension worker wishes to promote, or the exhibition of the results of good farming practices. Examples of methods discussed include the method demonstration, the result demonstration, field days and field trips. Contests, important in youth-oriented programmes, are also discussed.

Informational and innovative meetings form another important set of group techniques. Types of meeting discussed include the more traditional lecture, panel, seminar and symposium in which, for the most part, the audience plays a passive role by listening to one or more presenters. While this type of meeting can present a large quantity of information to the clientele, learning is sometimes hindered by audience boredom and lack of participation. It was to help solve these problems that the next grouping of methods, the innovative meeting techniques, was developed. The purpose of the innovative techniques, such as brain-storming and buzz sessions, is to improve the effectiveness of group discussion and to stimulate audience participation. These techniques can be used alone; however, they are often incorporated with the more formal information meetings to enhance interest and learning.

The final set of group methods are the simulation techniques. Simulation techniques stimulate learning by actively involving the participants in exercises modelled on reality. Individuals using this method look at the society (reality) in which their clientele live. Various elements are selected from that reality and a simplified model is put together which represents (simulates) that reality (Evans, 1979). The model, or simulation, can be put into a number of forms such as board games, open-ended scenarios for role playing, and descriptions of life-like situations or critical incidents requiring decisions or answers to problems. The purpose is to provide the participants with experience in real-life situations, which will ultimately lead to a greater understanding of their society and the role they play in it. Experience can be gained without the real-life penalties exacted if a wrong decision is made.

Simulation techniques have been used successfully in rural areas. Simulations were used for five years in Ecuador in a rural adult education programme, and the methods have also been used in Asia and Africa. Simulation methods rely heavily on the training and ability of the instructor, who must not only choose and/or develop the techniques, but must handle the discussion following the exercise in a way to extract all the various lessons generated by the exercise.

Method 6:   The Method Demonstration

The method demonstration shows a group or class how something is done step-by-step for the purpose of teaching new techniques and practices to extension clientele. A method demonstration could show how to use a tool, a new planting technique to prevent erosion, or how to cook a newly introduced vegetable. Ideally, each individual attending the demonstration would have an opportunity to practise the new skill during the session. Usually, however, time does not permit more than a few to participate. The effectiveness of the demonstration depends, in great part, on the amount of preparation and planning. Because the results of the demonstration can be observed and sometimes practised, this method is very effective in persuading clientele to try something new. A number of suggestions are given below on how to plan and conduct a method demonstration.

Planning the Demonstration

1. Identify the problem to be solved. The focus of the demonstration should be on solving a problem that exists locally. The problem can be identified through the observation and knowledge of the area, by a local individual or the extension worker. The extension worker should consider involving local clientele in problem identification.

2. Identify the skill to be taught. Once the problem has been identified, the extension worker must decide what skill to teach, and what educational objectives he or she wishes to attain. When going through the skill identification process, each proposed skill or method should be subjected to the following queries: (a) Is the skill important? (b) Can the people afford to adopt the skill? (c) Are there enough supplies and equipment available to permit its widespread use?

3. Gather information about the skill and study it thoroughly. The demonstrator (most likely the extension worker) must be familiar enough with the method to break it down into teachable segments.

4. Involve clientele in the planning and presentation of the demonstration. By seeking the people's advice and assistance, local interest and level of adoption will be increased.

5.   Assemble material required to conduct the demonstration. The material should include everything the farmer would need to apply the practice on the farm.

6.   Plan the presentation step-by-step, including an introduction and summary.

7.   Rehearse the presentation until it can be given with confidence.

Holding the Demonstration

8.   Schedule the demonstration at the most convenient time and place for clients. In some cases this may necessitate two demonstrations, one for men and one for women. In other situations, both men and women may attend the same demonstration if care is taken in scheduling.

9.   Arrange the audience so that everyone can see and hear the demonstration clearly.

10.   Introduce the demonstration by explaining why the new practice is important and what will take place at the gathering.

11.   Ask for assistance from the audience to help in demonstrating the various steps.

12.   Proceed with the demonstration step-by-step, answering questions and repeating difficult steps.

13.   Encourage members of the audience to attempt the new method during the meeting. If time allows, each individual should have an opportunity to practice the skill.

14.   Summarize the importance of the skill, the steps, and the supplies and equipment needed. Distribute illustrated literature, if available, showing each step.

Follow-up

15.   Evaluate the demonstration carefully, noting where the strengths and weaknesses appeared.

16.   Visit those clients who indicated an interest in the demonstration.

Method 7:   The Result Demonstration

The result demonstration teaches why a practice or input should be adopted by physically showing how a new or different practice compares with a commonly-used local practice. This technique is often used in crop farming. The purpose of using the result demonstration is to prove that the new practice is superior to the one currently being used, to persuade extension clientele to try the new practice, and to set up a long-term teaching situation.

An example of a result demonstration would be the comparison of two wheat crops, one with an application of fertilizer and one without. In co-operation with a farmer, two demonstration plots would be laid out in a field, side-by-side. In one, wheat would be planted using the local seed and the local cultivation practices. Because the purpose of the project is to demonstrate the effect of using fertilizer, no fertilizer would be used. In the second plot, wheat would also be planted, and subjected to the same practices and conditions, but fertilizer would be applied. Both crops would be harvested at the same time. If all goes well, the fertilized plot will produce a substantially greater yield than the unfertilized plot, visibly demonstrating the tangible benefits of fertilizer application.

A successful demonstration can produce positive results for extension workers by creating confidence in their judgement and ability. It is a technique extension workers new to an area might want to use to establish themselves in the community. Workers should consider the consequences if the new technique fails, however, and choose the demonstration carefully. A result demonstration is costly in terms of time, but if it is successful, it is an effective way to promote the new practice locally, and can open the way for further interaction with the clientele. The result demonstration requires careful and detailed planning to be successful and to obtain full educational benefit. The following suggestions may serve as a guide when planning to use this method.

Planning the Demonstration

1. Identify the problem to be solved by the demonstration.

2. Decide upon the objectives to be accomplished.

3. Gather complete information about the proposed practice and study it thoroughly.

4. Seek the assistance of the clientele in planning and carrying out the demonstration. Endorsement by the local leaders will lend credence to the effort. Seek their recommendations about good possible host farmers.

5. Develop a complete plan of work, clearly delineating each individual step and showing who has responsibility for each task.

6. Select accessible demonstration plots that are centrally located and near a road so people can easily visit the site.

7. Visit the host farmer and work with him or her to make sure that he or she understands the purpose of the demonstration and how it will be implemented. This step is essential because the host farmer will be answering most of the questions about the project, and will play a prominent and visible role during the entire project.

During the Demonstration

8. Ask area leaders to co-operate in encouraging people to attend the start of the demonstration.

9. Visit the demonstration site often.

10. Use the demonstration site for meetings and tours during the life of the demonstration. Encourage the host farmer to describe the process.

11. Keep records on both sites so the results at harvest may be compared. Detailed records maintained during the life of the demonstration will facilitate speaking about the process in meetings and elsewhere.

12. Publicize the demonstration by the best methods available. Use newspapers and radio if available; if not, talk about the demonstration on market day, to informal contacts, and in meetings held for other purposes.

After the Demonstration

13. Evaluate the process and results of the demonstration. This will be useful when using the method in the future.

14. Provide follow-up information and training to interested clientele.

Method 8: Contests

Contests are based upon the principle of competition and community-oriented activities, to encourage participation and heighten the practical agricultural skills. The purpose of holding contests is to provide farmers, especially young farmers, with powerful motivational forces and to offer opportunities to excel in specialized subjects and skill areas.

After periods of learning and practice in the community, individuals and teams demonstrate their proficiencies and talents before the public in contests involving such skills as crop and livestock judging, tractor driving and weed identification, to name but a few. Entering contests, shows, exhibits and demonstrations is encouraged by offering awards and rewards. Extension workers assume coaching roles. These include offering instruction in the fundamentals of the contest and suggesting strategies for winning. Coaching implies total involvement. It requires many hours, special kinds of resource materials, practice sessions, and extra travel to attend the competitive events. Preparation for contests causes coaches and farmers to become a team with a close relationship.

Method 9: The Field Trip

On a field trip, a group travels to another location to observe agricultural practices, projects or demonstrations not available locally. The trip may consist of one or more stops. The purpose of the field trip is (a) to provide first-hand observation of practices that might be of benefit to the farmer or householder and farm worker, (b) to enable the group to interact with individuals knowledgeable about the practices, and (c) to present a fresh and different learning environment for both the extension worker and the clientele.

A field trip's destination may be an agricultural experimental station, a farm, a home, or a community organization, such as a women's marketing co-operative. It is essential that the extension worker should work with the group beforehand so they know what they are going to see, why it is important, and how much time they will have at each stop. If the host is going to discuss the practices, make sure he or she clearly understands the objectives of the visit, what role the visit will play in the overall educational programme, and the time available for the visit. It is also recommended that the extension worker assign a task to the group to be completed during the trip. For example, each group member could be given a few questions relevant to the site visit to answer. Finally, time should be set aside for a group discussion at the end of the trip. This will provide the leader with an opportunity to highlight important aspects of the experience, and give the group members a chance to compare notes and ask questions (Laird, 1972).

A successful field trip takes a great deal of time and trouble to plan. However, it is one of the most effective teaching methods, combining, as it does, discussion, study, demonstration, and first-hand experience in a real-life situation. The field trip can be a very effective tool in persuading clientele to try a new practice. Some suggestions in planning a field trip are given in the section below.

Planning the Field Trip

1. <u>Identify the objective of the trip</u>. A field trip taken only to add variety in teaching methods will be of little value, and a waste of time and effort.

2. <u>Set aside adequate time to plan trip</u>. The trip necessitates the co-ordination of many factors, people, transportation, distance, and limited time. Allow enough time to complete all the necessary tasks.

3. <u>Contact site operators to obtain permission to visit</u>. Identification of hosts and destination may be decided upon with assistance of local leaders.

4. <u>Secure transportation</u>. Decide early in the planning process what kind of transportation will be used, how many people to plan for, and whether it will be available for the dates in mind.

5. <u>Draw up a tentative schedule</u>. Work with local area leaders to: (a) make out a time table, (b) decide who will go on the trip (farmers, farm workers, local leaders), and (c) learn what they believe should be gained from the trip.

6. <u>Finalize a detailed schedule, and assign tasks</u>. Steps 1-6 should be completed early in the planning period.

7. <u>Co-ordinate with site hosts</u> to be sure that everyone clearly understands the time of arrival, the number of visitors, the purpose of trip and their own role.

8. <u>Make a 'dry run'</u> to make sure the proposed itinerary can be completed in the allotted time.

9. <u>During the trip, keep the group moving</u> to maintain interest and keep to the time table. Provide for the comfort of the group to the best possible degree.

Method 10: The Field Day

A field day is a day or days on which an area containing successful farming or other practices is open for people to visit. Exhibits of a related nature such as tools, seed samples and educational material are often displayed. The purpose of the field day is to

permit extension clientele to observe personally, and ask about successful farming practices, and to create a situation in which informal contacts and learning can take place.

Field days are normally held once or twice a year, usually in each crop season.
They are held on farms, experimental stations or goverment centres to demonstrate successful farming techniques or research. This method helps to promote better farming by
putting the best agricultural exhibits on display, transmitting research results, and
providing an opportunity for farmers to see and discuss the demonstrations with one
another and with technical specialists. Exhibits and displays add another dimension to
the experience. If the field day is held on a farm, it is recommended that the host
farmer should play a prominent role in discussing crops and practices. The extension
worker should be available to clarify technical points. The following section contains
some suggestions for planning and holding a field day.

Planning a Field Day

1. Identify the objective to be achieved.

2. Select a demonstration site.

3. Work with area leaders and the host farmer to decide on the date, and essential
details of the event.

4. Publicize the field day well in advance.

5. Display sign boards at the field day site.

6. Arrange for an exhibition of related materials of interest to the visitors. Where
possible, include items which can be carried away by those attending.

7. Arrange for transportation to carry the farmers and other visitors around the
site. As with the field trip, arrange this well in advance.

Holding the Field Day

8. Distribute literature describing the farming practices being demonstrated.

9. Take visiting farmers around the plot. Conducted tours will not only show
consideration of the visitors, but will help ensure they see the important points of the
demonstrations.

10. Let the host farmer explain the practices being demonstrated. This will show
appreciation to the farmer for his or her extra efforts, as well as making visiting farmers
feel more comfortable about asking questions.

11. Emphasize distinctive features of the crops and/or practices being shown.

12. Hold group discussions with participants about the practices being shown.
Record their comments and reactions for use in the subsequent evaluation, and in planning future field days. The comments could be especially valuable in identifying what is,
or is not, acceptable to the farming community.

Follow-up

13. Evaluate the field day to determine its success or failure. Use the evaluation
to help guide future field days and programme planning.

14. Contact farmers who indicated interest in the new practices and assist them in
adopting it on their farms.

Method 11: Informal Discussion

Informal discussion in a small group is another type of group technique. The neighbours get together in a certain house at a certain time period once a month, or perhaps
once a week, to consider and communicate the common public problems and to get
acquainted with the neighbours, to exchange farming information and ideas, and to share
common problems, in order to help each other and the community.

The informal discussion is carried out at the villagers' monthly meeting. This discussion group cannot have professional leadership. The idea is that responsible villagers should get together to consider and talk about common problems. It is most successful in its pure form of government-assisted village and town work in developing countries, where the extension worker helps the community improve not only its own economic literacy levels, but also health, sanitation and family planning. Through this type of discussion, extension workers can encourage grass-roots discussion of public issues, improve the quality of the discussion, and form a common thought pattern.

Method 12: Lecture

The lecture is a formal, verbal presentation by a single speaker to a group of listeners. Visual aids may illustrate the lecture, and a question-answer period may follow the talk. The purpose of the lecture is to provide a body of organized information to an audience. While the lecture is a systematic way to present information, a major drawback is the passive role of the listener (Laird, 1972). Unless the speaker is gifted not only in his or her grasp of the subject matter but also in wit and style, attention tends to wander. Lecturers should be well-organized, deal with one or two central themes, and make every attempt to obtain and keep the attention of the audience. High-quality visual aids are often useful. A lecture series, with lectures being presented over a period of time, can sometimes be used.

Method 13: Panel

The panel is a moderated meeting in which a limited number of experts or specialists give short presentations on the same subject. Ideally, each panellist represents a different field or discipline. A panel provides a group with a series of informed opinions on a given topic. Those attending a panel session have an opportunity to receive an in-depth view of a subject by hearing the various experts give their opinions, by observing the interplay among the panellists, and by having an opportunity to question them. A panel should be guided by a strong moderator with a gift for summation, to ensure that each panellist has enough time to give his or her presentation, and to oversee the question and answer period.

Method 14: Colloquy

This is a modified version of the panel, in which three or four resource people discuss a specific topic. The audience is expected to express opinions, raise issues, and ask questions. The primary difference between a panel and colloquy is the degree of audience participation.

Method 15: Symposium

A symposium is a meeting in which 2-5 resource people give short, prepared papers on a given topic. Interaction with the audience is not expected. The symposium is used primarily for information gathering at the professional level.

Method 16: Seminar

A seminar comprises a small group of students or trainees engaged in specialized study under the leadership of an expert. The leader may give a brief, opening presentation, often on provocative issues, and guide general discussion. Research is sometimes conducted by members of the seminar. The seminar is normally reserved for advanced study and provides an opportunity for in-depth study of an issue or series of issues with an expert.

Method 17: Modified Conference Method

This method is a procedure in which a group of people, each of whom has had some experience in connection with the job or problem at hand, come together to discuss situations they are facing. This method provides those attending with an opportunity for constructive thinking under the stimulus of contributions offered by other participants.

Steps in the Conference Procedure

1. Define the job or problem under consideration.

2. Analyze the job or problem.

3. Select the functioning facts from the experiences given by members of the class.

4. Assemble these functioning facts either on a blackboard or a chart.

5. Evaluate the facts obtained in light of the problem.

6. Submit any additional facts to be considered, such as data from experiment stations.

7. Have each member of the group make a decision concerning the changes, if any, he or she can perform on their farm.

8. Have definite working plans prepared for implementing the desired, improved farming practices.

9. Supervise the making of the plans; at a later date visit each farm to check the results.

Method 18:   Clinic

A clinic is a meeting or series of meetings involving analysis and treatment of specific problems. This method gives the participants an opportunity to examine a problem or problems, with the goal of finding a solution. Those attending are also exposed to a process of analytical problem solving.

Method 19:   Workshop

A workshop is a co-operative gathering of individuals who discuss, learn, and apply practical skills. Participants are trained in a skill, procedure or practice which can be immediately utilized. Those attending are expected to produce a product, such as a visual aid, by the end of the meeting. The workshop normally involves between 15 and 30 people. Workshops can be a very effective teaching tool because every participant spends one or more days intensively working on a specific product. This method, properly conducted, gives practical experience ("hands on"), and is a widely used technique.

Method 20:   Brain-storming

Brain-storming is a group discussion technique in which the members generate as many ideas as possible on a specific topic without restraint or consideration of practical application. Spontaneity and creativity are an important part of the process. The purpose of utilizing this technique is to promote group creativity, so that all aspects of a problem are considered. Input from each member is encouraged with the intention of a better product or solution being produced than might have been possible by an individual. Brain-storming gatherings should be restricted to under twenty people per group. As ideas are presented, a recorder should write them down so everyone can see them. No judgement is made about the ideas during the session. Evaluation is made at a later session.

Method 21:   Buzz Session

This technique involves dividing a large group into much smaller ones, in which a topic is discussed within a limited period of time. Generally, a buzz group is expected to produce a product (such as a list of ideas, an opinion, or a group of questions) within about 5 minutes and to give an oral report to all groups involved in the exercise or meeting. The purpose of the buzz session is to facilitate the involvement of every member attending the meeting. This method is sometimes used to break up a larger meeting, and to add variety and interest. It can also be used to solicit solutions to problems, or to gather opinion.

Figure 9.2   Diagram of participant interaction using the brainstorming (left) and the
               buzz session (right) techniques

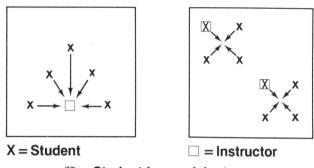

X = Student   □ = Instructor

⊠ = Student in special role

Note: From   Teaching Adult Education Courses: The Business Management Model   by
J. P. Bail & H. R. Cushman, (1979), Information Bulletin 98, Ithaca, N.Y.:   The New
York State College of Agriculture and Life Sciences, Cornell University.

Method 22:   Discussion Group

     A discussion group a is meeting or conversation between two or more people discuss-
ing a topic of mutual concern.   Members of the group normally share a common background, based on assigned readings or shared educational experiences.   The purpose of
this technique is to provide members of a group with an opportunity to gain a firmer
grasp upon knowledge by setting up a situation in which questions, problems, and feelings about a topic may be discussed.   Discussion is a versatile technique in that the
group can be assigned a moderator, be given goals, or have an agenda to follow.   To be
a true discussion group, one of these elements must be presented to provide structure
(Laird, 1972).

Method 23:   Listening Team

     A listening team consists of individuals within a group who listen, take notes, form
questions, or summarize a meeting such as a lecture, panel or a colloquy.   The purpose
of the listening team is to facilitate interaction between a speaker and the audience.   This
technique is especially useful when the speaker is not knowledgeable about a group's or
institution's specific problems.

Method 24:   Simulation Games

     A simulation game is a learning exercise in the form of a game which contains ele-
ments of a real life situation.   Its purpose is to give the participants practice in dealing
with a real life situation, in making decisions about it, and in observing and understanding any consequences of those decisions.   The exercise seeks to promote discussion about
the participant's own reality.

     An example of a simulation game is "Shantytown", which was developed to model the
situation facing recent rural immigrants to a city (Evans, 1979).   The game can be played
by 2-6 people.   Each player is given a certain amount of money at the start.   Play takes
place on a board marked with areas such as rent, salary, employment, food, bar, and
other factors which direct the player's action upon landing on the area.   Dice, cards
(marked with instructions and consequences) and tokens are used.   As the play pro-
ceeds, the dice are cast and the players advance around the board, encountering various
circumstances, for example, "buy food", "pay rent", "go to jail", "employment", and so
forth.   The game is over when all but one (or all) players are forced to return to their
village.   The game is designed to inform rural people of factors they may never have
considered when thinking about leaving their village.   For example, the need to pay
money for housing and food, needs which may have been fulfilled without the use of cash
in their village.   A game like "Shantytown" discloses many points for discussion, and
gives the players an opportunity to face real problems.

It is important to understand the culture or society before initiating game playing, because what is permissible in one culture, such as gambling, may not be so in another. Once a game has been chosen or put together, play it a few times and change it, if

necessary. Flexibility is important. Should a game need to be adapted, involve the clientele.

Method 25:   Role-Playing

Role playing is a simulation in which a simple, open-ended scenario is described, and participants are assigned roles to act out the situation or problem. There is no script to follow, and participants play the roles as they see fit, drawing on their own experiences. The purpose of using this technique is to involve participants in real life situations, to stimulate thought and learning, and to encourage discussion about factors involved in the drama.

The scenario used to set up a role-playing exercise may be simple, but it must be based on the local reality. In planning this exercise, the extension worker should (a) identify a setting, for example the landlord's estate, (b) identify and assign roles, describing each to the players (for example, the landlord, two tenant farmers), and (c) specify a task, or problem to be solved; for example, the tenant farmers are trying to convince the landlord to help them buy a new high-yield seed to increase production (Evans, 1979). Special emphasis should be put on the discussion following the exercise, and in helping the audience and the actors themselves to assess the points raised by the drama.

Method 26:   Critical Incident

The critical incident is an open-ended simulation technique in which each participant gives his or her own opinion about the outcome of a described situation. As in all the simulation techniques, the incident should be based on local reality. In this exercise a number of incidents can be used to promote discussion in which the proposed solutions can be compared and discussed.

PLANNING MEETINGS

Some general guidelines on planning and conducting a meeting are listed below.

1. <u>Purpose</u>   The initial step in planning is to identify the purpose of the meeting. Once the purpose has been decided upon, the extension worker should review all the information that he or she has at hand about the subject.

2. <u>Type</u>   Choose the type of meeting to best accomplish the purpose.

3. <u>Plan</u>   Once the purpose and type of meeting have been established, planning for the meeting can begin. Such matters as date, time, place, speaker (if any), publicity, number expected to attend, the desired set up (number of chairs, tables, visual aides, <u>etc</u>.), and how the meeting will be conducted should all be considered. Finally, determine how the meeting shall be evaluated.

4. <u>Publicity</u>   Publicize the meeting well in advance. Include the following in any announcements: subject of meeting; where, when, and why it is being held; the speaker's name (if appropriate).

5. <u>Holding the meeting</u>   Bear in mind the following facts (a) the subject should be of interest to the audience, (b) start the meeting with a challenge, (c) present ideas in a logical sequence with ample opportunity for questions and answers, (d) use visual aids whenever possible, and (e) after general discussion, a summary should be presented and conclusions drawn.

REFERENCES

Bail, J. P. & Cushman, H. R. (9176). <u>Teaching Adult Education Courses: The Business Management Model</u> (Information Bulletin 98). Ithaca, N.Y.: The New York State College of Agriculture and Life Sciences, Cornell University.

Evans, D. R. (1979). <u>Games and simulations in literacy training</u>. Tehran, Iran: International Institute for Adult Literacy Methods.

Jenkins, J. (1981). <u>Materials for learning: How to teach adults at a distance</u>. London: Routledge & Kegan Paul.

Krishan, R. (1965). <u>Agricultural demonstration and extension communication</u>. New Delhi, India: Asia Publishing House.

Laird, D. H. (1972). <u>Training methods for skills acquisition</u>. Washington, D.C.: American Society for Training and Development.

Morgan, B., & Holmes, G. E., & Bundy, C. E. (1976). <u>Methods in adult education</u>. 3rd ed. Danville, Ill.: Interstate.

Ruddle, K., & Chesterfield, R. (1974). The Venezuelan 'Demonstradora del Hogar': An example of women in nonformal rural education. <u>Community Development Journal</u>, <u>9</u>, 140-144.

Chapter 10
Using Mass Media for Extension Teaching

J. H. Behrens and J. F. Evans

Personal, face-to-face methods cannot reach everyone who wants and needs information. So mass media methods such as radio, newspapers, magazines, television, motion pictures, slide shows, exhibits and printed materials are used to reach large numbers of people quickly.

These methods are particularly useful in making large numbers of people aware of new ideas and practices, or alerting them to sudden emergencies. While the amount of detailed information that can be transmitted by mass media is limited, they will serve an important and valuable function in stimulating farmers' interest in new ideas. Once stimulated or made aware through mass media, farmers will seek additional information from neighbours, friends, extension workers or progressive farmers in the area.

Some mass communication techniques that may be developed as part of a campaign to support your other extension efforts are discussed below. They may be used singly or in combination, as needed, to meet the proposed objectives.

PRINTED MEDIA

The term, printed media, is used to cover those communication techniques that rely principally on combinations of printed words and pictures. They are our oldest formal combination. To use them effectively the educational levels and literacy rate of the audience must be considered. Extension programmes can take a broad and creative approach to ways in which to use print methods for conveying news to specific audiences. Newspapers may come to mind first, but they are only one of several print mass media available to convey extension news and information.

Newspapers

Newspapers vary greatly in their audiences and coverage, from the large urban daily newspaper to the small community paper. They are published by governmental, private, and other organizations (sometimes including extension) and can provide valuable channels for extension news.

How can you get news accepted and used by newspapers? Mainly by knowing what editors want and by being able to judge the news-worthiness of your information. Here are six ingredients that newspaper editors often use to determine what they print and how they use it.

1. Timeliness. The more timely the information, the greater the news value.

2. Nearness. The closer the information seems to the reader (geographically and psychologically), the greater is its news value. That is why local newspapers prefer local news.

3. Consequence. The more the readers are affected by the information, the greater is the news value.

4. Prominence. Prominent people, places, and things carry more news value.

5. Human interest. Readers are attracted by human interest elements such as unusualness, conflict, progress, emotion, and others.

144

6. <u>Newspaper policy</u>. Newspapers have editorial policies that influence the kinds and amounts of information they publish. So the use of various kinds of extension information may vary from paper to paper, and period to period, based on editorial policies.

Here are some of the main kinds of articles that extension workers submit to newspapers.

<u>Advance event articles</u> include announcements of approaching extension meetings, tours, speeches, and other events. Such articles are often brief, but should include details that would permit a reader to attend if interested i.e. the date, starting time, location, sponsor, nature of the event, agenda, and possibly the name of a contact person who could answer further questions.

<u>Follow-up event articles</u> report to readers about recent meetings, tours, speeches, or other extension events. Their main purpose is to report results, so they are often longer than advance event articles. They should include the date, location, sponsor, and nature of the event, to provide the background for the reader, but the greatest emphasis should be placed upon the outcome of the specific event. For example, an article about a speech should report what the speaker said. An article about a field tour should summarize what the participants saw and heard. An article about a business meeting should summarize the decisions that were made.

<u>Information articles</u> are used widely in extension to provide helpful information of various kinds: timely advice, "how-to" descriptions, reports of research findings, market news, relevant statistics, and others. Such articles are not tied directly to events.

<u>Feature articles</u> are informational and sometimes involve news, but are distinct in several ways from the types of articles mentioned earlier. Feature articles often interpret the news and provide background for readers. Often they are intended to entertain or inspire as much as to inform. They may feature ideas, places, techniques, persons, organizations, goals, successes, challenges, and almost any other aspect of human activity. They often involve more human interest than do news or information articles.

<u>News-writing style and format</u>. News-writing styles differ throughout the world, so the best approach is to use the styles and formats that local news editors prefer. Work closely with the local editors to learn their style rules, deadlines and other preferences.

Regardless of specific requirements, editors probably want news copy that is neat, readable, and well-spaced on the page. They probably prefer writing that is clear, simple, active, and concise, because such writing permits easy reading, which is your primary goal. And they want copy that is accurate. One of your most successful techniques may be to keep your readers in your mind as you write.

Wall Newspapers

Wall newspapers are used successfully by extension in many countries. The wall newspaper is known for being basically pictorial, using drawings and/or photographs, with a text as brief and vivid as possible. It is similar in size and appearance to a poster, but often contains more written material and a wider variety of information.

For example, a typical extension wall newspaper might use pictures and text to

1. announce the appointment of a new livestock specialist,
2. give a progress report on a current fertilizer campaign,
3. urge the use of vaccine to prevent fowl cholera in poultry flocks, or
4. report the results of experiments with new grain varieties.

In most countries where wall newspapers are used in extension programmes, they are produced in quantity by a central office and distributed by mail or through the extension organization. Some wall newspapers are printed by letter-press or offset methods, but methods such as silkscreen printing also work.

Distribution varies according to the requirements of each country. Mailings may be made directly to village leaders, school teachers, religious leaders and others. Sometimes the local extension worker handdelivers, and even posts, issues of the paper.

Walls of buildings at busy intersections are excellent posting places. Papers also may be posted effectively on village bulletin boards, in reading centres, in schools, and inside public buildings.

Blackboard News

Actually, extension workers need no typesetting or duplicating equipment to reach mass audiences with print news at the local level. Chalk and a blackboard, or felt-tip pen, or crayon with a newsprint pad, can provide a valuable kind of wall newspaper. In the absence of plain newsprint sheets, extension workers in India have even hand-printed news in large print over classified advertising pages of discarded newspapers. The classified advertising provided a suitably neutral background for readable, hand-written extension news posted in the village.

Newsletters

Newsletters can be an effective, low-cost way to reach readers. The content of a newsletter can be more localized and specialized than is possible with a general newspaper. Like the wall newspaper, the newsletter is well adapted to using local languages and dialects. And a newsletter can include hand-written, type-written or type-set copy. Duplication methods also can vary greatly.

A newsletter usually contains a larger share of text-to-visual than does a wall newspaper, but not necessarily. Page size is smaller than for newspapers, so space is often limited and brevity is vital. In fact, brevity is one of the benefits that readers find in a newsletter. Newsletter writers try to get into each subject quickly and use short sentences and energetic words.

The newsletter can be directed more selectively than newspapers. For example, a newsletter might be distributed only to new mothers in a village; content could provide news and advice about feeding and caring for infants. The newsletter can, therefore, be newsy, localized and specialized in what it covers.

Extension personnel might publish their own newsletter, or newsletters, or they often submit news to newsletters published by other organizations, such as co-operatives, that reach readers of interest to extension.

Folders, Leaflets and Pamphlets

Simple folders, leaflets and pamphlets can be used in many ways in extension programmes. They may be used singly, for example to explain the advantages of testing soil. They may be used in series of broader subjects like swine raising, with separate leaflets on feeding, housing, and breeding. They may be used as reminders of when to plant crops or what chemicals to use to control different insects.

Folders, leaflets, and pamphlets may also be used in co-ordination with other visual methods in long-range campaigns. Because of their low cost, they can be given away at meetings and fairs and offered on radio programmes. They are useful to supplement larger publications when new information is available and when reprinting the whole publication is not practical. An experimental campaign is being tried in the Northwest Frontier Province of Pakistan by air-dropping timely one-page leaflets on insect pests and vegetable growing practices from spray planes in village areas.

Besides the advantages of low cost and short preparation time, folders, leaflets and pamphlets take less time to get their message across. Their smaller size makes it necessary for the author to eliminate non-essentials from the message.

Fact Sheets

Fact sheets are "boiled down" treatments of subject matter. They usually cover a single topic, and often they are limited to a single page.

Most fact sheets are illustrated with drawings or photographs, or both. The illustrations are used to show details or steps in a process, to make the information clearer and more understandable. One of the important uses for fact sheets is to provide current subject matter to field workers. Field workers often complain that needed technical information is slow in reaching them. Much agricultural information is carried in technical bulletins and other lengthy publications. These take considerable time to process and distribute.

On the other hand, the essential facts can be put down and combined with drawings and/or photographs to make an effective summary which can be reproduced quickly and inexpensively in fact sheet form. This puts current information into the hands of local extension workers enabling them to give better service to farm families. Extension administrators who are concerned with the problem of speeding up intra-staff communication of subject matter should study the advantages offered by fact sheets.

Preparing the Material

1. When preparing these printed materials, keep the audience constantly in mind. Write with words people understand. Write about things that interest people. Change the method of presentation as the proposed reader changes (young people, farmers, women). Eliminate difficult scientific and technical terms.

2. The importance of illustrations cannot be over-emphasized. Even where literacy is not a problem, people interpret words differently because of differences in past experience.

Almost every extension service over-estimates the ability of its audience to read a printed message and understand it clearly. Almost every extension service over-estimates the extent to which people will be attracted to and read a printed message.

3. Illustrations reduce the risk of misunderstandings: they help make your message clear and more attractive, and they increase learning.

4. Good lay-out arranges material in a logical, easy-to-follow manner and makes it attractive to the reader.

5. Realistic illustrations are usually the most effective in extension work, although humourous drawings have a definite place. Use humour carefully so as not to offend anyone. Good pictures make any publication easier to understand and more interesting to read, but crop unnecessary details from photographs, and keep drawings simple.

6. Folders generally have more appeal when a colour or ink is used other than black. Choose colours that are legible as well as appropriate--dark green ink for pastures, dark brown for soils. Two or more colours can be used if the extra cost is justified. Coloured papers can also be used for interesting effects and are available at very little extra cost.

7. Attractive, effective publications can be prepared on spirit duplicators, mimeograph machines, offset duplicators or letterpress. It is not difficult to train an artist to produce illustrations, headings, or even copy in the local dialect for each of these methods of printing.

8. Recognize that the cover of illustrated literature has a function different from the pages within. The cover should be attractive, colourful and impelling. The audience should feel an urge to look inside. A bulletin that never reaches the hands cannot possibly reach the brain where judgements and decisions are made.

AUDIO-VISUAL MEDIA

Communication methods that rely on the audio or visual senses, either alone or in combination, help overcome the barrier of illiteracy and offer special advantages. They also have disadvantages, which will be discussed later.

Radio

Radio can be one of the most useful mass communication tools for extension workers, for several reasons. It offers immediacy, as radio programming can be changed quickly to meet new conditions. It reaches large numbers of people, especially as transistor radios become used more widely. They permit listeners to take their radio wherever they go, even where electrical power is not available. Radio provides the warmth of the human voice. It can tie into the strong oral traditions of communities and overcome the literacy barriers that face print media.

Extension workers find that radio works most successfully at the local level, to communicate local problems, solutions, and activities. They use local names, voices, and activities in programming. Farmer success stories and other kinds of neighbour-teach-neighbour approaches have been found to work well. Radio is most effective at the awareness and interest stages of the adoption process.

However, listeners cannot refer back to what they have heard on the radio, nor can they see what is being described. So radio is limited in its ability to convey detailed, complex information and, used alone, is limited as a teaching method.

Types of Use for Broadcast Radio

Two types of broadcast radio are commonly used for extension programming.

1. <u>Open broadcast</u>. The extension worker provides programming for the station's broadcasts, such as spot announcements to be taped and repeated at intervals during the day, or longer programmes, presented in person or taped for use on scheduled pro-grammes. Stations also invite printed news releases that can be read in newscasts and other programmes.

2. <u>Group listening</u> to open broadcasts. Extension services in many countries use approaches commonly called listening clubs, radio schools or farm forums. Local partici-pants gather at a certain time, listen to a programme broadcast by a certain station, then discuss the programme in terms of their own situations.

Organized group listening can produce good results because it involves the listeners more than individual listening does. However, listening groups are difficult to maintain and may require more of the extension worker's time than can be justified. Instead of organizing an extension listening group, it may be possible instead to promote collective listening in existing groups, such as co-operatives or farmers' associations. If extension wishes to organize its own group, it might be created for a limited period, perhaps keyed to a timely topic that is important to local listeners. When the series of broadcasts ends, the group disbands.

In general, group listening declines as transistor radios become more widely used by individuals and families. In some communities, participants in a radio group listen indi-vidually to the specified broadcast, then discuss the programme at a meeting later. As radio ownership grows in local communities, it seems likely that extension workers will use listening groups mainly for major topics that can arouse wide-spread local interest.

3. <u>Audio cassettes</u>. Low-cost, battery-powered cassette recorders are permitting extension workers to use recorders in some effective new ways.

1. A radio listening group can record its reactions, conclusions, questions and suggestions about a given programme or topic, then send the cassette to the station or sponsoring organization for information and follow-up.

2. A regional or central extension office can produce and record instructional programming on audio cassettes for use at local levels. Such cassettes might be used in group meetings or made available for individual listening. Local listeners can, in turn, record their own reactions or questions on cassettes which are returned to the regional or central extension office as valuable feedback.

3. A local extension worker can make field recordings for use in group meetings. The recorded material can add interest to meetings, yet is simple and cheap to produce.

4. Cassettes can permit multiple use of information aired on open broadcasts, by simply recording the programme as it is aired. The resulting recorded programme can be used later in group meetings. Conversely, the remarks of a guest speaker at a meeting can be recorded, and perhaps used in a later broadcast.

Producing and Presenting Radio Broadcasts. Here are some tips for preparing and presenting radio programmes for broadcast.

1. Try to localize the content and match it to the interests of listeners. Emphasize local matters and involve local people.

2. Take advantage of the timelines of radio by emphasizing current activities, trends, issues, developments, and so on.

3. Use sounds in creative ways. The voice is one kind of sound, but you can use many other kinds effectively.

4. Attract the listener's attention quickly, through a compelling remark, catchy introductory sound, or other technique. The first 10 seconds of a programme are especially important.

5. Give information a flowing quality that makes it personal and easy to follow. Good radio copy is written for the ear and uses simple, understandable words. Test your radio copy by reading it aloud and revising it until it reads easily and flows smoothly.

6. Speak in a normal conversational voice, at a natural speed. Speak as if you were conversing with one person. Sound interested.

7. Use changes of pace in your presentation, to hold interest. You can do so by varying your reading speed, for example, or varying the kinds and volumes of sound.

8. Repeat important facts, such as dates, times and places of meetings. Listeners cannot refer back, as they can with printed material, so they must rely on you to repeat important information.

9. Invite listeners to take part. You can involve them mentally by asking questions, posing problems, and otherwise encouraging interaction as if you were in conversation. You can even involve them physically, sometimes, by inviting them to carry out certain actions as you speak.

Television

Two types of television media are available for teaching purposes. The first and most familiar is broadcast television, in which programmes are aired over a large geographical area. The second type is sometimes referred to as closed-circuit television. This usage takes a video signal from a tape or cassette and carries it over a cable to one or more monitors. The monitors may be in several locations or next to the video player.

Broadcast television offers exciting possibilities for extension workers. The agricultural officer can demonstrate as well as talk. The home economist can demonstrate how to make a dress. The agricultural extension worker can present useful method demonstrations as well as show a whole series of result demonstrations through pictures which show change over time. All types of visual aids such as charts, graphs, live objects, and blackboards can be used to increase teaching effectiveness on television.

However, caution should be exercised before launching into using television. Television programmes require meticulous preparation. Every piece of equipment must be in place and the dialogue must be well thought out. Most important, study the geographical area that the transmitting station covers. Second, determine the number of receivers that are available to the intended audiences. It is useless to programme for rural audiences if they do not have the necessary receiving equipment or they live outside the range of the transmitter.

Organization is an essential ingredient of a television programme. The method used to arrange the sequence of related words and pictures that make up the story is called a "run down sheet." This sheet is divided into two columns, one headed 'video' for pictures, the other 'audio,' holding an outline of what is to be said. The run down sheet is not a script but an outline to guide the television crew as well as the performers. Start using this medium with programmes with all the action in one place. Then test out the ability of extension personnel to perform more involved sequences as experience is gained.

149

The basic rules are simple. Move deliberately to allow the camera to follow. Operate within a small area. Hold any material steadily on target for camera viewing. Avoid the use of complicated demonstration material. Time the presentation before going to the studio to make sure the programme fits into the allotted time. Have some extra points ready to present in case the material runs short. Colour combinations and light contrasts are important; the television engineer can specify which combinations are best.

In spite of the relatively high cost of receiving sets, television occupies an increasingly important role in developing countries. Many governments have installed sets in each village so villagers can receive official broadcasts. Extension administrators have only to convince authorities of the value of their educational broadcasts to open up this useful channel of communication and education to the masses of people.

Instructional Television

Instructional television can be an excellent tool for extension workers. Instructional television is distinguished from broadcast television in that materials are not designed for distribution by the mass approach of broadcasting. Productions need not be tied to the specific time constraints of broadcast requirements and can be as specific in length as needed. The medium of instructional television had its beginning in cable television, with programming distributed from a central source to outlets in various centres, such as classrooms or conference rooms. With the advent of new formats of videotape such as 3/4" U-Matic, ½" Beta and ½" VHS, the possibilities of using instructional videotapes or television have expanded dramatically within the last five years. These systems feature colour images and can produce programmes with lightweight portable equipment powered from battery sources. These videotapes can then be edited in production centres and duplicated to provide current materials for extension leaders and officers. This gives immediacy to a medium for transmitting information in times such as an insect or disease emergency. The ability to prepare timely topics in the field using identifiable farmers as subject matter carries an enormous credibility factor. Videotapes can be stored for use at later dates or erased and re-used when the subject matter is no longer timely.

Another value of this medium is as a training tool for extension workers enrolled in "in-service" education courses or refresher courses. Workers can be videotaped in presentation sessions and the tapes viewed to help individuals evaluate their own delivery styles, their presentation strengths and weaknesses.

Videotapes can be mailed, sent by messenger or carried by extension personnel to wherever extension audiences may be located. A video-player and a television receiver or monitor are needed for delivery. These units can be powered by batteries provided the right selections of equipment are made. Viewing stations should preferably have access to main power.

Videotape technology is the most expensive of all of the mass media methods. Users should be aware of this fact and financing and maintenance must be adequately provided before adopting the technique. Personnel who plan to use the equipment in productions will need to receive adequate training.

The basic rules for planning and production of broadcast television also apply to the production of materials for instructional television.

Projected Visuals

Motion pictures, slides, filmstrips and overhead transparencies have much appeal, and are among the most effective of the visual teaching aids. It is as well to remember that they have important limitations as well as advantages.

The main advantages of each type of projected visual are discussed in the sections that follow, but in general, the disadvantages or limitations are similar, that is, that special equipment is required both to produce and show the visuals. This equipment tends to be relatively expensive, and some sort of electrical power is required to operate the projectors. Transportation, maintenance and storage of equipment and materials require special consideration. If these limitations do not present a problem, projected visuals should be used as much as possible in your extension programmes.

Motion Pictures

Motion pictures are really not 'motion' pictures at all. They are a series of still pictures on a long strip of film. Each picture is flashed momentarily on the screen, and the rapid succession of still pictures, each showing the subject in a slightly different position, gives an illusion of movement.

Films have the potential to create powerful emotions and urges, and thus can be a tremendously effective tool in teaching. This means that selected and used properly, they can intensify the interest of an audience in the subject. Films are also excellent for showing the steps necessary in doing a task, or for showing a continuous action.

They can reproduce events long since past. They can record a demonstration that can be shown over and over again to many different people in many different places. They can slow or accelerate motion for better analysis of action and growth. They can magnify on a screen action that normally would be too small to be seen easily or clearly by an individual or group. They can condense or stretch time.

Many other strong points for using motion pictures could be mentioned, but the reasons already given are among the most important and help explain why films are a potent teaching tool. For motivating an audience, for appealing to the emotions, for a clear concise portrayal of action, few media approach the motion picture. It portrays reality.

The size of film most commonly used for educational motion pictures is 16 mm. All 16 mm. films are not alike however. Those made for viewing silently or with comment by a leader are made with sprocket holes on both sides of the film.

Films made by professional laboratories to which sound is added have sprocket holes on only one side of the film. You should not attempt to project sound film on a silent projector because the teeth of the drive mechanism will punch holes in the sound track. Another difference is that silent films are made to operate at 16 frames per second or somewhat slower than sound film, that runs at 24 frames per second. If a silent film is run at the speed for sound, an increase in the speed of action will take place.

In selecting a film to use in a given teaching situation, the same judgement must be exercised as in selecting other teaching aids and materials. In addition to the objective, any previous experience of the audience must be considered, with such factors as age, education, interests and customs.

A film should be used only as a teaching aid. Leaders frequently make the mistake of showing a film without preparing the audience or following up. Because movies sometimes cover too much ground or include too much detail (considering the experience of the audience) viewers often fail to fully understand the ideas presented.

To do the best kind of a job, leaders must first be thoroughly familiar with the subject to be taught. Leaders must know exactly how the film supports the ideas to be presented. Before showing the film, leaders should explain the lesson, tell why it is important and stimulate viewers to look for certain things in the film. When this procedure is followed, the end of the film should be the signal for the beginning of a lively discussion and question period.

A successful film showing depends on looking after a number of details. An adequate power supply that matches the requirements of the projector is necessary; this means checking on details such as extension cords and electrical connections. Some means should be available to darken the room without cutting off ventilation. Spare projection lamps should be on hand.

Before the audience arrives, the machine should be set up, threaded with film, focused on the screen and tested. The projector should be placed high enough to project over the heads of the audience and the screen high enough from the floor for everyone to see the bottom of the picture easily. A good rule is 4 feet (1.2 m.) from the floor to the bottom of the screen.

Slides and Filmstrips

The slide is one of the most popular and versatile visuals that can be used in extension education. There are two types of slides. The first is referred to as a "lantern slide." These were used in very early days and are almost never used today. The lantern slide measured 3" x 4¼" and is mounted in glass. The second, almost universal slide is the 35 mm or 2" x 2" (50 mm x 50 mm) slide. When 35 mm colour film is used, a direct colour positive transparency is the result. Some types of this film can be home-processed, others require commercial processing. When processed, the film is cut into individual pictures and mounted in cardboard or glass ready for projecting.

Filmstrips are visuals that have been photographed on a continuous length or strip of film and are projected in a special projector, one image at a time. Filmstrips require specialized production techniques which usually offset the production economies for use by extension groups when small numbers of the material are needed. Filmstrips have the same advantages and disadvantages as slides with the exception that image sequences cannot be changed.

Slides and filmstrips have the following advantages:

1.   they can be made by the individual worker at low cost;

2.   they can be made either in natural colour or in black and white;

3.   both the slides and the projection equipment are relatively light and can be easily transported;

4.   slide sequences can be readily changed to keep them timely and localized; and

5.   slide sequences can be changed in length to fit local needs.

Slides and filmstrips have these limitations:

1.   they do not show action;

2.   they normally require "live" narration, unless synchronized with a tape recorder;

3.   they require close co-operation with a projectionist throughout the presentation if the speaker desires to be in front of an audience, unless remote control equipment is available;

4.   most important, they require a dependable source of mains power or generating equipment, and maintenance. A supply of spare projection lamps is a must for the effective utilization of projected visuals.

The same attention to meeting room preparation (screen placement, location and set-up of the projector and room darkening) should be provided for showing slides and filmstrips as discussed in the section on motion pictures.

Overhead Transparencies

The overhead projector is the most recent development in projected visual techniques. It derives its name from the technique, whereby pictures or illustrations are projected over the head of the presentor. This medium also has advantages and disadvantages. The most important advantage is that the projector may be used in normal daylight conditions and presentation rooms do not require darkening. A second factor is that presentation transparencies are easy to prepare. The presentor or an artist/illustrator can draw or write directly on clear or coloured acetate sheets using a variety of writing implements.

Wax marking pencils, various types of felt markers and specially prepared pens with inks that project in colour can be used to prepare the subject matter. Overhead projectors cannot normally be used to project colour transparencies, as 35 mm slides are too small for the projection optics. In addition, the optics are not designed for projection of continuous tones and work best with lines and solid areas. Full colour transparencies similar to 35 mm slides also require darkened projection conditions. However, overhead projection transparencies can be produced from black and white copy that is transferred to special films for duplication in office copiers.

Overhead projection is a very versatile medium. Probably the most important factor is that the presentor maintains eye contact with the audience at all times. He or she can use techniques such as progressive disclosure to control audience attention. The teacher can point to parts of the transparency for attention. Shapes, such as arrows, cut from cardstock can be used as add-on features to attract attention and emphasize. The presentor can also write directly on the film to add material or emphasize points.

The overhead projector uses electricity for the projection lamp. Its usage then is also linked to a dependable source of main power or a portable generator. Spare lamps are also vital for continued and efficient use.

STATIC MEDIA

This group of media derives its name from the fact that the material does not involve motion or sound. Examples are posters, flip charts, wall charts, maps, chalkboards (black or coloured), magnetic boards and flannel boards.

All of these techniques require the use of some form of printed material. All can be effective when used properly. There are several drawbacks to their use, mainly due to the bulk of the materials which makes transport and storage difficult. Static media are often best used with small or intimate groups for maximum visibility.

Extension workers should keep the following in mind when designing any printed visual. Legibility of a letter is determined by letter height, line width, and letter style. Letter height is most important. A lower case letter, such as an "e," that is one inch in height is visible from a distance of 32 feet; a two-inch letter is visible for 64 feet, and a one-half inch letter is visible only at a distance closer than 16 feet. Letters should be bold and simple; fancy type styles should be avoided.

When using static visuals, make sure they are displayed prominently and are well-lit, so that members of your audience may see them clearly. Make sure they are secure, but do not be embarrassed by materials that fall to the floor.

Posters

A poster is a sheet of paper or cardboard with an illustration and, usually, a few simple words. It is designed to catch the attention of the passer-by, emphasize a fact or an idea and stimulate him or her either to support an idea, to obtain more information, or take some kind of action.

People do not walk around studying posters. They look at posters in the same way as they look at trees, birds, houses, cows, or other people. A brief glance is usually as much as the average person gives an ordinary object, long enough only to identify it. If something about the object catches the attention or stimulates interest, the passer-by will look at it longer. The design and use of posters as visuals in extension teaching are based on this principle.

Since a single glance may be all any poster will get, the message must be simple and clear. Details and wordy sentences have no place. Here are a few suggestions for designing more attractive, effective posters.

1.   Decide exactly who the audience is. Decide exactly what the poster must tell them. Decide what the audience should do.

2.   Put down on a sheet of paper words and rough pictures that express the message simply and clearly.

3.   Try to put the message into a few words, a concise, striking slogan. Visualize or put into picture form the most important central idea in the message. Remember that words and picture must be seen at a glance and must stimulate a response by the viewer.

4.   Rough out the poster in small scale, 1/3 or 1/4 actual size. If the services of an artist are available, he or she can produce an excellent finished poster from an original rough sketch.

5.   Use plain, bold lettering and lines; use colour to attract attention and for contrast (but remember that too many colours add confusion); allow plenty of space, do not crowd letters, words, or illustrations.

Posters should supplement, not replace, other communication methods. They are often used to "spearhead" or introduce a campaign, or they may be used to reinforce an educational effort after it has been launched. In general, the greater the number of posters used in an area, the greater the impact, up to a certain point. Most people find it annoying to be bombarded at every turn by the same poster. Over-use of posters defeats their purpose and may actually turn people against the idea they are trying to put over. Discretion and good taste will suggest the number to use in a given situation.

Posters may be produced in quantity by letterpress, by offset printing, or by silk screens. Where only a small number are required, they may be produced by the individual, by an artist or by other people such as school children.

Posters are put up on walls of buildings, fences, trees, poles, bulletin boards, store windows, trucks, automobiles and any other places where they are likely to be seen by people passing by.

Exhibits and Displays

Exhibits and displays have some of the same characteristics as posters, covered in the preceding section. The main differences are that exhibits and displays usually are larger and more detailed.

As with the poster, the job of the exhibit or display is to catch the attention of the passer-by, impress on him or her a fact or an idea, stimulate interest in the subject matter presented, and possibly urge him or her to take some sort of action. Differing from a poster, however, the exhibit is larger, may have three dimensions and, most important, imparts more detailed information than is possible with a poster.

Because of their larger size and because they usually are placed in the market place or other areas where people move slowly, exhibits and displays attract and hold attention for longer periods than posters. Even so, the periods are not long. The viewing time will depend on whether the exhibit is in an open area or in a separate enclosed room.

Viewing time may be as short as one minute or as long as ten minutes. On average, one should aim at telling the complete story in about three minutes. This means that whatever you can do to increase the attention getting power of your exhibit, increase its attractiveness and personal appeal, and keep its content simple and clear, the greater are the chances that the viewer will receive and understand your message.

Again, as is true of all other visuals, planning is the first step in preparing exhibits and displays. Decide who the audience is, what the message is, what the audience to do. Answering these questions will help to plan the scope of the exhibit, the appeal to use and the content.

The most effective exhibits are built around a single idea with a minimum of supporting information. In a few simple words and pictures, to tell farmers that a new seed variety is better and why, that is all. Make a miniature of the exhibit from paper or cardboard. This will help to visualize it at full scale. Experiment with colours and design; an artist's help in planning the arrangement would be helpful.

To attract attention and get people to stop and look at the exhibit, something that will catch the eye should be included. This might be a live object, such as a sheep in an exhibit about sheep, or it might be colour, movement, light, or any number of things suggested by a lively imagination.

One could try to define the "something" that causes people to stop and look for more detailed information. This "something" must produce a "mental shock." Once a person stops in front of an exhibit he or she is susceptible to the rest of the message or messages in the exhibit. Incidentally, a walking person passes by an exhibit in about the same number of seconds as there are lineal feet in front of the booth. In other words if the booth is 10 feet (3 metres) wide in front, a person has about 10 seconds to absorb enough of the intended message to be persuaded to stop and discover the rest. Thus it is obvious that the attention of passers-by must be attracted in a very short time.

Make sure that the central idea of the exhibit stands out. The lesson taught must be clear at once. A combination of real objects, models or illustrative material plus a bold sign will usually get the point across.

EMERGING METHODS

Extension organizations are experimenting with other mass communications systems, many of them related to computers, examples of which are listed below.

1.   Videotex, a two-way inter-active system that links computer data bases to television sets, through telephone or cable television lines. Agricultural uses of videotex are being tested in many countries, including the United Kingdom, France, Germany, Denmark, the Netherlands, Canada and the United States.

2.   Broadcast teletext, a one-way, non-inter-active system that transmits text and graphics through broadcast signals to television sets with special decoders. Teletext is in various stages of development in more than 15 countries.

3.   Slow-scan television makes picture communications possible through channels such as telephone, satellite, microwave and FM radio. Slow-scan television is being used in the South Pacific and other areas.

4.   Communications satellites offer special potential for reaching remote areas. Agricultural uses of satellites are being tested in India, Peru, Indonesia, the Pacific nations, Canada, Alaska, Colombia and the USSR, for example.

5.   Intelligent telephone, a system which combines the telephone with the computer to provide many possible uses of interest to agricultural extension.

REFERENCES

Agricultural Communicators in Education (1983). Communications handbook. 4th ed. Danville, Ill.: Interstate.

Brown, J. W., Lewis, R. B. & Harcleroad, F. F. (1977). AV instruction: Technology, media and methods. 5th ed. New York: McGraw-Hill.

Burnett, C., Powers, R. & Ross, J. (1959). Agricultural news writing. Dubuque, Iowa: Kendall/Hunt.

Eastman Kodak (1975). Planning and producing slide programs (Kodak publication no. S-30). Rochester, N.Y.: Eastman Kodak.

Eastman Kodak (1979). Speechmaking...More than words alone (Kodak publication no. ES-25). Rochester, N.Y.: Eastman Kodak.

Eastman Kodak (Ed.) (1982). Presenting yourself. New York: Wiley.

Hall, M. (1978). Broadcast journalism: An introduction to news writing. 2nd ed. New York: Hastings House.

Izard, R. S., Culbertson, H. M. & Lambert, D. A. (1973). Fundamentals of news reporting. 2nd ed. Dubuque, Iowa: Kendall/Hunt.

Kemp, J. E. (1975). Planning and producing audiovisual materials. 3rd ed. New York: Thomas Y. Crowell.

Minor, E. O. & Frye, H. R. (1977). Techniques for producing visual instructional media. 2nd ed. New York: McGraw-Hill.

Read, H. (1972). Communication: Methods for all media. Urbana, Ill.: University of Illinois Press.

Ryan, M. & Tankard, J. W. (1977). Basic news reporting. Palo Alto, Calif.: Mayfield.

Tiffin, J. & Combes, P. (1978). Television production for education. London: Focal Press.

Chapter 11
Organizing for Extension Communications

J. F. Evans and D. T. Dahl

World attention has focused on the important research contributions made in developing countries. While much remains to be done, the scientific communities in many developing countries harbour feelings of satisfaction as they consider what they have accomplished.

Extension colleagues in those same countries have also had success. However, they often speak more of their frustrations, and the difficulties they face in implementing extension programmes that have the potential to generate the same satisfaction levels.

Extension workers are concerned about inadequate methods and media for communicating knowledge to farmers, homemakers, and other clientele. They often do not have the latest scientific information readily available in reference form to help farmers solve practical production problems.

Often, the materials they have are research reports, which are designed and written for the scientific community. They often do not relate closely to the needs of the clientele extension is trying to serve. The extension workers say they need simple publications, fact sheets, slide sets, illustrated posters, voice recordings, and other kinds of communication materials that will help them inform farmers of recommended practices for increasing production.

Extension administrators, and programme and subject-matter specialists, often face difficulties when they try to develop reports that will effectively communicate the importance of their work to government leaders and other key audiences. They often have difficulty publishing appropriate annual reports and programme accomplishments. Even when they are successful in getting reports prepared, they are sometimes concerned that they have missed the opportunity to get their stories told by using films, audio-visual presentations, or national radio and television.

Such concerns have led many ministries, and extension leaders, to recognize that one link is missing in the agricultural knowledge-sharing system. That link is an extension communications unit, comprising professional men and women who have education and experience in the twin disciplines of agriculture and communications.

Today, widespread recognition exists that agricultural communications professionals have significant roles to play in the task of sharing agricultural knowledge at all levels. It is the extension communications specialists who help extension workers produce publications for communications to the farmer. In co-operation with extension leaders, they can produce radio and television programmes, they can plan and produce audio-visual presentations for farmer audiences, and help plan educational campaigns to transfer knowledge quickly and efficiently.

In addition, they can help extension administrators develop publications and other types of presentations to communicate the effectiveness of extension's programmes to other officials within the ministry or other branches of government. Such efforts can lead to new levels of understanding and appreciation that can greatly influence the support which extension will have from the government in the future.

Without question, extension in most developing countries has a crying need for the talents of professional agricultural communicators. This chapter will consider alternative ways a unit might be organized to make the most efficient use of such talent. The remainder of the chapter will focus on functions that a communications unit might undertake, ways to organize such a group, and finally, methods of staffing the unit to best meet the needs of extension, its staff, and its clientele.

FUNCTION OF THE EXTENSION COMMUNICATIONS UNIT

With research and extension efforts existing in most developing countries, it is probable that most countries also have a unit that is undertaking some responsibility for extension communications efforts. For example, a group may exist to produce "publications", the library may have been developed, or a unit may exist to conduct staff training.

The existence of such units, their effectiveness, and the willingness or hesitancy of administrators to make allowances and adjustments, may partly determine the breadth of functions undertaken by a communications unit. But where a function is dealt with is far less a concern than that the function is somehow being carried out. If existing arrangements continue to be ineffective or inefficient, astute administrators will realise the need to make adjustments. For that reason, this section will deal with a broad array of functions that must be carried out within extension, and that may very well fit best into an extension communications unit.

Certain key functions must be present to implement an extension programme effectively. These functions normally consist of the following.

Planning

This function is carried out by any extension personnel who are involved with, and concerned about, the well-being of a programme. The intention is to make effective and efficient use of all programme resources, to assure that reasonable goals are defined, that a logical approach is designed to reach them, and that a method for evaluating the success of the effort is determined.

Information Generation

Early in the planning process, it will be necessary to generate information about the audience to be reached, and about the media available to reach them. That function can most effectively be carried out by personnel within the communications unit with input from key extension personnel who have had extensive contact with the clientele or audience members.

Staff Training

This function assures that all extension personnel involved with the programme or campaign have a clear understanding about it, have the knowledge necessary to perform their roles, and understand how their efforts mesh with the work of others. Professional communicators may play a wide range of roles in carrying out this function, and they may share those roles with other extension personnel, depending upon the nature of the programme.

Professional communicators can support training for extension personnel by providing them with the following.

1. Technical subject-matter training and teaching materials. The intention is to help extension personnel at all levels to improve their knowledge of specific technical-subjects. The materials produced might range from simple flyers or fact sheets to complex self-study guides with audio-visual support.

2. Extension methods training and teaching materials to help extension personnel improve their own communications and teaching skills.

Evaluation

This function may be shared with other extension personnel. Communications unit personnel might be totally responsible for evaluating media use, audience reach, and pre-testing the design, or appeal used in printed materials. Other extension personnel may be completely responsible for other components of the evaluation effort.

157

Communication Services

Three communications services that may logically be provided by a communications unit are listed below. Details of exactly what is required, production details and deadlines, and any other requirements would be determined in the planning process discussed earlier.

1. Providing training materials for the clientele. Such materials would include publications, flyers, posters, and other such teaching materials which are distributed to the clientele by extension workers as follow-up to teaching, or to reinforce face-to-face teaching.

2. Providing mass media releases directly to extension's clientele. Work in this area would include magazines, news releases, radio and television programming, that is, the material which is developed with extension personnel but released directly through the media.

3. Developing and maintaining media relationships. The media are often bombarded with requests from organizations wishing to send messages to the vast audiences served by the media. Consequently, media personnel must be selective; they cannot accept all messages from every organization. Professional agricultural communicators can help the media to understand the value of releasing objective extension information. They can also learn of specific requirements, and the format preferred by the medium involved. The end result can be a much higher level of use of extension information by the media.

ORGANIZATION

Organizational structures for extension communications units will vary considerably. The best structure for a given unit will depend on the location of the unit within the extension system (regional, ministry, or other level), and the functions assigned to the unit, among other factors. However, experiences of extension services in various countries suggest several guidelines for planners.

1. At ministry or national levels, the extension communications unit should be represented at top administrative levels, to ensure that communications support is woven into the total work of the extension organization. The extension communications administrator should report directly to the extension director, and should serve on administrative bodies within the organization.

2. The extension communications unit should be connected, as appropriate, with related units, such as the extension training unit, the extension library, and the field-worker operations unit. These connections should be indicated on organizational charts and maintained through programming operations.

3. The organization of an extension communications unit should reflect a conviction that effective output of publications, training materials, radio programmes, posters, and so on, depends on effective analysis and planning. From a functional standpoint, analysis and planning processes are as vital as the processes of communication.

Table 11.1 shows some organizational relationships for extension communications activity although no given unit is probably organized exactly in this manner. For example, (a) a given unit may not be involved in certain activities, such as exhibits, television or video-tape programme production, or (b) the activities of certain staff members may cut across lines of the structure, e.g., staff members who are involved in communications planning may also produce communications materials.

However, the sample structure helps express some ways in which to organize related functions of an extension communications unit. From an operational standpoint, activity will tend to flow from analysis, to planning, to production, to distribution, to analysis of results. Smooth inter-action within the system is vital because no function is isolated from the others. Those who organize and manage extension communications programmes should resist the temptation to emphasize activities on the right-hand side of the table at the expense of those on the left-hand side.

Table 11.1

Some Organizational Relationships in an Extension Communications Unit

---

Director

Programming	Production/Distribution
Programming resources	Print Services

-Providing
    analysis and
    evaluation of
    audiences, messages,
    media, budgets, other
	-Editorial
	   -Printing
	   -Distribution

	Graphic arts
Communications programme planning	Photography
-Consulting with specialists -Conceiving plans	Radio
	Video
Communications programme management	

-Executing planned programmes
   -Evaluating progress and results
	-Television
	   -Videotape
	   -Film

	Exhibits, displays
	Maintenance
	-Equipment -Facilities

STAFFING FOR COMMUNICATIONS SUPPORT

The skills and capabilities needed in an extension communications unit are implied in the previous discussion about functions and organization. Staff members in such a unit should operate as extension personnel whose specialty is communications, a complex and often technical area. As a result, extension communications units are usually staffed according to media skills which the unit requires, such as printing, radio, graphic arts. Sometimes (especially in larger units) the staff members may also divide responsibilities according to topical expertise. For example, an editorial staff member may specialize in agricultural publications, or an extension communicator may specialize in projects related to family planning.

What is the best size for an extension communications staff? Obviously, sizes of staff vary greatly, depending on many factors including the functions assigned, and budgets. An international study of national extension systems in 1980 revealed that among 62 systems that reported having agricultural information units, the number of staff members ranged from 1 to 167 (Swanson & Rassi, 1981). The arithmetic average size was about 14 staff members, while the median size was 5 staff members.

The study also showed a median ratio of 1 agricultural information staff member to 48 extension personnel (in the total system) The arithmetic average was considerably higher, with 1 agricultural information staff member to 139 extension personnel. These staffing levels for extension communications are widely believed to be lower than the optimum. However, results of the survey are useful in revealing current ranges and averages.

A severe shortage of qualified extension communicators exists in many countries, especially those in nations which have limited programmes of professional education in communications. Results of various studies have shown that current staff members in extension communications units often have inadequate communications training. This situation poses a serious challenge to extension organizations as they work to improve their communications programmes.

159

REFERENCES

Bueno, P. B. & Frio, A. S. (Eds.) (1982, August). Development support communication for rural development. Proceedings of a regional workshop at the Southeast Asian Regional Center for Graduate Study and Research in Agriculture, Los Banos, Philippines.

Carpenter, W. L. (1975). States vary in information support for extension and research programs. American Association of Agricultural College Editors Quarterly, 58(4), 16-21.

Hussain, M. (1980, March). Evolution of agricultural information system in Pakistan and its impact on farm productivity. In Agricultural information to hasten development. Proceedings of the VIth World Congress of the International Association of Agricultural Librarians and Documentalists, Manila, Philippines.

Swanson, B. E. & Rassi, J. (1981). International directory of national extension systems. Urbana, Ill.: Bureau of Educational Research, College of Education, University of Illinois at Urbana-Champaign.

Webster, R. L. (Ed.) (1974). Integrated communication: Bringing people and rural development together. A report on the International Conference on Integrated Communication for Rural Development. Honolulu: East-West Communication Institute.

Whyte, W. F. (1975). Organizing for agricultural development. New Brunswick, N.J.: Transaction Books.

Woods, J. L. (1982). Making rural development projects more effective: A systems approach (Research bulletin RB no. 390). Bangkok, Thailand: Development Training and Communication Planning, UNDP Asia and Pacific Programme.

Chapter 12
Organizational Design
and Extension Administration

J. B. Claar and R. P. Bentz

The effective conduct of extension work usually requires a rather complex organization. The relatively large size of the typical extension service, the many relationships that must be maintained, the wide scope of subject-matter to be taught, and the large number of scattered client's to be reached, all affect the type of organization that is needed. Consequently, this chapter will outline the special features of extension's assignments, the functions to be performed, the conditions to be achieved, and include some conceptual guide-lines for use in decision making.

Effective extension work also requires management and operational procedures that reinforce the organizational structure. These must contribute to a favourable work environment, and result in systematic and expeditious handling of the many administrative tasks of the organization. Inadequacies in any of these areas can seriously impair the performance of an extension service.

It must also be recognized that in day-to-day management, extension administrators must deal with inter-relationships between organizational design and administrative decisions. For example, if local extension workers are well-trained (an administrative concern), they will require less supervision and the organizational span of control can be wide (organizational design). Therefore, even though organizational design and administration must necessarily be treated in separate sections, it is important to remember that they are not discrete topics.

ORGANIZATIONAL DESIGN

The Mission and Scope of Extension

The mission and scope of extension organizations vary widely in different countries. The nature of an organization, as well as its operation, is highly related to its mission and scope. A major reason that some extension services continue to have organizational problems is that their mission is unclear.

Mission statements should be broad enough to avoid frequent changing, but they need to be specific enough to make certain things clear. Some of the essential features of a good mission statement include the following.

1. The goals and objectives of the organization.

2. The expected outputs from the organization.

3. The scope or breadth of subjects to be taught and/or duties to be carried out.

4. The clientele to be reached, including criteria to decide priorities, if several different groups of clientele are to be served.

Carefully worded mission statements that cover these four points provide the basis, and starting point, for a successful extension service.

Special Features of an Extension Service

There are several factors that arise from the circumstances of an extension service that require attention when designing or reorganizing that service. They also affect how an extension service should be administered or operated. For example, research by Axinn and Thorat (1972) points out that

"The extent to which the goals of an agricultural extension program will be achieved tend to be directly related to the extent to which those toward whom the program is directed, have participated (possibly through representatives) in establishing the goals."

Client participation should be an important factor on the development of the programme related management procedures, and the structure of the organization. Some of the other major situations affecting the structure and approach in extension are as follows.

1. Because extension is educational in nature, the clientele is free to reject the information provided. Therefore, it is important for extension personnel to know and understand the local situation, and for extension to use a persuasive approach. Extension routinely relates new agricultural information and technology to the environment and conditions of rural people. This approach sets it apart from most other government agencies.

2. Most of the information that extension transmits to clients originates outside the extension organization; therefore, special linkages with the providers of information are required.

3. Extension generally functions at each level of government; therefore, extension should be organized and administered to maintain support at each level of government.

4. Extension workers need to have adequate and appropriate mobility. Research indicates (Rogers 1983) that extension's effectiveness is directly related to the number of contacts made by extension workers with given individuals, as well as the approach used by the worker. Roger's research (1983) also revealed that the use of demonstrations is an excellent technique to build trust and to gain acceptance of information. Consequently, mobility is essential for contacting clients frequently, and for carrying out field work.

5. Extension needs a two-way flow of regular communication with both research personnel and its clientele. Extension work involves circular communications from the researcher through extension to the clientele, with subsequent feedback to the researcher. In particular, farmer experience with technical recommendations and current farmer problems are two important kinds of information that need to be transmitted back to research through the extension system.

6. The credibility of extension with its clientele as a source of unbiased information is a major factor in explaining the relative effectiveness of local extension workers (Rogers, 1983).

7. Extension needs to relate to the entire agricultural sector. Farm policies, farm prices, input availability and dependability all have a critical bearing on extension's success. Therefore, extension needs to have continuing interaction with external agencies, including sources of credit and inputs, as well as with marketing agencies and policy makers. These organizations can then be involved in effective ways to support extension work.

Procedures and Requirements

Extension's operations in a country may be seriously affected by its overall requirements in a positively negative or a way. In the latter case, some may be so serious that extension administration may find it necessary to seek modification or relief from them. Three examples follow.

1. Extension should not commit more than 60-70 percent of its budgetary resources for personal emoluments, so that it can provide sufficient funds for programmed operations. A proportionately higher salary bill may be tolerable in agencies with largely office-bound personnel, but in extension, where the basic function requires extensive mobility and field work, it is essential to have sufficient funds and resources for extension workers to do their jobs.

2. The civil service systems of many countries classify field-level extension positions at a relatively low level, and this results in very unattractive employment conditions.

Capable people are difficult to recruit and then to retain in these field-level positions, because both salaries and status are low. On the other hand, incompetent personnel are often very difficult to dismiss, due to civil service procedures. Incompetent personnel may be manageable where the staff is large; but if there is a single employee, such as an extension worker at the village level, it can result in the frequent movement of extension personnel who simply cannot be left as the sole representative of extension at that location.

3. Extension organizations that require village extension workers (VEWs) to carry out administrative or regulatory functions, may effectively destroy the credibility of extension as an educational organization. Nevertheless, it is incumbent upon extension administrators to function within the framework that is decreed by the country; therefore, this may require extension to organize the unit in such a way that it can function as well as possible within the constraints that exist.

The following discussion deals with both the organizational and principal management problems facing extension administrators as they plan the educational mission of agricultural extension work.

<u>Organizing To Do Extension Work</u>

It is an old adage of organizational theory that "form should follow function". Therefore, the functions to be performed by extension and the conditions that are needed to facilitate their successful performance will be examined first. The basic functions which need to be performed are outlined in Figure 12.1. This diagram also indicates where the relationship is authoritative (supervisory) or a support function (advisory).

Figure 12.1 Functions of an extension organization

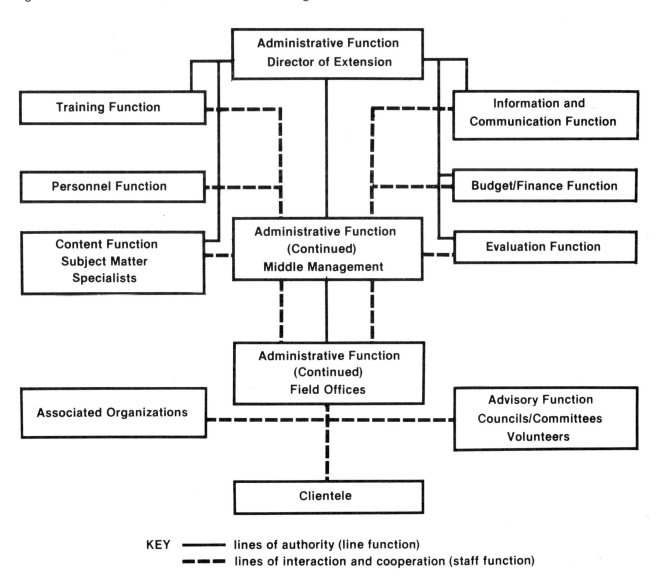

163

Extension, like all organizations, needs to have clearly understood lines of authority. The staff need to know who they are responsible to and for what, and how they are to be evaluated. Although these matters are normally a part of a job description, they should be rooted in, and consistent with, the organizational structure. In extension, as Figure 12.1 indicates, the lines of authority usually involve a hierarchy of personnel, starting with a director and proceeding through middle management, and finally to the field supervision of the village extension workers. The solid lines in Figure 12.1 indicate the lines of authority that need to exist, while the dotted lines indicate the linkages that require interaction and co-operation. Generally speaking, administration and supervision are "line" functions, while programme development, training, programme evaluation and communications are "staff" functions.

The extension organization needs close and open relationships with other organizations and agencies. Figure 12.2 indicates where some of these linkages and relationships are needed. It is, of course, especially important that extension has strong linkage to research agencies and other sources of reliable technical information. These relationships have already been examined in Chapter 2, but are mentioned here to highlight their importance in regards to organizational structure. The staff functions (represented by dotted lines) determine, to a large degree, the content and usefulness of extension's educational programme and are critical to extension's success.

Figure 12.2 Relationship of extension organization to other government entities

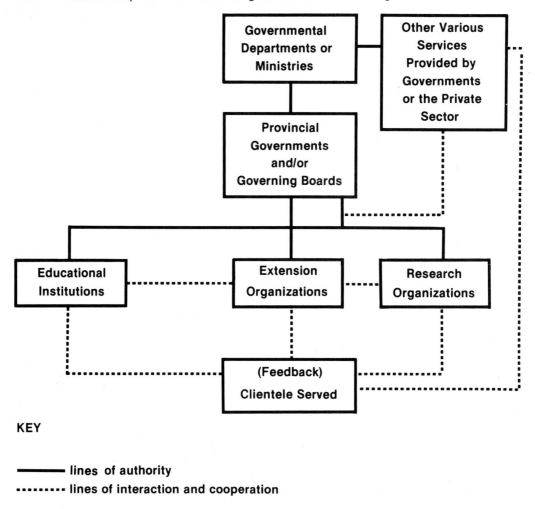

KEY

—————— **lines of authority**

•••••••• **lines of interaction and cooperation**

Importance of Client Participation

There are other special features of extension that are very important and that bear on how it should be operated. For example, the extension structure needs to provide two-way physical access between extension and its clientele. If this can be provided directly by travelling extension staff, then fewer office locations may be needed. But if mobility isn't practical, then the organization needs to provide access by locating the staff in close proximity with the clientele.

The organization also needs to provide for involvement of the clientele in several key areas. Some of these key areas are listed below.

1.   Problem identification; only farmers can explain why they act as they do. They can identify problems from their own perspective.

2.   Feedback on the effects of newly tried technology; only farmers can help to identify certain problems that may limit their adoption of new technology, which will provide guidance for future research and extension programmes.

3.   Clients' evaluation of the effectiveness of extension personnel and programmes is invaluable.

In short, the organization should delegate as much authority as possible to the local level to secure continued involvement of the clientele in programme planning, implementation, and evaluation.

Some factors that affect organizational design are: (a) the scope of the subject-matter to be covered, and the scope of client problems, (b) the size of the area to be covered, (c) the number, type, and characteristics of the clienteles to be served; (d) the duties to be performed by different personnel, (e) the extension methods that are relevant, and the relative emphasis to be placed on them, (f) the number and type of extension personnel to be employed and supervised, and (g) the distribution of knowledge centres, such as universities and especially field research stations, with which extension will need a special relationship. Therefore, the functions that need to be performed, the relationships that need to be cultivated, and the special conditions that facilitate effective performance, all need to be kept in mind.

There are a number of organizational guide-lines that can be helpful in setting up or modifying an extension organization.

1.   The organization should be hierarchial to accommodate the different administrative and geographical levels at which extension needs to function.

2.   Units with similar functions should be grouped together to provide for a reasonable span of control and workload for each administrator.

3.   Similar functions should be established at the same level in the organization, to avoid any perception of unequal access and treatment.

4.   Authority should be delegated so that it is commensurate with the responsibilities that are assigned throughout the organization.

5.   Lines of authority should be as short and direct as feasible, especially where communication is slow and knowledge of the local situation limited.

6.   Each individual staff member should have only one supervisor.

Structure of Extension Organizations

Extension is organized in many different ways around the world, because special circumstances and the personal preferences of administrators all affect organizational structure. From this wide array of organizational structures, it can be concluded that different organizational approaches can operate effectively, if the people involved understand them and want to make them function. There is no ideal model that can be recommended for all countries. The challenge to administrators is not to employ a specific model rigidly, but to create desirable conditions for effective extension work.

It is true that every organizational pattern has certain inherent strengths and weaknesses that must be given attention. For example, one of the classic issues in extension is whether subject matter specialists (SMS) should be housed with their research counterparts or separate from them. If the latter course is chosen, programme planning and esprit de corps in extension tends to benefit, but extension specialists' knowledge of new research results may suffer because there is less than day-to-day contact with researchers. The opposite tends to be the case when specialists and researchers are housed together.

Chapter 1 describes several different approaches to organizing extension. These models will not be repeated here; however, a few examples of different organizational

165

structures will be discussed, in order to indicate the diversity and problems. Figure 12.3 shows a conventional extension organization that has many typical components. In

Figure 12.3   Organization chart of a conventional extension service

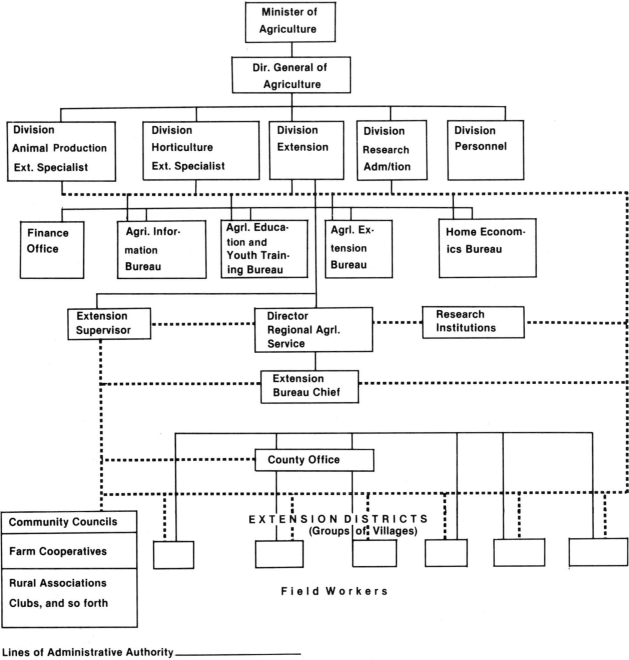

Lines of Administrative Authority _____

Lines of Staff •••••••••••••••••••••••••••••••••

this example, responsibility for livestock is included in the ministry of agriculture; it is not a part of extension's responsibilities. This pattern is common, where agricultural extension is synonymous with food and fibre crops. However, when extension is not concerned with all potential farm enterprises, its assistance in farm management for any integration of livestock and crops into a farming system will be limited.

The extension service depicted in Figure 12.3 is also organized at the same administrative level as the various subject matter units it depends on for training and support. In this model, extension is mainly a field operation, and it must depend on other units, which it does not control, for a variety of support services. In this context, it is highly unlikely that its programme development and its subject matter support needs will be adequately met.

Still another problem is that in many of these cases, universities and other sources of agricultural information and training are linked to the ministry of education.) Working relationships between different ministries are difficult, and it is rare to find any systematic lateral communication, thus research done in such settings may remain unused.

In Figure 12.3, the lines of authority seem clear but rather involved. Frequently there is a permanent secretary between the director-general and the minister of agriculture. Sometimes many critical decision-making areas are not delegated, especially for politically sensitive matters, and the line of decision-making authority for field units may be longer than it appears. In some cases, it may take up to six months to get ruling on a personnel decision.

Another area of concern is that it is not unusual for extension personnel to report to the governor's office at the provincial level or; if they do not formally report there, considerable ambiguity may result. For example, in one country extension personnel were frequently borrowed for other duties in the province, and vehicles were commandeered for other purposes. Except where the Training and Visit System has been introduced, the typical extension system is a multi-purpose organization with a variety of "nonextension" duties, and poorly defined lines of administration. (See Chapter I for a brief description of the T & V System).

Figure 12.4 depicts a government-operated extension service with a broad subject-matter scope and deliberate ties between extension and research. Subject-matter leadership and support is built in under the director's control. This organizational structure provides for good lateral communication. However, livestock is in another bureau in the government, at the same level as agriculture; hence, farm management issues can not be handled effectively.

Figure 12.4   Extension organization with a broad subject matter scope

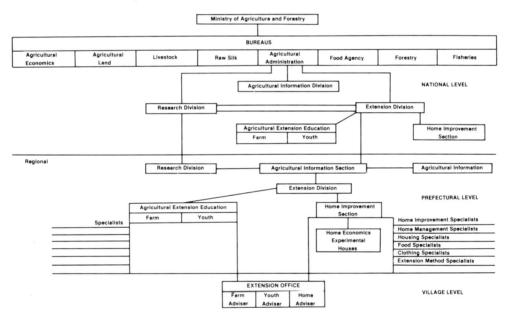

In the United States, the Smith-Lever Act of 1914 set up a decentralized system, so that the extension mission is carried out through one or more land-grant universities in each state. The federal government has an extension unit to administer the federal act and to work with the various state extension services. Relationships between the federal government and the states are handled by the Department of Agriculture, and the details of these relationships are spelled out in a memorandum of understanding with each land-grant university. The federal government provides matching funds to the states.

Figure 12.5 shows a typical state extension organization in the United States. The system features short, clear lines of authority with direct ties to research, and with subject-matter specialists under the control of the extension service at the department level. In this arrangement, the scope is broad and the client groups served are numerous. The subject matter areas covered include home horticulture, marketing and rural development, in addition to home economics, agricultures and youth programmes.

Figure 12.5   Typical extension structure in a state of the United States

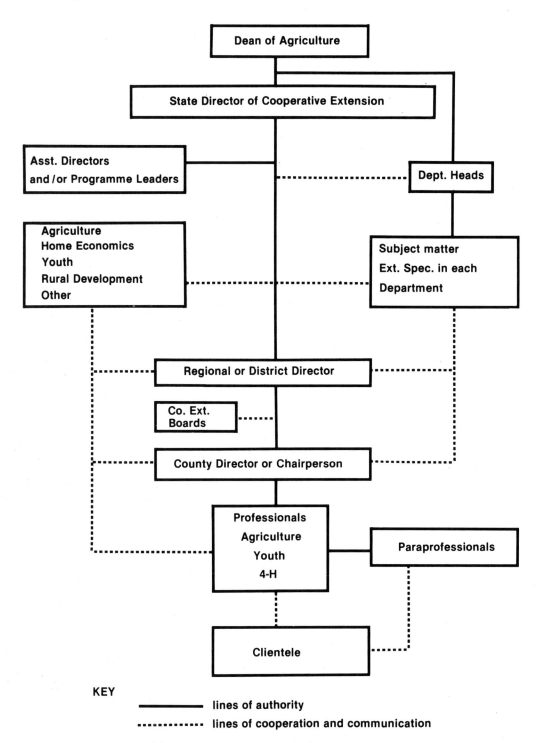

KEY

——————— lines of authority

••••••••••••• lines of cooperation and communication

Several other features of the extension service in Figure 12.5 should be mentioned. The subject-matter extension specialists are usually members of academic departments in which research and university teaching are also centred. Extension subject-matter specialists are normally housed with their research and teaching counterparts, and many have joint extension and research appointments.

County extension boards (or councils) are made up of representatives of the clientele served, and represent the clientele in programme development, budget development, and in programme and personnel evaluation. These councils also play a critical role in extension's public reporting, because they have first-hand knowledge about extension operations.

Figure 12.6 depicts an extension organization having a structure with many of the desirable alignments or relationships which are necessary for success in lesser developed countries. Some of the characteristics of this structure are that

1. the director has control of the staff offices, including subject-matter specialists.,

2. research and extension specialist divisions are organized so that it is easy for personnel to relate easily with each other,

3. university research is tied in through memoranda of understanding,

4. the extension service has clear lines of authority with district subject-matter staff to train and supervise field workers, and

Figure 12.6   Organization chart extension service indicating desirable relationships

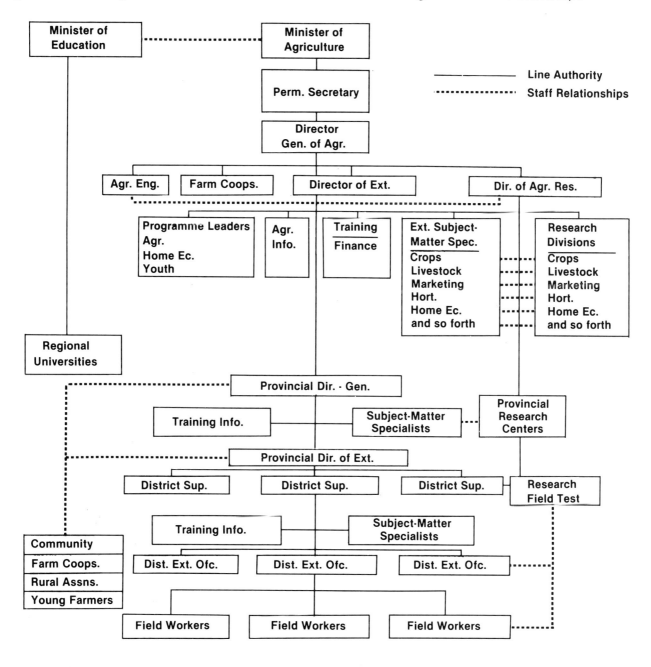

5. provincial and district extension specialists do not report to political subdivisions, but are a part of a professional extension service.

These four organizational charts were chosen to illustrate different conditions, but they do not cover all of the possible variations in organizational structure by any means.

It appears clear that extension structures inherited from the colonial past in developing countries are not well-suited to carry out a modern educational mission. Historically, the educational and communication functions were frequently tacked on to an existing unit, which carried out other non-extension functions.

There are many subtle organizational aspects that affect how extension is perceived. For example, if extension is tightly linked to an administrator who is known to deal with political matters, it may be very difficult for extension to avoid the image that it is a channel for political manipulation. If extension performs regulatory functions, or collects debts, all extension personnel may be viewed warily by the clientele. Because of these factors, it may be wise for a country wishing to utilize a persuasive approach to agricultural reform to establish some organizational distance between extension and the units that perform those non-educational functions that destroy the image that extension wishes to project.

One developing country has put extension and research together into a semi-autonomous unit that is not a functional part of the ministry. Through this organizational move, the country wishes to underscore the credibility of extension as a source of reliable knowledge and a "friend of the farmer." To do so, research and extension should be set up in a parallel, integrated fashion at all levels.

EXTENSION ADMINISTRATION

Staffing An Extension Service

The first concern in staffing an extension service should be the administrator or director's position. The type of administrator that should be chosen depends upon the type of organizational style one wishes to emphasize. Specific qualifications for the director or administrator will be covered later; however, a general overview is provided here.

Generally speaking, in most extension services, the administrator is advised to develop the approach known as "staff development leadership". This is very appropriate in remote locations, with the associated freedom of operation, and sometimes infrequent supervisory contact with, field personnel. The words leader and leadership are not necessarily synonymous with the term administrator. Leaders should be able to administer, and administrators should serve as leaders of an organization. But unfortunately, all administrators do not show evidence of the qualities necessary for good leadership, and some leaders prove to be poor administrators. A good administrator is one who is both an effective leader and an efficient administrator.

The administrative role is to plan, organize, direct and control the activities of an organization. These functions are normally detailed in the job descriptions of administrators. The successful administrator tends to emphasize means to obtain ends, whereas unsuccessful administrators often focus only on the ends.

Basil C. Moussoures, former Director of Extension in Greece, described the situation as follows.

"It is the group then which must work to accomplish certain goals. The process of administration requires that the group be taken into consideration in all activities designed to accomplish the ultimate goal of the group. This takes a special kind of individual who can utilize the resources of the group to carry out the necessary steps of planning, organizing, etc. Unfortunately it seems much easier for some administrators to do the jobs themselves. This is perhaps the clue as to where the dividing line occurs between successful administration and unsuccessful administration depends upon the ability of the administrator to lead others in the activities necessary to accomplish the objective of the organization." (FAO, p. 185, 1973).

The quality of leadership is dependent upon any individual's ability to

1. command respect and loyalty,
2. transfer responsibility,
3. include all levels in policy development,
4. instil confidence,
5. generate enthusiasm,
6. teach,
7. learn,
8. instil team spirit and action,

9. make prompt decisions,
10. assume responsibility,
11. listen,
12. recognize limitations in self and others,
13. judge fairly,
14. be honest, and
15. be objective.

These are the essential attributes of a good administrator and these are in those who truly lead. They are not qualities usually found in autocratic administrators.

Thus, the administrative styles usually sought when choosing extension administrators are those that support team-building and group involvement. This concern affects the type of staff to be chosen throughout the organization.

Three broad types of positions (local workers, specialists, and administrators) provide a general framework for staffing an extension service. But there are many combinations possible within these groups, and many factors to be taken into account. For example, what is the ideal mix among the three kinds of positions, what specialties should be covered, how much technical support staff should be provided, how many administrators and supervisors are required for effective management, and so forth.

Number of Extension Personnel Needed

The number of paid personnel required to operate an extension service cannot be stated with certainty. There are several factors to take into consideration in determining the number needed for different assignments. They are (a) the size of area served, (b) the diversity of agriculture, (c) the size and complexity of the farms, (d) the number and the educational level of the potential clientele, (e) the complexity and scope of the programme (e.g., crops, livestock, marketing), (f) the ease of communications between staff segments, (g) the mobility of the extension staff, (h) the educational level and experience of the extension staff, and (i) the major extension methods used to reach the clientele.

Although there are many factors which affect the size and staffing ratios of agricultural extension systems, it is instructive to examine some general staffing profiles of national extension systems. Swanson and Rassi (1981) developed Table 12.1, which shows a proportional breakdown of staff by function.

To understand the meaning of these data fully, a few explanatory comments are necessary. First, there is some ambiguity between the categories labelled "extension workers" (or officers) and "extension assistants" (or para-professionals). In some countries, extension workers or officers are the front-line extension personnel working with farmers. In other countries, agricultural extension officers (AEOs) play a supervisory role for extension assistants who carry out front-line extension activities. Therefore the proportion of extension staff carrying out supervisory functions may be understated.

Secondly, while the proportion of technical support personnel, or subject matter specialists (SMSs) is calculated to be only 8 percent world-wide, it is instructive to consider the European and North American data. In these countries, 18 percent and 19 percent respectively, or nearly one in every five extension workers, is a SMS. In Africa and Asia, comparable data were reported to be only 6 percent or one SMS for every 16-17 extension workers. Since SMSs provide the essential link between research and field-level extension workers, these figures for Africa and Asia suggest a serious bottleneck in providing adequate technical backstopping and training for field-level extension workers. Furthermore, SMS in many developing countries only hold a first university degree or equivalent, while in Europe and North America, SMSs are more highly trained; they are experienced professionals, mostly holding post-graduate degrees.

Another source of useful information that can be considered in projecting extension staffing requirements is the T & V System (see Figure 12.7). Under this proposed staffing arrangement, there would be approximately 12 percent administrative and supervisory staff (ZEOs, DEOs, SDEOs and AEOs), 6.5 percent SMSs, and 81.5 percent Village Extension Workers (VEW). As noted above, this proportion of SMSs appears low, and may partially reflect simply the current supply level of SMSs in Asia and Africa. However, as a proposed level, this proportion of SMSs appears to be too low. Rather, it would

Table 12.1

Summary of Extension Personnel Data

SUMMARY OF EXTENSION PERSONNEL DATA

Region	Type of Position				Total	Percent of Total	Distribution by sex		Total
	Administrative Personnel	Technical Support Personnel	Extension Agents	Extension Assistants			Percent* Male	Percent* Female	
AFRICA									
No. of Personnel	1,170	1,368	12,531	9,284	24,353				
Percentage	5%	6%	51%	38%	100%	8%	97%	3%	100%
ASIA & OCEANIA									
No. of Personnel	17,967	12,476	66,054	107,587	204,084				
Percentage	9%	6%	32%	53%	100%	70%	77%	23%	100%
LATIN AMERICA & CARIBBEAN									
No. of Personnel	1,217	2,369	12,271	2,631	18,488				
Percentage	7%	13%	66%	14%	100%	5.6%	86%	14%	100%
EUROPE									
No. of Personnel	2,409	3,416	11,316	1,446	18,587				
Percentage	13%	18%	61%	8%	100%	6.5%	86%	14%	100%
NORTH AMERICA									
No. of Personnel	1,448	4,790	12,678	6,164	25,080				
Percentage	6%	19%	50.5%	24.5%	100%	9%	81%	19%	100%
TOTAL	24,211	24,419	114,850	127,112	290,592				
PERCENTAGE	8%	8%	40%	44%	100%	100%	81%	19%**	100%

* These percentages are based only on those 57 countries where male-female data were available

** Approximately 41% of the female extension agents (or 7.8% of the total) are engaged in home economics related programs

Note: From International directory of national extension systems by B. E. Swanson and J. Rassi, 1981, Urbana, Illinois: University of Illinois at Urbana-Champaign.

172

seem more appropriate for extension organizations in Asia and Africa, over the next 10-15 years, to work toward a staffing level of SMSs of about twice the current rate, or about 13 percent, as well as to up-grade the training level of this group.

Figure 12.7   Organization pattern of intensive extension service in one state of India

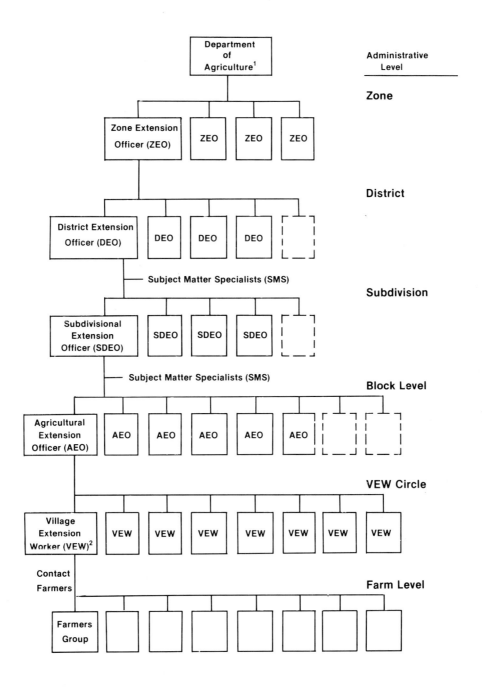

1/ A Director of Extension and Deputy Directors for (a) administration, personnel, and finance (b) implementation and monitoring and (c) technical and professional services are recommended at the central level.

2/ The ratio of VEW's to farmers average about 800 but will vary widely depending on such factors as the dispersal of farms, cropping intensity, number of differ crops, and so forth.

Note: From Agricultural extension: The training and visit system by D. Benor and J. Q. Harrison, 1977, Washington, D.C.: The World Bank.

The Swanson & Rassi study shows that the average for communications staff in an agricultural information unit is about 1 communication specialist for every 50 extension personnel (for details, see Chapter 11).

In considering an appropriate extension ratio between extension workers and farmers, it must be said that these ratios found in countries around the world vary considerably from several thousand farmers for every extension worker in some African nations to less than 1:300 in some special donor-assisted projects. The T & V System has a range between 1:1600 and 1:280 farmers in South and South East Asia (Judd, 1983). The average ratio is about 800 farmers or farm households per village extension worker. In conclusion, while there cannot be a single standard to use in determining extension staff size and ratios, these data do provide some general parameters for extension directors to use in planning future personnel requirements.

Qualifications and Functions of Extension Personnel

The qualifications and functions of personnel assigned to an extension organization depend on the mission, and the organizational structure. If the extension organization has responsibility for enforcing regulations and/or providing direct governmental services, then these extension personnel will function quite differently from those who are responsible only for providing problem-solving education. Likewise, if the extension organization is mainly responsible for disseminating information through publications, exhibits and the mass media, it too will have different personnel requirements.

Most extension services have created several categories of paid staff to carry out the extension mission. Typically paid staff are separated into the following three broad categories (a) Local or Village Extension Workers (VEWs) who work directly with the clientele, (b) Subject Matter Specialists (SMSs) who are responsible for training and providing technical expertise for VEWs and (c) administrative and supervisory staff, who are responsible for supervising VEWs and other staff, as well as carrying out the usual functions of administration (including staff selection, salary administration, budgeting, financial management, programme evaluation, and policy formation).

Village Extension Workers (VEWs)

The village extension worker under ideal conditions is expected to be a teacher, facilitator, organizer and leader. The VEW helps farmers to become aware of new research findings and technology, and to adapt this new technology to the local situation, and conditions. The goal is to help farmers move from the awareness stage in the adoption process to the point where they have gained sufficient knowledge, skills, and, possibly, changed attitudes to make a decision about the value of the new technology. The intent is to bring about improvements in farm practices that help to increase productivity and incomes, and to enhance the quality of life. Some guidelines to consider when implementing this position include the following.

1. The VEW should live and work in the geographic area where he or she is assigned. This practice may limit the pool of people interested in extension work, but it is an important factor in the VEW being accepted by his or her clientele.

2. The VEW should be assigned only to educational and communications duties. The mixing of extension and non-extension functions at the local level leads to confusion in the minds of the clientele, and can contribute to role conflicts. It can, consequently, have a negative impact on how the local extension worker is viewed by the community.

Some of the duties a local extension worker is expected to carry out include the following.

1. To develop a working relationship with all groups of farmers and other clientele in the community. To accomplish this, the VEW should

    (a) visit villages, farms, demonstration sites, and other installation in his or her area regularly,

    (b) co-operate with farmers, community leaders, dealers, and other business people (particularly those related to agriculture), and

(c) establish a liaison with people in the general service area assigned to other public agencies (government, research agencies, and educational institutions).

2.   To develop appropriate advisory mechanisms; for example,

(a) establish committees, councils or other mechanisms to obtain local-level input and feedback into the process of problem identification (programme development), programme execution, and programme evaluation, and

(b) participate in committees or other structures and organizations, that are concerned with agricultural development in the area

3.   Develop plans for, and prepare reports on, extension activities carried out in the service area.

4.   Conduct educational activities such as demonstrations, meetings, surveys, field days, workshops, and so forth for all farmers, including women and young farmers. The local extension worker should co-operate and assist as appropriate in farming systems research projects carried out in his or her service area, and also to organize exhibits and campaigns, and conduct other special projects.

5.   Provide accurate information to all groups of farmers, based on an authoritative and unbiased source. Local extension workers should make use of one-to-one instruction, face-to-face group discussion, distribution of single concept fact sheets (impact points) and other publications, as well as to assist as needed in preparing releases for the press, radio, and/or television.

6.   Co-operate fully with staff members in the extension organization in planning, conducting, and evaluating extension programmes.

In summary, the local extension worker serves as an educator and communicator in the process of social and economic change in agriculture, home economics and rural life. The VEW should be considered a respected community leader with proficiency in rural-oriented subject-matter. The VEW is a teacher of adults and youth and is an individual who helps clientele to identify their own problems and to find practical solutions. VEWs have the primary responsibility of working directly with farmers, the farm family, and/or the farm household.

The minimum qualifications for local-level extension workers include the following.

1.   Sufficient formal training in relevant subject-matter, to be able to make a positive contribution to the agricultural development of his or her service area.

2.   Practical skills and experience in farming and/or closely related activities.

3.   Ability and desire to work co-operatively with rural people and to help them reach their goals.

4.   Understanding of how adults and youth learn, and familiarity with the techniques, processes, and methodologies of extension education.

Each national extension organization, based on the availability of trained agricultural personnel, will have to determine how to apply the above requirements. Most countries do not have a sufficient supply of trained personnel beyond those with intermediate-level agricultural training. Some countries have only secondary school diploma holders available. Agricultural graduates from universities may not be available for hiring, except for the higher level administrative positions. Consequently, extension organizations should establish realistic minimum qualifications for local extension worker positions based on what is obtainable. If this level is low, then appropriate in-service training will be needed to resolve these deficiencies.

Support Staff

Local extension workers are, out of necessity, generalists. However, to be effective, they need a continuing flow of new technology and other information to make farmers' methods more productive, as well as to improve the quality of life of the farm family or household. The primary support staff are the subject matter specialists (SMS) who are expected to provide a continuing flow of new technology and other relevant information to VEWs. In addition, there are other smaller groups of specialists, such as agricultural communicators and trainers, who provide an essential support function.

Subject Matter Specialists (SMSs)

If a national extension system is to be effective in transferring improved technology to farmers, then SMSs play an essential role within the extension service. Their task is essentially one of linking VEWs with sources of new technology and information (primarily the agricultural research system), as well as feeding back farmer problems to research personnel, for analysis and solution.

Some of the duties assigned to these positions are the following.

1.   Maintain contact with research agencies and researchers to know current recommendations, and subjects being researched, and to provide input into research activities by bringing to researchers the insights of the "real world" situation.

2.   Study and analyse trends and research data related to their particular field of interest.

3.   Develop training plans and other activities that will convey knowledge and the results of research to VEWs who can transfer it to farmers for the resolution of their problems.

4.   Co-operate with agricultural information specialists to prepare bulletins, circulars, news-letters, single-concept fact sheets, and appropriate visual aids to be used to convey information to VEWs.

5.   Participate in field activities (as and when possible), working directly with researchers, VEWs, and farmers on on-farm research demonstrations, workshops, and so forth.

The minimum qualifications for subject-matter specialists should include the following.

1.   Completion of a first university degree, or equivalent, with specialization in an appropriate agricultural discipline. Post-graduate education would be highly desirable.

2.   Ability and desire to translate research findings and materials into appropriate technology that can be used in practical situations to meet farmer problems.

The role of the subject-matter specialists is to help bridge the gap between research findings and practices/found in farms and homes. The link between research and extension is critical, because extension cannot exist for long without an unbiased source of knowledge. Research, on the other hand, will not be useful to a country's agricultural development if there are no means of putting new technology into practice, or a means for feedback. In addition, the subject-matter specialists are expected to help modify research findings to realistic recommendations that will fit the local situation.

Other Support Staff

The other kinds of support staff include persons who are specialists in communications, editing, programme development, training, and evaluation. Most of these support positions require a high level of specialization. All of them require an understanding of extension methods and a dedication to the purposes of the agricultural extension organization.

Administration and Supervision

The first two categories of staff (village extension workers and support staff) will not be successful without the third category of extension personnel, those responsible for administration and supervision. Administrators and supervisors are responsible for managing the organization and for giving it leadership and direction.

The Extension Director

The director of extension is the organization's chief executive officer. As is true for all administration, the function of the director is to plan, organize, direct and control the activities of the organization. More specifically the position include responsibility for supervision, personnel management, salary administration, facilities management, fiscal management, programme development and co-ordination, programme execution and programme evaluation. In short, the director is ultimately responsible for all aspects of the extension organization.

The above functions, almost by definition, must be delegated to some extent to others in the organization. Because of the need for delegation, there is also great need for effective communications within the organization.

The person who is to head an extension organization should have the following minimum qualifications (a) a university degree in a relevant field and usually, a post-graduate degree, and (b) several years of professional experience, preferably at different levels in the extension organization.

Beyond these minimum qualifications, there are a number of other important considerations, including (a) the ability to make decisions, and to follow through on those decisions, (b) to have a positive attitude in dealing with superiors, subordinates, and the general public, (c) a highly developed sense of co-operation, (d) the ability to inspire in others a sense of responsibility and duty to the people being served, (e) a basic understanding of the educational philosophy and methods of an effective extension organization, (f) the technical skills of management and (g) an understanding of agriculture and the ability to relate to rural people.

Middle Management Personnel

A team of managers and supervisors are required for the effective operation of an extension organization. The composition of the team, the numbers involved and the qualifications required for the different positions will vary from country to country depending on the size, complexity and functions of the organization. One of the key features of the T and V System is to install a clear-cut system of management.

The example from India (see Figure 12.7) had four managerial levels (zonal, district, sub-divisional, and block) below the state level. The extension officers at each level are clearly responsible, and in charge of supervising those officers and/or field personnel (VEWs) in their respective service areas. The Subject Matter Specialists (SMSs) are also clearly designated to provide a "staff" function within this system of management.

The supervision of field extension personnel in most developing countries is of critical importance. Firstly, VEWs are distributed widely, with many living in their service area; therefore, supervision is difficult to carry out. Secondly, because the front-line extension staff have only secondary, or at best intermediate level, education, they may need closer supervision than a professionally trained staff with university degrees. While supervision is a key function at each level of the extension organization, the agricultural extension officers (AEOs) who supervise the VEWs have a particularly important responsibility.

Proper supervision of staff is a key element in developing effective extension workers. It is a process by which workers are helped to do their jobs with increasing satisfaction to themselves, to the people with whom they work, and to the organization. Supervision of extension staff, therefore, is concerned with the improvement of extension personnel as individuals and as educational leaders. Individual guidance, encouragement, and on the job counseling constitute important aspects of supervision.

The qualifications of middle-level administrators and supervisors should generally be the following (a) a university degree, or its equivalent, in an appropriate field of agriculture; and (b) special training in extension education. In addition, it would be desirable for extension officers to have special training in programme administration and supervision, a farm or rural background, and field experience as a front-line extension worker.

Pre-service and In-service Education as an Administrative Function

The performance of extension staff must be a major continuing concern of extension administrators. Although the pre-service and in-service education of extension personnel are covered in other chapters (see Chapters 14 and 15), the role of extension administrators to perform these services deserves brief treatment here. Pre-service and in-service education for extension staff is so important that a training officer should be employed by extension. In the case of the T & V System, a training officer is a key member of each three-person SMS team operating at the subdivisional level.

177

Pre-service Education

Research in extension methods indicates that change agents are most effective when there is minimal "social gap" between the extension worker and clientele. The system of selecting people for extension positions may result in new staff not having suitable backgrounds for the work. In addition, their training will probably have been theoretical, with little practical experience. Extension administrators should be concerned about the type of candidates available for employment and they should work with -the deans, directors, and principals of pre-service training institutions to include areas of present serious deficiency in the curriculum and the methodology of instruction in the future (see Chapter 14).

In-service Training

Extension administrators need to ensure that field staff maintain their competencies throughout their careers. It is imperative that field-level extension workers stay up-to-date, so that they may function effectively as educational leaders. The major types of in-service training can be described as follows.

1. Training which is designed to correct deficiencies of new or promoted staff in their appropriate subject matter areas.

2. Bringing staff regularly up-to-date on new developments in their respective subject matter areas.

3. Training in the extension teaching methods that are necessary for success on the job.

The in-service training function is generally of such size and scope that a full-time training officer is required to co-ordinate it. The training officer should have a degree of visibility in the organization, and should report at a high level in the administrative structure. In-service training stresses the necessity of staff always evaluating the relevance of their clients' needs. In-service training plays an important role in organizational survival.

Extension organizations can utilize several different methods to meet this need. Part of in-service training may be accomplished via the mechanism of providing "released time" for staff to participate in credit or non-credit progammes at an educational institution. It is also essential that some in-service training be provided by the organization itself, and be tailored to specific needs. For example, the T & V System provides in-service training to all field-level extension personnel on a regular basis (one day every fortnight for VEWs; in some countries, VEWs have one-year training programmes at extension-operated training centres).

PERSONNEL MANAGEMENT

Personnel management consumes a significant portion of administrative time and attention. The details of the various aspects of the personnel function are beyond treatment here. They include the policies, procedures, and practices that result in the hiring, firing, training, and promoting of staff. How these matters are dealt with affect morale, loyalty to the organization, and pride of accomplishment. Clearly, the personnel management function needs to be well-handled if extension is to fulfil its mission.

There are a number of factors or considerations that can serve as useful guide-lines for effective personnel management.

1. Personnel policies should be designed to facilitate effective extension work. Policies or procedures that hamper this goal should be avoided, or modified.

2. Personnel policies should be designed to facilitate employee co-operation. Positive approaches give better results than do restrictive approaches.

3. If the organization is large, it is more efficient to centralize the personnel management function. However, when this is done, consideration should be given to the need for flexibility and the recognition of differences among staff members.

4.   Personnel decisions on important matters such as recruitment, placement, performance appraisal and promotions, should be made bearing in mind the needs of the organization and should be based on merit, not external factors.

5.   Personnel management should be conducted in a fair and impartial manner. If staff are treated equitably, their responsiveness to the goals and objectives of the organization will be enhanced.

6.   The extension staff should be permitted, and encouraged, to participate in the decision-making process. Group consultations tend to build morale, and permit staff to share in the credit for any improvements. While extension administration must eventually shoulder the responsibility for decision-making, it is likely that group participation in decision making can help to legitimize actions and to gain early acceptance throughout the organization. The group decision-making process often results in better decisions.

UTILIZING VOLUNTEERS

A corps of trained and dedicated volunteers can increase the reach of an extension organization. Volunteers can add an important extra dimension to the extension staff, and this can have considerable impact. For example, the Illinois Cooperative Extension Service (1982) in the United States recently reported that it has approximately 500 paid professionals, but that more than 30,000 different volunteers assisted in extension activities in a one-year period. Approximately 800,000 different individuals received direct instruction, and many more people were reached through indirect means. The impact of extension in Illinois would be far less if the volunteer component were removed.

Volunteers give of their time for a variety of reasons, such as (a) the prestige that accrues to them because of the role they play in the organization, (b) specific benefits from their participation, (e.g., they may be given special training, or early access to new information), (c) the larger benefit to their community, because of their participation, and (d) the desire to be involved and to have a voice in matters that are important to the future welfare of their family and community.

The kinds of tasks performed by volunteers are many and varied. Typically, they can be categorized as follows.

1.   Advisory groups. They can help identify important problems, and then help set programme priorities. Advisory groups are often used to help assess needs, contribute to programme development, assist in programme execution, and participate in programme evaluation.

2.   Information Transfer. Volunteers, such as contact farmers, can help to disseminate information to others. This role may be performed quite informally, and it may entail nothing more than passing information to a neighbour. Volunteers often manage demonstrations on their own land and allow their neighbours to see the results; they may also be formally involved in on-farm research.

3.   Organizational Maintenance. Volunteers also help carry out many tasks that help to maintain and promote the organization. They may serve as leaders of young adult or youth groups. They can provide liaison with other organizations, or help raise funds for extension programmes. In some extension organizations, selected volunteers participate in a variety of administrative functions, (such as fund-raising, financial management, personnel selection, and facilities management.

FACILITIES AND EQUIPMENT MANAGEMENT

The physical facilities required for doing effective extension work are not as extensive as those required at an agricultural research station. Subject matter specialists can make effective use of small plot equipment, such as those found at research stations. However, the typical extension worker needs only basic physical facilities and equipment. It is a mistake to assume that an extension organization can function well without adequate equipment and physical facilities; extension workers need the necessary tools for them to succeed. It is a common complaint in many developing countries that most operating funds are spent on personnel emoluments, leaving almost no budget for extension programmes. Such a practice severely reduces the effectiveness of extension, because of the high mobility and training requirements.

Office Facilities

The primary function of an extension organization is to transmit information and skills about new agricultural technology. Therefore, the first need of an extension organization is to be able to communicate with its own staff. Extension workers, consequently, need to have a place where they can be reached. They need a place where clientele can meet them. Most extension organizations have found that these needs are met most effectively by creating extension offices. Determining the number and location of extension offices requires careful planning. Obtaining and maintaining office facilities requires constant attention. The questions of office configuration, necessary office equipment, the need for clerical staff, and office procedures must be resolved within the resource constraints of each country. Some common errors associated with facilities management include (a) inadequate clerical support, resulting in extension professionals sometimes performing clerical and other routine tasks, to the detriment of their educational roles; (b) offices which are poorly equipped and laid out (Carefully selected furniture, equipment, and office machines make effective work possible, and staff are stimulated to work better and to increase their efficiency in congenial surroundings), and (c) offices are often poorly located and lack visibility and/or accessibility.

Resource Materials

Extension staff, especially AEOs and VEWs who work out of field offices, need to have access to resource materials. They need relevant bulletins, manuals and reference guides. They also need equipment and supplies (seed, fertilizers, pesticides) for use in demonstrations. The kinds and amounts will vary depending on the programme area and the situation concerned. Extension workers cannot be successful in demonstration work without access to necessary supplies and equipment.

Transportation

Adequate transportation is a prerequisite for effective extension work and presents special problems for extension organizations. All categories of extension staff must be able to travel throughout their respective service areas if the organization is to fulfil its mission. How an extension organization can best address this need will depend on local circumstances.

Some solutions call for major investments. However, if the staff is centralized in urban areas to reduce the need for too many vehicles, there are negative impacts on programme delivery. Like so many aspects of extension administration, providing for adequate transportation involves "trade-offs".

Housing

Generally extension workers should live in their service area to increase their availability to clients. Because suitable housing is frequently unavailable in villages, it may be necessary for extension to build and maintain housing for its field level staff.

FINANCIAL MANAGEMENT

Extension organizations are financed in different ways throughout the world. The laws under which the extension oganization operates determine to a large extent how extension is to be financed. In most countries, agricultural extension is a public agency under the ministry of agriculture. Budgets are created based on appropriations and/or grants and income. Expenditures are then incurred for salaries, wages, transportation, facilities, supplies, equipment and so forth. Reports are made as to how funds were expended, and requests are made and justified for subsequent funding. When extension is a governmental entity, budgeting, disbursing, and control of funds must be managed by extension administrators in accordance with established policies and procedures. Some general guidelines are listed here.

1. Extension administrators should portray the financial needs of the organization honestly and realistically and be prepared to explain both when the resources are needed and how they will be used.

2.   Sources of financial support should be sought from all levels of government, as well as from those who benefit directly from extension services. Active, local-level participation in the financing of extension's efforts is viewed by many as an indication of the interest in, and need for, the extension organization. Local-level financial involvement may not represent a significant fraction of the total resources required. However, it appears that a feeling of ownership and heightened level of interest tend to follow a financial commitment to a programme. As positive attitudes of support, and reports of benefits from the programme, filter back to the national level, they in turn will have an impact on financial support at that level.

3.   Budgeting should be handled in such a way as to facilitate the management of money allocated to the organization. The budget should be broken down into logical categories so that budget managers may know how the money is to be used. Procedures need to be developed and followed so the manager will know how the money was actually used. There also needs to be some means by which budgets and expenditures can be adjusted to meet changing situations.

4.   Authority to expend money within given parameters should be delegated to the extent possible. Persons responsible for an educational activity need some freedom to make decisions on how to go about the task. Generally speaking, it is better to keep the "distance" between the budget manager and the person who is delivering the educational programme as short as possible.

5.   Adequate control procedures must be developed and followed. There must be adequate safeguards to guarantee that public financial support is not redirected for private gain. The extension organization must be in a position to account for all funds used and to be able to say that these funds were expended in accordance with relevant law, rules and procedures. This is an absolute necessity in all public organizations.

REPORTING ACCOMPLISHMENTS

Reports provide a major means for staff to maintain communications with extension administration, and to explain accomplishments. If used properly, they can provide a basis for decision-making, and can serve as input to other agencies. Extension staff tend not to like paperwork; they tend to be "doers", concerned with the here and now. They also tend to under-estimate the value of making plans, and then preparing a written report on outcomes. However, it is the planning and reporting functions that introduce structure into an unstructured environment. If developed and implemented carefully, these functions can serve useful purposes.

Most extension organizations make use of a whole array of planning and reporting mechanisms. There is considerable diversity in the systems used. Again, there probably is no single approach that fits all situations. However, there are several general principles to be considered.

1.   Plans and reports should be useful to those who prepare them, as well as to those who read them. Resistance to the whole process will be a continuing problem if those who prepare plans and write reports see no value in them.

2.   Plans and reports should be used as input by decision makers. A planning and reporting system is far too expensive in terms of staff time if its only use is to show that everyone is busy. The corollary to this is that plans and reports should not be required any more than is absolutely necessary.

3.   Plans and reports by extension staff should be used locally with advisory groups and programme development committees. This use will help such groups to know what worked or didn't work previously. It should help in the targeting of educational activities on current problems.

4.   Plans and reports should be developed in light of realistic expectations. Goals of the extension organization might be phrased in lofty terms (e.g., to do away with hunger and poverty in rural areas), but plans and reports should be set in a framework of activities and goals that can be accomplished within given time frames and for identified clientele groups.

5.   Plans and reports should be used to help extension staff improve their performance. They can serve as input for counselling by supervisors. They are useful for

the training of new staff. They should be used by supervisors in the process of measuring performance and determining merit.

6. In nearly all countries, plans and reports are used to justify public expenditures and to support requests for future financial support. Without an adequate reporting system to collect the required information, administrators are unable to give creditable responses.

Some additional guidelines that might be helpful include the following.

1. Reporting should be standardized as much as possible. Comparable data can then be collected from all staff and summarized for the organization.

2. Only the information really needed should be collected. Routinely gathering data without a specific purpose in mind merely clutters up the system.

3. Staff should be urged to prepare concise reports. They should identify as simply as possible (a) the problem, (b) the activity or activities conducted, (c) the number of clientele who participated or who benefited in some way, and (d) the change, if any, that resulted from the effort.

4. The review and analysis of plans and reports should be accomplished in such a way as to facilitate the retrieval of information for comparison purposes-over time and between units. There is little value in a reporting system if the information it contains cannot be retrieved and used.

5. Plans and reports should be accurate. It is useful to record data at the time it becomes known, so that subsequent plans and reports are based on facts rather than recollections.

LIAISON WITH OTHER ORGANIZATIONS

The need for liaison and interaction between the extension organization and other agencies was covered in Chapter 2. There are often a number of public, semi-public, and private organizations concerned with rural development in a country. In addition to national organizations, there can be a number of local agencies such as farm co-operatives, community councils, farmer associations, and commodity groups. Many of these local and national groups have goals that are similar in essence to those of the extension organization. Where similarities occur, the opportunity exists for extension to integrate its activities with those other organizations. The resulting co-ordination tends to increase the impact of these efforts to mutual advantage.

Extension's credibility is based on the fact that it provides unbiased knowledge, technology and research findings, although other groups may promote a particular viewpoint. In these situations, extension staff need to refrain from taking sides. This does not mean that extension should avoid the appearance or the reality of conflict of interest; relationships should be established with other organizations in ways that protect the integrity of interests of all parties. The opportunities for benefits to rural people that come from inter-organizational co-operation may be great. Potential hazards are real, but they can be overcome.

REFERENCES CITED

Axinn, G. H. & Thorat, S. (1972). Modernizing world agriculture: A Comparative study of agricultural extension education systems. New York: Praeger.

Barker, R. & Sinha, R. P. (1983). The Chinese agricultural economy. Boulder, Colo.: Westview Press.

Benor, D. & Harrison, J. Q. (1977). Agricultural extension: The training and visit system. Washington, D.C.: The World Bank.

Judd, P. (1982, May). The training and visit extension system in South East Asia. Paper presented at the Regional Seminar on Extension and Rural Development Strategies, Universiti Pertanian Malaysia, Serdang, Selangor, Malaysia.

Maunder, A. H. (Ed.) (1973). Agricultural extension: A reference manual. Abridged edition. Rome: Food and Agriculture Organization of the United Nations.

Rogers, E. (1983). Diffusion of innovations. 3rd ed. New York: Free Press.

Swanson, B. E. & Rassi, J. (1981). International directory of national extension systems. Urbana, Ill.: Bureau of Educational Research, College of Education, University of Illinois at Urbana-Champaign.

183

Chapter 13
Evaluating Extension Programmes

J. Seepersad and T. H. Henderson

THE MEANING OF EVALUATION

Evaluation is an activity we engage in every day because we are always making judgements relating to the value or worth of things we do or experience. For example, we are constantly evaluating the food we eat, the jobs we do, the programmes we listen to on the radio, and so forth.

The following sequence of steps are usually involved in all evaluations.

1. Evaluations are usually prompted by the need to make a decision about the value or potential value of something. For example, if we are listening to a programme on the radio for entertainment, we may need to decide whether such a programme is likely to provide the type of entertainment we are looking for. Or, at the end of the programme we may want to decide whether we would listen to similar programmes in the future.

2. We define criteria as to what constitutes an entertaining programme for us (type of music, amount of a certain type, etc.).

3. We make observations or collect evidence relating to the criteria (what type of music is being played, and how often?).

4. We form judgements relating to the value or potential value of the programme (not valuable, or not likely to be valuable because the music we like is hardly being played).

In our day-to-day activities we may hardly be aware of these steps. However, in systematically evaluating extension programmes, explicit attention must be given to each step in the process.

Definition of Evaluation

Extension evaluation can be defined as a continuous and systematic process of assessing the value or potential value of extension programmes. This process includes developing criteria from the concerns of the relevant audiences for the evaluation, the collection of data relating to the criteria, and the provision of information that adequately addresses the concerns.

The definition attempts to be as broad as possible, and emphasizes the continuous and systematic nature of evaluation. With regard to the role of objectives in evaluation, a major concern is usually the extent to which the programme met some or all of its goals or objectives. However, different evaluation audiences may want to focus on different aspects. Additional concerns may also become prominent during the execution of the programme. Finally, extension evaluations conducted regularly during programme implementation will indicate the potential value of programmes.

TYPES OF EVALUATION

Informal and Formal Evaluations

One of the earliest writers on extension evaluation (Frutchey, 1967) pointed out that there are several degrees of evaluation. This can be illustrated by means of a continuum

(Figure 13.1). At one end of the continuum there are "casual every day evaluations," or informal evaluations, and at the opposite end, "extensive formal studies".

Figure 13.1 Degrees of evaluation

Casual every day evaluations can be equated with first impressions of a particular experience. According to Frutchey (1967, p. 2), "They are the ones we ordinarily make without much consideration of the principles of evaluation in the decisions we make about simple problems". On the other hand, extensive formal studies involve the use of sophisticated research procedures and are often conducted by teams of evaluation specialists.

Informal evaluations are unsystematic; the criteria and evidence used in making judgements are implicit. They can, therefore, be biased and misleading. The more systematic the evaluation, the more likely will it contribute to making useful decisions about an extension programme. Thus, we should at all times attempt to make our evaluations more systematic and more formal. This is not to imply that the only good evaluations are those which approximate the extensive formal study. Such studies may only be justified where a major extension programme is involved.

Formative and Summative Evaluations

The terms formative and summative are associated with two important roles of evaluation. Taylor (1976, p. 355) provides the following definitions of these types of evaluations:

> "Formative evaluation attempts to identify and remedy shortcomings during the developmental state of a program. Summative evaluation assesses the worth of the final version when it is offered as an alternative to other programs."

In the past, the emphasis has been on summative evaluations that were conducted after the completion of the programme to assess the accomplishments and whether intended objectives were achieved. Nowadays, more and more attention is being paid to formative evaluations that are conducted before programme completion, more particularly, during programme implementation. Such evaluations provide early feedback on programme weaknesses, which can then be used to modify or adjust the remaining stages of a programme.

Monitoring and Evaluation

Conceptually, monitoring and evaluation correspond in many respects to formative and summative evaluation. However, in extension, the former has been most extensively used in conjunction with a specific monitoring system developed for the Training and Visit System (Benor and Harrison, 1977). According to Cernea and Tepping (1977, p. ii), the system is designed "as a management tool to ensure the extension organization is operating efficiently, to enable management to take corrective action when necessary and to provide policy makers with appropriate information". Cernea and Tepping defined monitoring as follows (1977, p. 11);

> "It is the gathering of information on utilization of project inputs, on unfolding of project activities, on timely generation of project outputs, and on circumstances that are critical to the effective implementation of the project."

185

Indicators used for monitoring are the number of contact farmers reached by the village extension worker (VEW), the number of visits made by the VEW, and so on.

With reference to evaluation, Cernea and Tepping distinguish between on-going evaluation and ex-post evaluation as follows:

On-going evaluation is an action-oriented analysis of project effects and impacts, compared to anticipations, to be carried out during implementation.

Ex-post evaluation would resume this effort several years after completion of the investment, to review comprehensively the experience and impact of a project as a basis for future policy formulation and project design.

Indicators used for evaluations include yields of major crops, and changes in cropping intensity and patterns.

Monitoring and evaluation will be discussed in greater detail later on in the chapter.

Reasons for Doing Formal Evaluations

The discussions in the previous sections have already touched on some important reasons for doing formal evaluations. An expanded discussion of these and other reasons is provided below.

1. Evaluations guide and direct future action. It is important that the merit or worth of programmes be determined as accurately as possible through formal evaluations which are less subject to the effects of personal biases.

2. Evaluations can help to improve on-going programmes. Information provided from evaluations conducted during programme implementation (formative, monitoring and on-going evaluations) can be used to modify or adjust subsequent stages of a programme.

3. Evaluations can provide the basis for planning future programmes. For example, summative evaluations can provide answers to questions relating to whether a programme should be continued, expanded or terminated. All evaluations should also describe the environment, resources and constraints that could affect a programme, and factors that contribute significantly to the success or failure of a programme should be identified. Such information will be of immense value in planning and conducting future programmes.

4. Formal evaluations are indispensable where accountability is an important concern. Sponsors of special extension projects for example, will be interested in whether funding has served the purposes for which it has been disbursed, and they may require that further funding be contingent upon formal evaluations conducted at certain points during the project. Agricultural extension organizations also have a responsibility to account for public funds: accountability for programmes should be an important concern, although the mandate may not be as explicit.

5. Formal evaluations can serve important public relations functions. Outside the immediate relevant audiences of administrative and technical staff directly connected with a programme, the information from evaluations can be presented to other persons and organizations who are linked to extension or concerned about its effectiveness.

6. Since formal evaluations necessitate a systematic approach to decision-making, they can contribute to the development of professional attitudes in the extension worker. Also, feedback information on programmes from evaluations can help to improve the morale of the extension staff.

BASIC STEPS IN EVALUATING EXTENSION PROGRAMMES

The actual procedures used in evaluating extension programmes may vary considerably depending on the nature, scope and complexity of the programmes, and the resources available for conducting the evaluations. However, a series of steps can be identified which are basic to all formal evaluations. These are discussed below.

Develop An Evaluation Plan

A plan should be drawn up indicating what will be done, why it needs to be done, and how it will be done. The other steps that will be discussed subsequently will constitute the major components of the plan.

An evaluation plan is important for three major reasons. First of all, resources to conduct an evaluation are always limited, and adjustments will constantly have to be made between what is the best or ideal way to conduct such an exercise and what is possible, given the limitations existing in the situation. Planning will help to ensure that the exercise is conducted and completed within the existing constraints.

Secondly, planning will also help to focus the evaluation on questions that are of concern to relevant audiences. One issue that has arisen in the past is that the findings and recommendations of evaluations have only been used to a limited extent in making subsequent decisions about programmes. An evaluation plan will give the relevant audiences an opportunity to be involved so that the evaluation can address their questions and concerns. Planning will help identify which evidence would be considered "credible" by such audiences.

Thirdly, the existence of an evaluation plan will facilitate getting useful input from everyone concerned. Specialists, for example, will be better able to guide extension staff if a blueprint exists linking the intended procedures with the objectives of the evaluation.

Consider the Need for the Evaluation

"The major purpose of evaluations is to assist in program decisions. Formal evaluations are worth doing only if they have a chance of affecting such decisions" (Bennett, 1973 p. 21). An additional consideration is the significance of the need. A major extension effort involving large numbers of farmers and substantial resources will (with other things being equal) be more significant than smaller programmes centred around a few farmers. Signficance can also be gauged by looking at the issues with which the evaluation is concerned, and assessing the likely impact of the evaluation on the resolution of these issues.

List the Reasons For Wanting to Evaluate the Programme

As mentioned before, there can be several reasons for evaluating a programme. The evaluator should decide which reasons are most important and focus the evaluation accordingly.

List the Audiences for the Evaluation Report

The potential audiences for the evaluation should be considered; they may consist of the evaluator, fellow extension staff, supervisors, advisory councils and other support groups, programme participants (men and women farmers, farm families, members of youth organizations), programme sponsors, and the general public. The evaluator will need to select the primary audience for a specific evaluation effort because different audiences will have different concerns about the programme.

State the Criteria for Evaluating the Programme

Criteria are the yardsticks used to measure the merit or worth of a programme. For example, a criterion for an extension programme may be the number of women farmers who adopt a particular practice. If an evaluation indicates that the specified number did, in fact, adopt the practice, the programme can be considered a success as far as this criterion is concerned. For example, where programme emphasis is on increasing the output of cash crops, an unintended outcome may be that land formerly used to grow food crops changes to cash cropping land. This has particular effects on women farmers who frequently grow food crops. Unintended outcomes such as these should be a part of the evaluation.

The main source of criteria should be the basic intent and objectives of a programme. If a programme was developed in response to a particular need, a major concern of the evaluation should be whether the programme is meeting the need, or to what extent it met the need.

However, several problems may emerge in attempting to develop criteria from programme objectives. First, the objectives may not be explicitly stated. Steele (1972, p. 5) pointed out that in such cases "this guide to effectiveness has to be set retrospectively by asking the question, 'What results can be expected to have occurred given these particular participants, this programming situation, and this programming input?'"

Many times, too, objectives may be vague, consisting of very general statements of expected outcomes. Much difficulty will be encountered in attempting to translate such objectives into specific ones that can be identified, examined or measured.

Another problem is that even though objectives may be carefully developed and properly stated, the programme emphasis may shift as it progresses, and the statements of objectives may not adequately reflect these later shifts. Additionally, stated objectives may not be sufficiently inclusive to identify all the major results of a programme, because these are difficult to predict beforehand.

As a result of these problems, many writers advise against focusing solely or too heavily on programme objectives in conducting evaluations. The evaluator should, therefore, discuss the issues and concerns to be addressed with the primary audiences. In addition, the resulting consequences, both intended and unintended, should be considered. Of course, the extent to which the programme met its objectives is quite likely to emerge as the significant concern, especially if programme planning procedures were adequate in the first place. However, during the process of discussion, refinements will have occurred. Objectives of special concern may be identified; also standards for judging the value of programmes may be raised or lowered based on experience obtained during the conduct of the programme. Other important concerns will also be given due consideration.

List the Resources that will be Available for the Evaluation

Evidence consists of information related to a particular criterion. In deciding on which type of evidence to use, adjustments will almost always have to be made between what is the best or ideal type to use and what it is possible to obtain.

There are various ways of classifying evidence that can be used in extension evaluations. Sabrosky (1967) distinguished two major types; evidence in terms of changes in the behaviour of people, and evidence in terms of opportunity. In the former case the major consideration is whether extension's audiences have changed their attitudes or practices as a result of the extension method or activity. In the latter case, Sabrosky pointed out;

> "When it is difficult or impossible to measure progress at the level or original status or change in people themselves, it is desirable to measure work in terms of the learning situation we have set up. If no learning situations are set up (no written materials go out, no talks are given, no demonstrations are put on, no visits are made), we cannot expect the people to learn anything as a result of extension work" (p. 26).

What can be considered an expanded version of Sabrosky's classification of types of evidence has been given by Bennett (1977). He proposes seven levels of evidence for programme evaluations that can be arranged in a hierarchy. The levels of evidence, and examples of evidence at each level, are shown in Table 13.1. At each level what was planned or anticipated can be compared to what was actually achieved. For example at the "inputs" level, the actual time spent by extension staff on a programme, or aspect of a programme, can be compared with the amount of time such staff had planned to spend. In many extension programmes, such sophistication in planning may be rare; however, evidence obtained at each level can still be useful in aiding programme decisions.

Bennett (1977, p. 8) further proposed the following guidelines to assist in deciding which level of evidence to use.

1. "Evidence of program impact becomes stronger as the hierarchy is ascended." Levels 1 to 3 provide ways of measuring possible opportunities for education to occur. He also pointed out that, "Ascending to the fourth level, reactions, can provide somewhat better confirmation of whether given activities are helpful as intended. But such evidence

indicates less satisfactorily than evidence of KASA [Knowledge, Attitudes, Skills, Aspirations] changes the extent of progress towards ultimate program objective." The ideal assessment of impact would be obtained at the highest level in the hierarchy, in terms of whether the desired end results have been achieved, and the assessment of any significant side effects.

2. "The difficulty and cost of obtaining evidence on program accomplishments generally increases as the hierarchy is ascended" (Bennett, 1977, p. 9). Although evidence at the lower levels do not provide as strong an indication of impact as those at the higher levels, it is relatively more difficult and costly to obtain evidence at the higher levels.

3. "Evaluations are strengthened by assessing extension programs at several levels of the hierarchy including the inputs level" (Bennett, 1977, p. 9).

Table 13.1

Hierarchy of Evidence for Programme Evaluation

Criteria Categories	Examples of Types of Evidence
7. End Results	Attainment of ultimate objectives. Changes in the quality of life and standard of living of farmers.
6. Practice Change	Number of farmers adopting improved agricultural practices.
5. KASA Change	Changes in knowledge, attitudes, skills, and aspirations of target audience.
4. Reactions	Number of persons indicating whether extension program is useful.
3. People Involvement	Percentage of target audience participating in program (attending meetings, etc.).
2. Activities	Learning situations set up. Subject matter taught.
1. Inputs	Number of visits, meetings, etc.

Note: From Analyzing impacts of extension programs, by C.F. Bennett, 1977, Washington D.C.: Extension Service, U.S. Department of Agriculture.

4. "Evaluation is strengthened to the extent the specific criteria for evaluation are defined prior to conduct of the Extension program" (Bennett, 1977, p. 11). The basic point here is that early clarification of programme objectives will assist in the subsequent conduct of evaluations. Evidence obtained prior to programme execution (e.g., level of knowledge, attitudes and skills of programme participants) will provide a benchmark against which progress (as a result of participating in a programme) can be judged.

5. "The harder the evidence for evaluation, the more an evaluation may be relied upon in program decision making" (Bennett, 1977, p. 12). Examples of hard and soft data are given in Table 13.2. Here again the decision as to which type of data to use rests on careful consideration of what is ideal and what is possible.

Designs for Evaluation Studies

A variety of designs can be used in collecting evidence for evaluation studies. Bennett (1977) provides a list of these in order of their potential ability to provide strong

Table 13.2

Examples of Hard and Soft Data in a Hierarchy of Evidence for Program Evaluation

		Examples	
		Hard Data	Soft Data
7.	End Results	Trends in profit-loss state ments, life expectancies, pollution, indexes, and satisfaction with health.	Casual perception of changes in quality of health, economy and environment.
6.	Practice Change	Direct observation of use of recommended farm practices over a series of years.	Retrospective reports by farmers of their use of recommended farm practices.
5.	KASA[1] Change	Changes in scores on vali- dated measures of knowl- edge, attitudes, skills and aspirations.	Opinions of extent of change in participants' knowledge, attitudes, skills and aspira- tions.
4.	Reactions	Extent to which a random sample of viewers can be distracted from watching a demonstration.	Recording the views of only those who volunteer to express feelings about demonstration.
3.	People Involvement	Use of social participation scales based on recorded observations of attendance, holding of leadership position, etc.	Casual observation of atten- dance and leadership by participants.
2.	Activities	Pre-structured observation of activities and social processes, through par- ticipant observation, use of video and audio tapes, etc.	Staff recall of how activities were conducted and the extent to which they were completed.
1.	Inputs	Special observation of staff time expenditures, as in "time and motion" study.	Staff's subjective report regarding time allocation.

Note: From Analyzing impacts of extension programs by C.F. Bennett, 1977, Washington, D.C.: Extension Service, U.S. Department of Agriculture.

[1]KASA stands for Knowledge, Attitudes, Skills and Aspirations.

scientific evidence of the degree to which observed change is produced through extension programmes. A modified list of these designs is as follows:

1. The Field Experiment
2. Matched Set Design
3. "Before-After" Study
4. The Survey
5. The Case Study

The field experiment provides the strongest scientific evidence and the case study the weakest, for the purposes of evaluation; some evaluation studies may incorporate elements of several of the designs listed above. Generally, the first two designs are hardly used in the regular conduct of evaluations, because they are expensive and dif- ficult to handle. The last three designs listed will be described briefly below.

"Before-after" study. In this type of study, observations are made before and after participation in an extension programme. The changes in the status of participants can be attributed to the programme after other competitive explanations (for example, unusual weather affecting crop yields, other programmes) have been logically ruled out.

The survey. This design is perhaps the one most often used in conducting extension evaluations. It does not require observations before a programme is implemented, and is generally easier to carry out and less expensive than the "before-after" design. However, according to Bennett (1977, p. 19) it "generally provides rather weak conclusions about the extent to which extension, rather than other forces, produces any observed differences between extension clientele and non-clientele."

Surveys can be used to collect data on people's perceptions and opinions about programme activities, and the results of programmes. Surveys can also seek information on the status of participants prior to their participation in a programme.

The survey design usually requires use of questionnaires sent through the mail, or administered through personal interviews. Sampling techniques are generally used to select the study population.

The case study. According to Bennett (1977, p. 20), "Case studies observe intensively one or only a few selected individuals, groups, or communities. Observation may involve examination of existing records, interviewing, or participant observation". Although the evidence provided by this design is not as strong as those from other designs, case studies can reveal information about a programme which is not accessible by other means. It is usually most effectively used as a supplement to other evaluation designs.

Conduct the Evaluation

Analyze the Evidence

Various kinds of data analyses can be used ranging in complexity from percentages to statistical techniques. The important considerations are the expertise possessed by the evaluator, and whether the analyses can provide the information for answering the questions with which the evaluation is concerned. Good data analysis relies on emphasis on those aspects that are related to the particular issues addressed by the evaluation.

Report the Findings

The style and content of the evaluation report should be tailored to the audiences being addressed. A variety of reporting procedures may be used (written reports, audio-visuals, question-answer reports, etc.).

With regard to implications and recommendations, the evaluator should clearly state the reasons behind any recommendations made. As far as extension evaluations are concerned, only in a few cases will the audiences for the evaluation prefer to draw their own conclusions without any input from the evaluator. Audience expectations should be determined in advance in terms of which concerns may need more definitive judgements.

Apply and Use the Findings

The evaluation should not be regarded as complete until the findings have been used to improve the on-going programme, and/or in the planning of future programmes. In cases where the evaluators are extension workers evaluating their own programmes, there should be little difficulty about incorporating the findings into the programme. However, depending on the nature and extent of the changes, approval may have to be obtained from supervisors. Where the evaluators are not persons conducting the programme, the likelihood of evaluation findings being ignored is greater. However, if the initial steps in planning and conducting evaluations (identifying concerns of relevant audiences, etc.) have been adequately pursued, the task of applying and using evaluation findings is easier.

EXAMPLES OF EXTENSION EVALUATIONS

Monitoring and Evaluation

This section will briefly discuss the major features of a system for the monitoring and evaluation of extension projects under the Training and Visit System, as proposed by Cernea and Tepping (1977). The proposed system, though not applicable in its entirety outside of the T and V System, provides some useful insights into how some of the evaluation concepts could be applied.

The Training and Visit System

Before looking at the proposed procedures for monitoring and evaluation, it will be useful to review some of the basic features of the T and V System. This system, developed by Benor and Harrison (1977), has been introduced in extension projects assisted by the World Bank, in a number of countries. The following are some of the essential characteristics of this system.

1. The village extension workers (VEWs) are assigned purely educational responsibilities.

2. The total number of farm families to be visited by each VEW is clearly defined.

3. At each level in the extension organization, the span of control allows close guidance and supervision of the level below.

4. Extension programmes concentrate on the most important crops and on improving farming practices which have the greatest potential for increasing yields, and which do not generally involve large cash inputs.

5. Specific recommendations for improving farming practices are carried to selected contact farmers, who will assist in spreading the new practices to surrounding farmers.

6. The contact farmers are visited every fortnight at a set date and time.

Procedures for Monitoring and Evaluation

A synopsis of the recommended procedures for monitoring and evaluation is presented below. The basic steps are discussed in the order in which they were presented previously in the chapter (although this may not have been the way in which the proposed system was originally conceptualized).

The Need for Monitoring and Evaluation

With specific reference to India, the number of state extension organizations using the T and V System has expanded rapidly to include all states since it was first introduced in 1974. Large numbers of extension staff, as well as farmers, are involved. Obviously then, such a major programme needs to be critically examined from its earliest stages to ensure that extension's efforts are not being wasted and that important ends are being served.

This new extension approach depends to a large extent on the efficient organization and management of the extension services. Because of the rigid time-table for visits, training sessions, and so on, all the components of the system need to function smoothly. Thus, it is important to detect any malfunctioning or breakdown in the system quickly as possible through monitoring and evaluation.

Purposes of Monitoring and Evaluation

The major purpose of monitoring and evaluation is to provide the information management needs on how efficiently the extension organization is operating. If deficiencies occur in certain regions or districts, then corrective action can be taken quickly.

The basic consideration in monitoring is whether scheduled visits are being made, appropriate recommendations are being given, farmers are adopting the improved practices, and the like. Evaluations at a later stage will indicate whether results are satisfactory, and whether corrective action needs to be taken. Since a major concern is whether staff are completing assigned tasks, the monitoring and evaluation exercises are therefore divorced from the responsibilities of such staff, and are conducted by a separate evaluation unit. There are, of course, advantages and disadvantages in having a separate unit for conducting evaluations, but these will not be discussed here.

Another important consideration is the question of accountability. The T and V System entails the disbursement of considerable funds and the deployment of large numbers of agricultural extension staff on purely educational duties. Justification of such an approach, especially in the initial stages, thus becomes a critical issue.

Audiences for the Evaluation

The primary audiences for the monitoring and evaluation exercise are the policy-makers and the management personnel in the extension organization.

Criteria and Standards for Evaluating the Programme

The criteria for monitoring and evaluating T and V System agricultural extension projects were developed from the goal or basic intents of such projects. Cernea and Tepping (1977, p. 7) stated the goals as follows.

> "The new agricultural extension projects put a definite emphasis on reaching quickly the mass of small farmers, tenants and sharecroppers. Their goal is to increase the productivity of large numbers of small and marginal farmers, help them to meet their basic human needs and contribute to an overall increase in food production."

These increases in food production were to be attained by getting farmers to use low- and medium-cost labour-intensive technology in improving their productivity, rather than through major changes in the infra-structure for agricultural development.

The ultimate goal of these projects can be said to be the improvement of the social and economic welfare of the farmers, and economic wealth of the country. A series of intermediate goals can also be identified. For example, an initial goal is visiting the farmer with specific recommendations every fortnight; a subsequent goal is that an acceptable number of farmers adopt the recommended practices. Criteria for monitoring and evaluation are developed from these intermediate goals and consequently, a hierarchy of evidence is established.

The difficulty in using objectives strictly is also illustrated here. While specific objectives and standards can be set beforehand on initial goals (for example, number of visits) considerably more difficulty will be encountered in doing this for subsequent goals. For example, what will constitute an acceptable number for farmers adopting the recommend practices? This number may be affected by unforeseen circumstances (weather, etc.) outside those allowed for by the programme.

The specific criteria used and the evidence to be collected for monitoring and evaluation are outlined in a later section.

Resources Available for the Evaluation

The usual constraints of time and money also consititute a limitation in this case. Thus the emphasis in monitoring and evaluation is in obtaining the absolute minimum information that will aid the policymakers to make decisions about the programme.

Collection of Evidence

This section will list the criteria for monitoring and evaluation and indicate what evidence will be collected and how it will be collected relative to the criteria.

Table 13.3

Monitoring

	Criteria	Data Collection Procedures
(1)	Degree of exposure to extension - Farmers reached directly - Farmers reached indirectly	Monitoring sample survey
(2)	Quality of visits	Monitoring sample survey
(3)	Farmers' evaluation of programme	Monitoring and harvest survey
(4)	Adoption of farm practices	Monitoring sample survey Harvesting study
(5)	Role behaviour (VEWs, AEOs)	Monitoring survey

The data collection procedures outlined above are intended to complement the reporting procedures built into the system. Monitoring data are obtained primarily from interviews with a sample of contact farmers with whom the VEWs work conducted during the pre-harvest stage. The questionnaire for the monitoring survey includes questions relating to the following (Cernea and Tepping, 1977):

1. Name of the VEW.

2. The frequency of visits by the VEW.

3. Attendance at group meetings.

4. Amount of land operated, and the proportion that is irrigated.

5. The practices recommended by the VEW, the area adopted, the area to be adopted, the extent of adoption, or reasons for non-adoption.

6. Whether any increased yields are expected on areas on which the recommended practices have been applied and, if so, how much.

7. The number of farmers they know who have learned the recommended practices from them.

8. A rating of the usefulness of the agricultural extension programme.

9. Any other comments or suggestions.

Table 13.4

Evaluation

Criteria	Data Collection Procedures
Yields of major crops	Harvest survey
Changes in cropping intensity and patterns	Harvest survey
Area under high yielding varieties	Reporting
Spread of key practices	Monitoring and harvest surveys

The evaluation data are obtained from sample surveys of the total farming population, conducted during the harvest stage. Crop-cutting is used to estimate yields of farmers included in the sample. The questionnaire for the evaluation survey includes questions relating to the following:

1. Gender of the respondent, whether or not the respondent is a contact farmer or was a contact farmer at any time.

2. Name of the VEW.

3. The frequency of visits by the VEW.

4. Attendance at group meetings.

5. Amount of land operated, and the proportion that is irrigated.

6   The practices recommended by the VEW, the area adopted, the extent of adoption, or reason for non-adoption.

7. · New practices used on the plot selected for the crop-cutting, and sources of information about the practice.

8. Estimated difference between present yield and what would have been obtained without following the recommended practices.

9. Total area planted in the crop, and the proportion under newly recommended practices.

10. A rating of the usefulness of the agricultural extension programme.

In linking the concepts previously discussed to the proposed procedures for monitoring and evaluation, the following points should be noted.

1. As previously mentioned, monitoring corresponds quite closely to formative evaluation, carried out during programme implementation to diagnose possible weaknesses. Thus, the information sought relates to the number and quality of visits and so on. The farmers' early reactions to the programme will also indicate whether or not the programme is on the right track, and what kinds of adjustments should be made to the programme. Evaluation examines the results of the programme (changes in yields, etc.), and this is obviously a summative type of evaluation.

2. The evidence collected for monitoring and evaluation corresponds to various levels in the hierarchy of evidence. Monitoring focuses primarily on the inputs and activities level. It also examines evidence relating to "people involvement" and "reactions".

Evaluation takes a closer look at "people involvement" and "reactions" in particular, women farmers and contact farmers. The focus, however, is on "KASA Change" and "Practice Change" (see Table 13.2).

3. The proposed procedures also incorporate both hard and soft data (see Table 13.2). At the "practice change" level, crop-cutting provides harder evidence of changes in yield than estimates of changes. Subjective ratings of the usefulness of the programme are, of course, soft data. Evidence relating to inputs and activities are not entirely soft since they are not based on staff reports, but on farmers' reports of the VEWs' visits. On the otherhand, such evidence cannot be regarded as being completely hard since it is not derived from direct observation of staff's activities.

4. The overall design is the sample survey. The main advantage here is that such a design allows the collection of "credible" and "acceptable" information, which can be quickly fed to the primary audiences for the evaluation (the policy makers).

With regard to benchline data, Cernea and Tepping (1977) recommend that information from the initial surveys be used as benchline information. Retrospective reports by farmers, about which practices are based on technical advice from the extension organization also provide evidence (primarily soft evidence) of the programme's impact.

Competitive explanations for changes must also be taken into account. However, since the programmes are closely monitored from their inception, the relationship between the results, and the extent to which such results are directly due to the extension programmes can be more clearly established.

5. Processing the date obtained involves a quick summary of the data by the field investigator; this is immediately sent to the District Extension Officer. The summary, together with the completed questionnaires, are then sent to the state evaluation unit for further processing and analysis. In the case of evaluations, the field investigator also enters preliminary data on the produce from crop-cutting. Data on the produce is also sent to the state evaluation unit.

6. Cernea and Tepping (1977, p. 59) also recommend special in-depth studies "to complement and/or deepen the information generated by the large-scale sample surveys". These include the following.

a. Study in the selection of contact farmers.

b. Sociological village case studies on the impact of extension programmes.

c. Study of the village extension workers and the agricultural extension officers.

d. Study on the quality of training sessions.

Conclusion

This chapter has presented some important evaluation concepts, identified basic steps in planning and conducting extension evaluations, and illustrated how these steps could be carried out, using specific examples. The point that has been emphasized throughout is that evaluations should be tailored to specific situations. In the extension setting, some form of evaluation effort is always possible and desirable. Evaluation studies, at whatever level they are planned and conducted, should be regarded as an integral part of the extension process.

REFERENCES CITED

Bennett, C. F. (1977). Analyzing impacts of extension programs. Washington, D.C.: Extension Service, U.S. Department of Agriculture.

Benor, D. & Harrison, J.Q. (1977). Agricultural extension: The training and visit system. Washington, D.C.: World Bank.

Cernea, M. M. & Tepping, B. J. (1977). A system of monitoring and evaluating agricultural extension projects (World Bank staff working paper no. 272). Washington, D.C.: The World Bank.

Frutchey, F. P. (1967). Evaluation--What it is. In D. Byrn (Ed.), Evaluation in extension (pp. 1-5). Topeka, Kansas: H. M. Ives and Sons.

Sabrosky, L. K. (1967). Evidence of progress towards objectives. In D. Byrn (Ed.), Evaluation in extension (pp. 25-29). Topeka, Kansas: H. M. Ives and Sons.

Steele, S. M. (1972). Six dimensions of program effectiveness (Evaluation series). Madison, Wisc.: Division of Program and Staff Development, University of Wisconsin - Extension.

Taylor, D. B. (1976). Eney meeny miney, meaux: Alternative evaluation models. North Central Association Quarterly, 50, 353-358.

Chapter 14
Pre-Service Training of Extension Personnel

B. E. Swanson

Extension personnel are the major resource of a successful extension service. Extension personnel who enter the extension service should be prepared to perform basic extension tasks. In addition, a successful extension organization makes provision to further strengthen the professional and technical competencies of its staff members once on-the-job, because staff must be trained and up-to-date to provide the essential link in the technology transfer process, between agricultural research and farmers. Therefore, attention should be paid both to the pre-service training of extension personnel and an overall staff development and in-service training plan.

All men and women who enter the extension service whether as specialists, administrators, supervisors or field-level agents should have basic skills in and an understanding of the following broad topics.

1. Technical subject matter in areas such as agriculture, animal/veterinary science, fisheries, forestry or home economics.

2. Extension service organization and operation, including its overall purpose, mission, policies and procedures.

3. Human resource development, including the participatory processes of involving people in programme planning and development, group behaviour, staff-clientele relationships, and personnel management.

4. Programme development process, from problem identification and needs assessment to programme design, implementation and appraisal, including programme administration.

5. Pedagogical skills, including the teaching-learning process of adults (men and women) as well as young people (aged 15-24), and instructional design and teaching strategies.

6. Communication strategies not only for programme delivery (utilizing both modern and indigenous systems of communications) but also for obtaining feedback information from client groups and feeding this information to research and other appropriate agencies or groups.

7. Evaluation techniques for the purpose for determining the effectiveness and value of extension programmes to users.

The purpose of pre-service training programmes is to prepare individuals with appropriate technical and pedagogical skills, so that they can enter careers as extension workers. Therefore, to determine the content of pre-service training programmes, it is very useful to have available job descriptions which set out the standards and criteria for successful employment in the extension organization. Within these broad guide-lines, specific training objectives should be identified in the following areas (a) technical subject matter skills and knowledge, (b) communication and education skills and knowledge, including methods and techniques, (c) knowledge of rural social systems, including cultural norms, community organization, and so forth, and (d) information about extension organization and operations (this area is more frequently covered in induction training).

Because many of these extension related topics have been covered in other chapters of this manual (i.e., communication, the adoption process, and educational strategies and methods, the rural social sciences, and extension organization and administration), the focus in this chapter will be largely devoted to how to organize a pre-service training programme, particularly to teach technical subject matter, with the agricultural sciences being used as the primary example. In addition, the use of extension internships and a social laboratory are discussed in terms of providing professional field experiences for future extension workers.

PRE-SERVICE TRAINING OF EXTENSION PERSONNEL

Pre-service Training Defined

Pre-service training may be defined as a programme of learning activities that pre-pares an individual for a career in extension, and usually leads to some type of diploma, certificate, degree, or other qualification in one or more of the following, agriculture, fisheries, forestry, animal and/or veterinary science or home economics. In most countries, the extension service has little direct responsibility for implementing pre-service educational programmes; these programmes are generally conducted by the intermediate and higher agricultural education institutions in a country. However, the extension service should work closely with these educational institutions to influence pre-service training objectives and the approach pursued by these institutions. This advisory input should come from knowledgeable people in extension, who can clearly state the perfor-mance objectives, criteria and standards that graduating students and other prospective applicants should have to be accepted for the employment in the extension organization.

Types of Pre-service Training

There are generally three or four main categories of extension personnel (a) admini-strative/supervisory, (b) technical, or subject matter specialists, (c) extension workers or officers, and, frequently (d) extension field assistants or para-professionals. In many countries, extension assistants are, in fact, the front-line extension cadre, and they are in turn supervised by extension officers.

The pre-service training required for each of these positions differs from country to country, depending in part on the type and relative capacity of each nation's agricultural education institutions. Generally administrative and supervisory personnel should hold the minimum of a university degree (e.g., B.Sc. degree in agriculture) and, preferably, some additional post-graduate training in extension education; or the qualification of Ingenieur Agronome or Ingeniero Agronomo, and possibly some post-graduate education as well. In addition to their technical and academic education, administrators and super-visors also need skills in areas such as personnel selection and counselling, personnel management, evaluation and training, programme development and administration, as well as supervision skills. Most extension personnel will not receive this type of training as part of their first university degree; therefore, special short courses, possibly at an administrative staff college or post-graduate education at a university, may be required.

Technical or subject matter specialists (SMS) also need a university degree in agri-cultural science, or a specialized field (such as entomology, plant pathology, agronomy, horticulture), and where possible, they should also have post-graduate education in their specialty. In addition, most will need basic crop production skills, so they can success-fully grow a crop on farmers' fields. Since SMS, will be largely responsible for providing some in-service training for field-level extension workers, it would be highly desirable if they have some basic training skills as well.

Extension officers in some countries may hold the first university degree or the equivalent. However, it is more common for field level extension personnel to hold a post-secondary qualification such as an intermediate level diploma in agriculture or other similar qualification, such as technicien superieur, Agronomo, and so forth, which would represent a two- or three-year course of study in agriculture or agricultural extension following secondary school. In some countries, extension assistants only hold a (tech-nical) secondary school qualification or the equivalent (i.e., certificate, perito agronomo, agent technique, moniteur, etc.).

University degree programmes in agricultural science or agronomy tend to be similar, in that the curriculum is organized around specific fields of study or disciplines, and subjects are generally taught in a theoretical rather than practical manner. Furthermore, because of the lack of locally (or regionally) produced textbooks, courses in developing countries are often taught using texts and references that are written for European or North American conditions. These factors may result in students having a relatively good academic foundation in their specialty, but university degree programmes generally do not provide the necessary practical skills and knowledge for extension officers and specialists to be effective on-the-job. They must acquire these practical, production skills either before entering the university or through on-the-job and in-service training.

APPROACHES TO PRE-SERVICE TRAINING

The typical method of instruction in secondary, intermediate, and higher agricultural education institutions is by lectures and other teacher-centred techniques of instruction. The teacher is viewed as the source of knowledge, and students are passive recipients in the "teaching-learning" process. Students learn to respect authority, and not to question the validity of the subject matter presented. The philosophy of education underlying this approach is "top-down" in nature, and is opposite to what students will need to become effective extension workers in the field. As Benor and Harrison (1977, p. 7) have observed, ". . . pre-service training . . . is often too theoretical and provides little opportunity to apply in practice what has been learned".

Farmers are, by their very nature, problem solvers. However, in traditional production systems, their options or production alternatives are very limited. The production methods and techniques they use may have been developed and tested over many years, perhaps even generations. However, with the movement to so-called "modern" or science-based systems of farming, there will be an increasing number of technological alternatives available for farmers to consider and possibly use.

The process of considering, trying, and perhaps adopting some of these new technological alternatives was described in Chapter 5. The importance of understanding the adoption process and its implication for extension methods, is that farmers must be active participants in the learning process. "Top-down", teacher-centred approaches to agricultural extension will seldom be as effective; in many situations, particularly in working with small, subsistence farmers, "top-down approaches" will actually be ineffective.

Instead, extension workers must use an inter-active method for programme planning and instruction that focuses on problem solving. Both the extension worker and the farmers he or she works with must be active participants in seeking ways to increase the productivity and income of farm families and households. Extension workers bring knowledge about new technology and related information to the learning environment, while farmers bring knowledge about their farming situation and of their current farming practices. Together, they must seek ways to overcome production problems and, in the process, to increase farm output and incomes.

If the premise is accepted that extension personnel need to utilize a problem-solving approach in their extension activities, then it seems appropriate to ask the rhetorical question, "Where are extension workers expected to learn these pedagogical skills"? They have little knowledge or experience of problem-solving skills if they are only taught using teacher-centred methods of instruction (see Table 14.1). What appears to be called for is a major revamping of the curriculum and method of instruction used in agricultural education institutions that prepare field-level extension personnel.

"What must be kept in mind is the importance of preparing extension agents to work effectively at the village level. Often there is too little emphasis placed upon having extension agents master the art of listening and reacting with a carefully measured problem-solving approach" (Maalouf, 1983, p. 8).

This conclusion is based on the assumption that extension personnel cannot teach or utilize teaching skills that they have never seen practised. Therefore, to suggest how this approach might be implemented, the following section has been developed as an example showing how a practical, problem-solving instructional programme might be organized for an intermediate-level school of agriculture. These institutions are often weak and traditionally organized, however, they offer the greatest potential in improving the quality of field-level personnel going into national or provincial extension organizations.

Table 14.1
Why Teachers (and Extension Workers) Should Use Problem-Solving Methods in Teaching
Agriculture

1. Problem-solving offers students the opportunity of learning a process (how to solve problems), as well as subject matter.

2. Problem-solving gives students the opportunity to help plan what they are to learn, and how they are to learn.

3. Problem-solving keeps the teacher's feet on the ground. By giving the class a share in planning the learning activities, the instruction is kept at a practical level.

4. In problem-solving teaching, students gather subject matter and use it in a practical way; hence, they learn the proper use and application of knowledge. This is far superior to learning knowledge for knowledge's sake, or learning knowledge to earn a grade.

5. Problem-solving is a flexible teaching approach.

6. Problem-solving that is properly executed generates much interest in learning on the part of students because the reasons for studying, and the applications of classroom learning, are clearly developed.

7. Problem-solving is sound pedagogy. It makes use of the following important learning principles.

    a. In order to learn, people need the opportunity to practise the activity which is being learned.

    b. Motivation for learning is highest when interest in the learning activity is high. Class interest in an activity depends largely on students seeing the relationship between what they are doing in the classroom and their own goals as future extension workers.

    c. People remember best when the subject matter can be practised (learning by doing) and related to familiar situations.

    d. Transfer of training occurs most readily when the application of principles to various situations are made clear in the classroom.

    e. Persons are most interested in an activity when they are actively participating in that activity, as opposed to watching or passive listening.

8. Problem-solving teaching enables teachers to allow for individual differences, because students may advance at varying rates and pursue individual objectives.

9. In problem-solving, classroom work and practicals in the field, combined with teacher visits to student plots, can be tied together into a meaningful whole.

10. In addition to learning "agriculture", students learn the following important tasks from problem-solving.

    a. set goals,

    b. identify their own problems,

    c. use books and other references to help solve problems ,

    d. discriminate between reliable and unreliable data,

    e. supplement their own experience and knowledge with other sources of information,

    f. make wise management decisions

    g. translate information and facts into approved farm practices,

    h. think systematically,

    i. relate school work to real life situations, and

    j. evaluate their progress, and make plans to correct their deficiencies.

ORGANIZING A PRE-SERVICE TRAINING PROGRAMME
BASED ON PROBLEM SOLVING

The philosophical basis for a problem-solving approach to agricultural education is to view new knowledge and technology as a means to solve problems, rather than an end in its own right (as is the case in most academically organized educational programmes). This difference in orientation has a profound effect on the teaching objectives pursued, the motivation of students, and the role of the teacher. It also has major implications for how the curriculum is organized, the teaching methods used, and the teaching resources required. Each of these major topics will be covered briefly in the following sections.

Teaching Objectives

Students who are preparing to be agricultural extension workers and who successfully complete an agricultural training programme using a problem-solving approach should know or be able to do the following

1. understand the role men, women and young farmers play in agricultural production, marketing and utilization, particularly in the division of (agriculturally-related) labour that is practiced in each region,

2. understand the predominant production practices and farming systems followed by the major groups of farmers in their district or province,

3. have the ability to actually produce the major food and cash crops grown in their district or province, using improved technology to maximize productivity and income,

4. have sufficient farm management skills and knowledge so that they can work with farmers in analysing production alternatives, and to select the optimal technology for their farming systems,

5. have the ability to use successfully problem-solving approaches to extension teaching in individual, group and mass media situations, and to know when each approach should be used, and

6. understand the social factors and/or characteristics of ethnic groups in the district or province, as well as the ways the agricultural roles of men, women and young people could affect extension programmes and success.

Curriculum

The curriculum of intermediate-level agricultural education institutions should be organized around the crop-growing season and/or livestock production cycles, so that each student gains practical experience growing the important food crops and/or livestock using recommended improved technology. Appropriate agricultural mechanization skills should be taught, as well as appropriate learning experiences in extension education, agricultural communications, rural sociology, farm management and record keeping.

If a three-year course is offered, the goal should be for each student to fully participate in on-farm experiences for two crop years. The first year would focus on learning production skills, while the second year would utilize these skills in the context of conducting demonstration plots. During the third year, students should spend time directly with farmers in a pilot project area (or social laboratory).

In the first crop year, students should have the opportunity to carry out farm practices on individual plots of approximately one-fourth acre (or one-tenth hectare). If individual plots are not possible, group projects can be set up. They should grow all of the important food crops (including vegetables) in proportions more or less similar to that of a local farm family operating a small, intensively cultivated farm. Each student should keep production records and accounts in a farm management record book.

At the beginning of the first school year, students should be assigned plots at random. One of the first tasks should be to collect soil samples and test the soil to determine fertility requirements. Classroom teaching and practicals should then be organized around the farm production process. For example, first land/seed-bed preparation should be discussed in the classroom, with students then developing a farm plan for their plot and preparing their land for planting. Different varieties of each food crop should be compared in the classroom, including factors such as their responsiveness to fertilizer, disease resistance, the nutritional and taste qualities, and preference for different varieties. Essential cultural practices (such as plant population, date of planting, timing of irrigation, etc.) should be discussed.

Following classroom instruction, students should then decide which varieties they will grow on their respective plots and determine the level and type of fertilizer to be applied (computing the amount needed for each crop in their plot). Students would then proceed through the growing season, carrying out recommended practices and finally harvesting each crop, keeping production records to compare results.

To follow this approach, students will need a locker or a similar storage area where they can keep work clothes, basic farm tools, and so forth. All inputs, including seed, fertilizer and pesticides, should be available for purchase (on credit) from the farm store (operated by the school). Students can be assessed a rental charge for their plots and for any power machinery used. All farm products should be sold back to the farm store at prevailing prices to repay the cost of inputs, plus interest. Any profits that they accrue should belong to the students so there will be a real incentive to maximize production and income. Alternatively, profits could be used to support a special educational activity, such as a class trip to an agricultural fair or to the ministry of agriculture in the nation's capital. In short, students will work harder and learn more if there is an adequate incentive.

Students should be evaluated on their production plots periodically throughout the season, including the appropriateness of their chosen production practices, as well as the productivity/profitability of their plots. Their farm management record book should also be evaluated. In addition, students should be evaluated on classroom instruction through periodic examinations.

At least one introductory course in agricultural extension should be included in the first-year curriculum. This course might cover a broad range of topics such as understanding the clientele of extension, including (a) the unique and inter-related contributions to agricultural productivity, marketing, and consumption made by men, women, and youth within farm households, and, (b) the extent to which women-headed households in the area are farming households, to determine the number and proportion of women farm managers.

Second-year students should also be in charge of individual plots if possible. However, the objective during the second year should be to conduct demonstrations, such as varietal trials, fertilizer demonstrations, and/or the effect of different cultural practices on productivity (e.g., plant population x different fertility levels). At the end of the growing season, trainees should learn how to measure yields accurately within their demonstration plots. A field day should be held for farmers near the school where each trainee would describe each demonstration, including the results obtained from individual treatments. Courses in agricultural communication and extension methods, including the effective use of audio-visual equipment, should be included, as well as courses in agriculture mechanization (maintenance and repair of small tillers, tractors, and implements), and possibly in animal husbandry and production.

Third-year students would not be expected to grow individual plots, because they should participate in off-campus activities working directly with farmers in a pilot project area near the school. They should also be assigned to work with experienced extension personnel in a 8-10 week professional internship that is periodically supervised by school officials. Third-year students also need to complete courses in complementary areas, such as rural youth organizations (including young farmers), extension programme planning and rural sociology, as well as other technical agriculture courses.

Teaching Methods

The implementation of the curriculum described above would be greatly enhanced by using a problem-solving method of instruction. Problem areas (instructional units) would be scheduled to follow the growing season and, therefore, be taught just prior to the students' need for this information in their plots. In this manner there would be direct and immediate application of knowledge and skills acquired in the classroom. Furthermore, students should determine themselves which practices they will follow in their plots, so they can see first-hand the consequences of their decisions and efforts. As particular production problems arise during the growing season (such as insect or disease problems), the class can immediately consider the problem, possible solutions, and actually implement a particular solution.

The problem-solving method of teaching involves the active participation of both the teacher and students, and is well suited to the curriculum being proposed here. The teacher guides the learning process; students identify specific problems or types of knowledge they will need to know, or skills that they should have in order to carry out a specific production practice. Examples of typical problem areas for a crop like wheat include selecting wheat varieties, fertilizing the wheat crop, irrigating wheat, controlling pests in wheat, harvesting wheat, and so forth. An outline of a teaching plan, using the problem-solving approach, can be found in Appendix 3 at the end of the manual.

After students identify the important problems in a particular instructional area, they should investigate different solutions to each problem, using extension bulletins, research papers and books. After students have investigated a problem in detail, class discussion should then be used to consider the range of alternative solutions, with the most appropriate ones being noted. Following this classroom activity, students will be expected to use their own best judgement as to which practices they would use on their own individual plots. This basic teaching method can be used for both technical and extension-related courses being covered in the curriculum. An example of steps used in the problem-solving method of teaching can also be found in Appendix 4.

Professional Internships

It is recommended that all students who are preparing to be extension workers have a professional internship, or practical experience, during their third year of study. Such an internship would involve placing students with well-trained and competent extension workers for some direct on-the-job training and experience.

The selection of "supervising" extension workers for this programme is very important as they will serve as role models for prospective extension workers. They should be both technically and professionally competent, and capable of performing their extension duties in an exemplary manner.

The length of a professional internship can vary from a few weeks to several months. Six weeks would be the minimum amount of time for the experience to be useful to the student, and 16 weeks would be about the maximum, without eliminating other important courses during the third year of the curriculum. In some countries, the 2-3 month summer recess (before the final year of courses) is used for the internship; others view the internship as an integral part of the final year of studies. In either case, the internship should be scheduled so that students can see a wide variety of extension activities being carried out.

Professional internships must be supervised by school personnel who visit each co-operating extension worker to make sure that the student is receiving appropriate experiences, and is participating in some extension teaching activities. These visits can also be used to discuss any problems or questions that might come up between the student and the co-operating extension worker. Finally, these visits provide the opportunity for the school supervisor to evaluate the performance of the student (in consultation with the extension worker) and to determine if the co-operating extension worker should be used in the future for the professional internship programme.

Students will likely not have sufficient financial resources to cover their food and housing costs while they are on the internship. It may be necessary to provide them with a modest living allowance while they are in the field on the internship phase of their training. In at least one country, the extension service considers students as temporary staff during their internship and provides them with a modest salary to cover their living costs while in the field. Additionally, prior planning by the school staff will be required to insure that there are adequate housing facilities at these placement sites for women, as well as men, students.

The professional internship is an excellent way to bridge theory and practice in the professional skills side of extension training. By having the opportunity to observe and work under the tutelage of a competent extension professional, this experience will provide students with an excellent example of how they should perform once they are working in the extension service.

Pilot Project or Social Laboratory

A second dimension of on-the-job experience can be provided through the use of a pilot project, or social laboratory, that is operated by the school as a training site for third-year students, and as a demonstration site for the school staff. In most cases, the (pilot) project area would be adjacent to, or would surround, the school. The size of the project area will vary according to the size of the school (i.e., number of students in the third-year class) and the physical characteristics of the district. Ideally the area should include at least two different agro-ecological zones, so students can experience different farming systems, and the effects of different technical recommendations.

The school will need to have at least two or three full-time extension professionals operating in the pilot area at all times. These extension workers might be on assignment from the extension service, or the school might be given full responsibility for carrying out extension programmes in the project area and be expected to staff it with their own personnel. It is expected that the technical backing of extension activities in the project area, would be the direct responsibility of the technical staff of the school, in consultation with researchers at the regional agricultural experimental station.

However, it is expected that, during one growing season, third-year students would actually plan (in consultation with the technical instructors), and help carry out, on-farm demonstrations and trials in the project area. These trials could be group or individual projects, but the purpose is to give students first-hand experience of conducting field demonstrations and trials on farmers' fields, and to observe first-hand the results of different recommendations. It is expected that students would also help organize farmers' field days, to show the results of different recommendations on productivity and income. This is most important; it gives the students the experience of working directly with farmers, under farmer conditions. Through these experiences students will learn why farmers follow certain practices, the difficulties farmers may face in obtaining inputs and/or credit on time, as well as some of their marketing problems. These learning experiences provide students with some contact with real world conditions.

Resources and Facilities Required

To implement a three-year training programme such as the one described above would require sufficient land for first- and second-year student plots, plus whatever animal husbandry facilities would be appropriate and necessary for group projects. First- and second-year students would require hand tools for field work plus appropriate small tillers, back-pack sprayers/dusters, harvesting equipment, and so forth, as necessary. Trainees could use the larger pieces of equipment as needed for a small rental charge.

There would need to be a school-operated farm store where trainees could purchase seeds and other inputs (on credit), check out equipment, and sell their farm products (which could alternatively be made available to the school cafeteria, or sold to the faculty or in the local market). If students suffer losses, they should not be expected to pay the difference; however, it is recommended that any net profits should go directly to students (either in direct compensation or to fund special education activities), as this would increase their motivation to use the optimal combination of inputs to maximize production and income.

In addition to the land requirement, animal husbandry facilities, the farm store, and a trainee work area, there should be an agricultural mechanization shop that could accommodate up to 25 students at one time. In this shop, problem areas such as agricultural equipment maintenance, adjustment, and repair should be taught, with students gaining as much practical experience as possible. Also, the use of basic wood- and metal-working hand tools should be taught.

In addition to the above facilities, adequate classroom and housing facilities would be needed to accommodate the student population, plus a school library, an audio-video centre and office space for the principal, instructors, and other personnel. The availability of appropriate housing facilities for women students is critical; housing will play an important role in supporting school efforts to recruit women students. Finally, an adequate number of bicycles and other appropriate vehicles will be needed to transport third-year students working with farmers in the pilot project area, as well as for the teaching staff when they supervise third-year students while they are on their professional internships.

CONCLUSION

The overall objective of using a problem-solving approach to organize the pre-service training of future extension personnel is to immerse them in the problem-solving process. The actual technical subject matter and skills students learn is of secondary importance to the process itself; The technical content is actually used as a means of teaching problem-solving skills and attitudes.

Teaching problem solving is far more difficult and complex than any technical subject matter students will encounter during their pre-service training. Only by repeating the process over and over again will students learn how to formulate, and then solve, problems themselves. In short, the problem-solving approach is the application of the scientific method to education and training. If extension agents are to be successful in assisting farmers to change from traditional to science-based systems of farming, using improved technologies of production, then it is essential that they develop problem solving skills. This is why problem solving is a very appropriate and necessary method to use in preparing extension workers to carry out the educational function of technology transfer.

REFERENCES

Bernado, F. A. (ed.) (1974). Organizing various schools/colleges of agricultural sciences into an efficient national system, AAACU Publication No. 4. Bangkok: Asian Association of Agricultural Colleges and Universities, Inc.

Binkley, H. R. and Tulloch, R. W. (1981). Teaching vocational agriculture/agribusiness. Danville, IL: Interstate.

Contado, T. E. (1978). The social laboratory, in Rural Development: The Philippine Experience. A. A. Gomez and P. Juliano (eds.). College, Laguna, Philippines: Philippine Training Centre for Rural Development, University of the Philippines at Los Baños.

Crunkilton, J. R. and Krebs, A. H.(1982). Teaching agriculture through problem solving (Third Edition). Danville, IL: Interstate.

Drawbaugh, C. C. and Hull, W. L. (1971). Agricultural education: approaches to learning and teaching. Columbus, OH: Charles E. Merrill.

Food and Agriculture Organization of the United Nations (1982). Guidelines for agricultural training curricula in Africa. Rome: Food and Agriculture Organization of the United Nations.

Maalouf, W. D. (1983). Basic and in-service training for agricultural extension. Paper presented at World Bank/DANIDA symposium on agricultural extension, Copenhagen, Denmark, 3-4 October, 1983.

Phipps, L. J. (1980). Handbook on agricultural education in public schools (Fourth Edition). Danville, IL: Interstate.

Rowat, R. (1979). Trained manpower for agricultural and rural development. Rome: Food and Agriculture Organization of the United Nations.

Saguiguit, G. F. (1982). The social laboratory: some experiences and perspectives, in Training for agriculture and rural development, 1981, FAO Economic and Social Development Series No. 24. Rome: Food and Agriculture Organization of the United Nations.

Swanson, B. E. (1976). Coordinating research, training and extension: a comparison of two projects, in Training for agriculture and rural development, 1976, FAO Economic and Social Development Series No. 2. Rome: Food and Agriculture Organization of the United Nations.

Chapter 15
In-Service Training and Staff Development

V. M. Malone

In this chapter, the major focus is placed on a description of how to develop and maintain a comprehensive plan for extension personnel staff development. Within this framework, some discussion is presented regarding definition of terms, purpose and responsibility for programming through (a) induction, (b) in-service, and (c) continuing education, including the importance of increasing professionalism through association with colleagues and peers.

STAFF DEVELOPMENT DEFINED

Staff development is a "program of activities designed to promote the professional growth of individuals" (Dejnozka and Kapel, 1982). Such a definition assumes, however, that "no deficiency exists in the individual." Staff development provides individuals with an opportunity (a) to develop a sense of purpose, (b) to broaden perception of clientele, and (c) to strengthen capacity to gain knowledge and mastery of techniques (Dejnozka and Kapel, 1982).

Staff development assumes individual extension workers to be responsible for engaging in the activities of an overall program of self development from the time of their preparation for the profession to the close of their career role. However, the extension service should provide an opportunity for workers to participate in as many of these staff development components as is possible within the organization structure.

STAFF TRAINING DEFINED

Staff training is a term used to describe the programmes and activities that are conducted by the organization for the purpose of maintaining and upgrading competencies of the staff to perform those tasks related to their jobs which aid the organization to reach its goals within its stated missions. Therefore, the extension organization is responsible for the design and implementation of staff training programmes which have the following general objectives:

1. To strengthen technical subject matter competencies.

2. To strengthen those educational process skills that aid in the delivery of programmes to appropriate audiences.

Staff training is designed, basically, to narrow the gap between the staff's current level of competence and that needed by the organization at any given time. Staff training, as opposed to staff development assumes that deficiencies do exist which need to be corrected.

COMPONENTS OF A STAFF DEVELOPMENT PROGRAMME

Individuals who view extension work as a career should be encouraged to participate in staff development programmes as early as possible during their pre-service period of study. In doing so, they commit themselves to project the image of a professional extension worker on a lifelong basis. Such workers tend to do a better job, have a longer tenure in the position, and find satisfaction in doing the extension work itself (Malone, 1973; Vroom, 1972). These individuals are the key to successful extension programmes from the field level to the ministry level.

The components of a successful staff development programme include pre-service training, induction training, in-service training, and continuing education. Each component is described in terms of its purpose, content, and suggested format, with a recommendation for appraisal of outcomes.

Induction Training

A major phase of staff development occurs at the time when new personnel are employed in the organization, to begin the work designed to assist the organization to meet its goals. This phase is called induction training (in some programmes it is called orientation). The purpose of induction training is to "make new employees familiar with the practices and procedures of the organization" (Dejnozka & Kapal, 1982, p. 173). These should be related to the stated job description.

Induction training has two basic assumptions. One is that everyone entering the organization needs to have his or her perceptions and expectations of the organization clarified so that they are similar to those understood by others in the organization. The second assumption is that new employees who have the necessary competencies to do the broad category of tasks, need to have a clarification of the specific tasks related to their position.

A good induction programme assists new employees (a) to develop a feeling that they are an important part of the organization, (b) to reduce the initial stress related to performance of their tasks, (c) to identify the resources available to support them in their role, and (d) to strengthen their personal commitment and dedication to the extension audience through the extension organization.

Induction training has as its focus the structure, policies and procedures of the extension organization. Therefore, induction training should occur the first day of employment, on (or as close as possible to) the site where the workers are to perform their daily tasks.

Content and Format

Induction training should be goal specific, learner-oriented and designed to move the employee to a status of independent work as quickly as possible. The degree that this can be done is based on the background of the new employee. The course should review the history of the organization, the economic, political and social influences that affect the people in the area, the programme development process, professional conduct and behaviour expected of the employee, and the extension organizational structures. It should also specify

1.   rules and regulations governing the job,

2.   resources available to conduct the specific tasks,

3.   administrative supports to achieve work goals,

4.   the social interaction process which will assist employees to find job satisfaction, and

5.   resource needs on- as well as off-the-job, for example, getting settled in the community, identifying key leaders.

The induction course should also specify the skill practice of the new employee. The employer should be able

1.   to manage programmes, monitor budgets, and other resources,

2.   to complete office and programme forms,

3.   to store and retrieve information,

4.   to utilize the mass media,

5.   to conduct demonstrations,

6.   to use various instructional strategies, and

7.   to Involve a variety of human and technical resources in programmes.

The suggested format for an induction programme might be similar to the one shown in Table 15.1.

Table 15.1
Suggested induction plan

Day 1 (First Day of Employment)	First Week	1st month - 1st year
Meet Programme Administrator to be welcomed to the organization	Visit research station, farm and home sites to observe activities of those sites	Attend classes related to the organizational structure and procedures
Meet with immediate supervisor to identify day-to-day activities, to complete any employment papers and to identify work station facilities	Meet with immediate supervisor to work on policy and procedures for extension work activity	Observe research work and practices at research stations, farm and home sites
Meet other resource people at the work site, to discuss selected rules and regulations	Visit community area to observe day to day activities of local people	Participate in field day under the supervision of an experienced worker
	Review resource materials related to the scope of extension work in the local community	

Time Frame for Session on Day 1		
Morning	Afternoon	Evening
Meet immediate supervisor to discuss general job description; complete any employment papers	Tour selected areas of community with a resource person, e.g., farms, households, etc.	Visit any community events, or a key family.
Visit work site and examine materials available to be used in support of programme	Visit one or two key leaders, or make appointments to visit them	
Review policy handbook on Office Procedure	Meet the immediate supervisor to clarify other extension policies, structures	
Set up 1-week calendar of events from current mail/visit lists		

Case Situation

     Mr. Siddique has been assigned to work in his home village as the new extension assistant. He has a diploma from the Agricultural Extension Training Centre with high grades in agronomy. He has been away from the village for three years. He is married

208

to a woman from another village. This is his first extension assignment. This village needs assistance in agronomy.

Task. Using the information in Table 15.1, describe a plan of activities for the first day for Mr. Siddique.

In-service Training

In-service Training Defined

In-service training may be described as any "planned program of learning opportunities afforded staff members...for purposes of improving the performance of the individual in already assigned positions" (Harris, 1980, p. 21). In-service education is also described as a planned programme of "systematic practice in the performance of a skill" (Dejnozka and Kapel, 1982, p. 346).

In general, inservice training is a program designed to strengthen competencies of extension workers while they are on the job. Therefore, it should be a (a) problem centred, (b) learner oriented, and (c) time defined series of activities.

Basic Assumptions and Rationale

Extension leaders planning in-service training programmes should have certain assumptions and a basic rationale in mind before the activities are developed. These assumptions should include the fact that men and women, alone and in groups can and do learn on the job, and learn best when they are actively involved in the learning process.

These assumptions, with a stated rationale, should provide a framework for the design of an effective in-service training programme. The rationale is based on two questions: To what extent does the proposed programme (a) fit into the goals, objectives and mission of the organization, and (b) use current research findings regarding accepted practice, based on the successful experiences of current users of the system?

Guide-lines for the Design of an In-service Training Programme

A programme planning process (Houle, 1980) should be used to design an effective in-service training programme. Figure 15.1 is a visual description of the one presented here. It includes the following components with appropriate questions to be answered by the planner.

Problem Identification. What are the incidents that indicate a problem exists or is anticipated (Rowat, 1980)? Is it one that can and should be changed by a training programme? How can the problem be changed into a need (need assessment)? Are the physical, human, and financial resources available to resolve the problem?

Learner Identification. Who are the target groups of learners? What competencies do they have now relative to the problem identified? What is the gap that exists between present level of competencies and needed level of competencies? To what extent do learners perceive the problem?

Goal and Objective Identification. What should be the overall purpose of the programme (goals) (Tyler, 1975). Are the intended learner outcomes (objectives) realistic (Mager, 1974)? What job performances will be improved?

Learning Opportunities and Selection of Instructional Strategies. (Bender, 1972; Knowles, 1980; Malone, 1982). To what extent are the instructional strategies compatible with the stated objectives and learner characteristics? To what extent can an appropriate climate for learning be arranged within the existing organizational constraints? What will be the subject matter for each session, and how appropriate are the learning opportunities for the selected subject matter?

Format and Scheduling of Learning Events. What will be the scope and sequence of the overall programme and of the individual sessions. To what extent is a sequencing of activities related to competencies built into the programme plan?

Figure 15.1 Evaluation and appraisal for impact on learner and organization

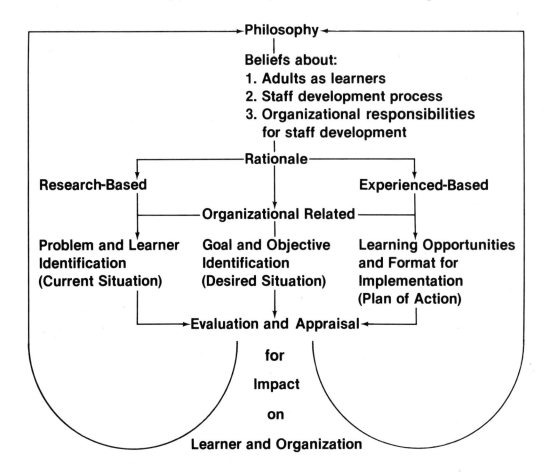

Evaluation and Appraisal. To what extent are activities included which will allow one to measure the worth of the learning experience for the planners, administrators, learners, and other significant people? What plans are there to monitor the on-going programme, so that adjustment can be made?

The in-service education programme is the responsibility of the organization. Such a programme is designed to assist staff to maintain and strengthen their competency to perform the tasks related to a specific job, role, and function in the organization. However, adults as learners are more likely to participate in such a programme for competency development if they have had an opportunity to share in the programme design. They tend to learn more effectively if they are actively involved in the learning experience and if the programme is connected to their experiences. Therefore, most in-service training programmes should be designed around the problem-solving approach to learning or experience (see Table 15.3).

Sample In-service Training Format Using an Experiential Learning Cycle

The planner for training programmes should use a plan that provides an outline for the design of the training programme. Such an outline might look like the one in Table 15.3.

In-service training on subject matter content should be planned as a series of experiences: for example, Course I: Introduction to Soybean in Various Cropping Systems, Course II: Selected Issues and Problems Using the System, Course III: Training Farmers to Use the System, Course IV: Other topics including "Updates".

A week-long, in-service training programme might have a format that provides an opportunity to practise the behaviour on a more intensified basis, if the course is related to one subject. If the training is designed to provide an overview of many subjects, then a week's course will provide fewer instances for practice. Training sessions that include active involvement of the learner should provide no less than one hour for participants to practise behaviour.

Table 15.2
The experiential learning cycle

---

1. Experiencing To generate data.
   - How do you feel about that?
   - What do you need to know to ...?
   - Would you say more about that?

2. Sharing To report data.
   - What went on/happened?
   - How do you feel about that?
   - What did you observe?
   - What were you aware of?

3. Interpreting To make sense of data.
   - What does that mean to you?
   - What struck you about that?
   - How do those fit together?
   - How might it have been different?

4. Generalizing To develop abstractions from data.
   - What might we draw/pull from that?
   - What does that suggest to you about the subject in general?
   - What principle/law do you see operating?
   - How does this relate to other experiences?

5. Applying To use data in a new place.
   - What could you do to hold on to that?
   - What are the options?
   - What would be the consequences of doing/not doing that?
   - What modifications can you make work for you?

Note: From the experiential learning model and its application to large groups by S. E. Marks and W. L. Davis, 1975, in The 1975 annual handbook for group facilitators, LaJolla, California: University Associates Publishers, Inc.

Week-long training should provide for breaks of 20 to 30 minutes in the morning and afternoon, with a 90-minute break for mid-day and/or evening meals. This time provides participants with an opportunity to take care of personal needs, refresh themselves and interact with others. Afternoon sessions should be designed to be very active. Avoid using long films, or other visuals on a continuous basis, in the afternoon or evening. Keep in-class sessions limited to no more than a two-hour period before participants are allowed to move around. Adult attention span is limited to about two-hours.

In designing in-service training for professional staff, it is important to answer the following questions first.

1. Who are the learners, and how much do they know about this subject?

2. Is this training to be an awareness, knowledge and/or skill building session, or a combination?

3. What visible evidence of learning will be acceptable?

4. What standards will be used to measure acceptable evidence of learning?

A combination of skills for staff with limited backgrounds cannot be taught in a one-session programme. Such skills must be planned into a training programme which allows for a gradual increase in competencies over a period of time. For example, the programme, "Use of Pesticides in Household Grain Storage Areas", might be offered in four, one-day workshops over one month, or two three-day workshops over a six-week period, or eight half day workshops over an eight-week period.

Table 15.3
In-service training outline

Title:	Soybeans in Various Cropping Systems
Location:	Experimental, Farms, University Agriculture Centre
Date:	Wednesday, Nov. 18, 1984, 8:00 a.m. - 11:45 a.m.

Audience/Expected Number: 34 Field-Level Workers , 22 Specialists

Instructor(s):	A. V. Phillips, P. W. Singh and M. Louaose, Research and Extension Specialists
Materials:	Hand-outs (Instructional Package) Visuals, Equipment for Visuals and for Field Demonstrations, Evaluation Forms, Enrolment Forms (for attendance count)
Purpose:	To provide an opportunity for participants to strengthen competencies concerning various cropping systems which have an affect on the productivity of selected soybean cultivars.

Goals: Participants
- to discuss the various cropping systems utilized by local people.
- to examine and identify soybean cultivars grown under various conditions.
- to analyze the results of various cropping systems on soybean production.
- to demonstrate ability to select soybean cultivars and cropping systems which provide best conditions for increasing production on selected farms.

Format:

8 a.m.	Introduction to Overall Programme and Resource People; Presentation of Major Concepts with Visuals
8:45	Group interaction (Structured) to discuss concepts presented as these relate to local situations (8 groups of 7 people) (Experiencing)
9:30	Comfort Break (Plan according to local customs)
9:45	Speakers provide demonstration in field on identification and selection of soybean cultivars and cropping systems (Questions and Answers during Demonstration). (Sharing)
10:45	Small groups work with a leader to test knowledge and skills in the selection of cropping systems which work well with selected soybean cultivars (Interpreting and Generalizing)
11:25	Summary with handouts to allow a follow-up for practice and evaluation. (Applying)

Note: A follow-up activity should be planned to occur within a 30-day period by mail, visit, telephone or other. A second course should be planned to reinforce learning, with a focus on demonstrated practices.

Other In-service Training Formats

In-service training may be offered in a variety of formats. Each should be selected based on the criteria listed earlier. All training should be well-designed to meet intended outcomes, and should fit into an overall plan for staff development. Most of these training methods were described in Chapters 9 and 10. During any one year, an extension service staff development programme might include a variety of programme formats to achieve overall staff development goals (see example in Table 15.4).

Table 15.4
Sample: One-year staff development programme (To be adjusted to fit local customs)

January	February	March	April
Annual Conference	Short Courses	Refresher Courses	
Updates	Updates	Updates	Updates
Induction	Field Trips	Induction	Symposium

May	June	July	August
Short Courses		Staff Retreat	
Updates	Updates	Updates	Updates
Induction	Seminars	Field Trips	Field Trips
		Induction	

September	October	November	December
Refresher Courses		Seminars	
Updates	Updates	Updates	Updates
Induction		Induction	

- Annual Conference, for all staff (specialists, field agents, administrators)
- Updates, specific subject matter for selected staff
- Induction--new staff
- Short Course, subject matter oriented to introduce new trends
- Refresher Courses, subject matter oriented for skill development
- Symposiums, specialists involvement
- Field Trips, observations of practices

Fortnightly Training

Another complementary approach that should be discussed briefly is the fortnightly training sessions associated with the Training and Visit System (T & V System). The actual teaching methods used (e.g., lecture-discussion, demonstration, etc.) under this approach, have been described in other chapters. What is unique about the T & V approach is the close integration of training and extension work itself. Fortnightly training sessions are the chief means of continuously up-grading the professional and technical skills of VEWs. The VEWs have the opportunity to actually use the new skills and knowledge by teaching them to farmers during the following fortnight. This close integration of training and extension work gives each VEW a high motivation to learn during the one-day training sessions, and then reinforces the application of these new skills and knowledge through repeated presentations and explanations to their eight groups of contact farmers during the next series of fortnightly visits.

The fortnightly training sessions have two main objectives. The first is to present the specific recommended practices (called impact points) that are to be disseminated during the following two weeks. The second objective is to report problems and other questions to Subject Matter Specialists (SMSs) that could not be solved or answered during the previous fortnightly visits. SMSs then are confronted with an ideal problem-solving situation, to enhance learning on the part of VEWs and AEOs, as well as the SMSs themselves. When SMSs do not have an appropriate solution to a problem, they, in turn, raise these problems with researchers for guidance. This feedback from farmers to

VEWs, to AEOs, to SMSs and then to researchers is precisely what must be established if both research and extension are to be responsive to farmer needs and problems. Therefore, the fortnightly training sessions not only improve the technical and professional skills of VEWs and AEOs, but, equally important, they form the link for the essential feedback system between farmers demanding new or additional information and research personnel, communicated through extension.

The operational details of organizing fortnightly training sessions are well-covered elsewhere (see Benor and Baxter, 1984). Therefore, only a training and visit schedule for a district or subdivision being served by a three-person SMS team is included here (see Table 15.5). It should be noted that one member of the SMS team is designated as the Training Officer (TO), and this individual is in charge of preparing for the fortnightly training sessions. Under the schedule as shown in Table 15.5, there would be approximately 3 or 4 AEOs and 20-30 VEWs in each group; therefore, each SMS team would train and backstop four groups, approximately 100-120 AEOs and VEWs.

Table 15.5
Fortnightly training and visit schedule

| | Week 1 | | | | | | | | Week 2 | | | | | | | |
	M	T	W	Th	F	Sat	Sun		M	T	W	Th	F	Sat	Sun	
Group 1	Trg	V	V	V	V	EV	H		Mtg	V	V	V	V	EV	H	
Group 2	V	Trg	V	V	V	EV	H		V	Mtg	V	V	V	EV	H	
Group 3	V	V	Trg	V	V	EV	H		V	V	Mtg	V	V	EV	H	
Group 4	V	V	V	Trg	V	EV	H		V	V	V	Mtg	V	EV	H	

Abbreviations: Trg = Fortnightly training of VEWs and AEOs given by SMSs.

V = Visit of VEWs to farmers' groups, and of AEOs to VEWs; also, field visits (and other professional activities) of SMSs, Training Officers, and SDEO.

EV = Extra visit by VEWs and AEOs.

H = Holidays.

Mtg = Meeting of AEO with the VEWs in his/her range

Note: From Training and visit Extension by D. Benor and M. Baxter, 1984, Washington, D.C.: The World Bank.

Continuing Education and Career Development

Successful extension workers, at all levels, will find that satisfaction with the work itself is enhanced if they take the time to design and develop a plan for their own continuing education and professional development. Extension services that provide the opportunity for all staff to prepare such a plan will receive the benefits of having longer tenured and more satisfied employees. Such employees increase both the effectiveness and efficiency of an extension service.

Continuing education is a process by which men and women, alone or in groups, consciously choose to engage in learning opportunities for the purposes of becoming more professional. This tends to be a lifelong process.

Therefore, extension services should be supportive of continuing education activities through release time, fellowship/scholarship support, salary adjustments and other formal/ informal recognition/reward systems.

Extension workers are responsible for the design of their own continuing education programme. Such a programme would include both formal and informal modes of inquiry, instruction, or performance. Using the form in Appendix 5, the extension worker could create an overall plan, and modify it on an annual basis. The typical continuing education plan might include the following categories of experiences.

1.   Academic course work beyond the entry-level degree programme.

2.   Self-directed learning experiences with his or herself or with others (study groups, etc.)

3.   Professional association involvement.

All three categories should have the support of the extension service as described above. Of the three categories of experiences for continuing education, the one which has as its major focus the professional growth of its member is the professional association. Almost every subject has a professional association to which members pay dues; members associate with other colleagues in the local area and at international levels. The associations set standards for the profession, sometimes including credentialing and the identification of ethical practice, with sanctions. Professional associations help members to build an image of the group of people who are the leaders of their field. The association also helps members to increase their level of morale for doing the tasks related to their jobs.

Members can achieve a sense of status and leadership when they are part of a professional association. Therefore extension workers should make every effort to become actively involved in the establishment of professional associations. If one does not exist, key members of extension should take steps to organize an association. The following guide is useful in establishing such an association.

The Professional Association

1.   Two or more individuals meet to determine what needs of professionals could be met through an association.

2.   Interest is shared with other colleagues, to determine the level of interest that exists for the establishment of an association.

3.   An organizational meeting is planned, and conducted, in which some tentative purposes and goals are set out. Participants are encouraged to refine these goals, and to identify activities to achieve goals. Such activities should be centered around the following areas (a) education, (b) leadership, (c) public information, (d) service, and (e) socialization

4.   Establish a formal structure for the association including the identification of leaders, governing policies and procedures, financial obligations, membership services, meeting times, and recognition and award procedures.

A good association exists because it provides member involvement, and it meets their needs to be professional. The size of the organizing group is not a problem. However, be sure to involve those who can sanction or legitimize the group's activities.

Houle, in a discussion of the professionalizing process identifies several characteristics of the process. Members of a professional association can

1.   help to clarify the function of the profession and ethical practices,

2.   identify the need to master theoretical knowledge, to use practical knowledge, and to increase capacity to solve problems,

3.   broaden the relationship of members to other vocations and to service users, and

4.   increase self-enhancement and strengthen public acceptance of the profession.

In one, small, Carribean island state, extension workers established an association; within a three-year period, membership had spread to other islands. This year, they recognized an outstanding extension worker through a well-defined plan of action. That worker was chosen by his peers to represent them at an international series of meetings. The recognition of that individual will have an educational impact on the members and on the public acceptance of all member of that profession at the local, national and international levels.

Finally, career development is the act of acquiring information and resources that will enable one to plan a programme of lifelong learning related to his or her worklife (Schein, 1978). This is a self-directed activity and may be geared to the notion of adult developmental stages. Such a plan (see Tables 15.3 and 15.4) assumes that certain events in one's career and personal life will trigger (Cross, 1982) the need to learn. Such triggers can be predicted in some cases.

Successful extension services can provide overall staff development programmes that are well-structured but flexible enough to accommodate the situations that are triggered in a person's extension career.

REFERENCES

Bender, R. E., Cunningham, C. J., McCormick, R. W., Wolfe, W. H., & Wooden, R. J. (1972). Adult education in agriculture. Columbus, Ohio: Charles Merrill.

Benor, D. and Baxter, M. (1984). Training and Visit Extension. Washington, D.C.: The World Bank.

Benor, D. & Harrison, J. Q. (1977). Agricultural extension: The training and visit system. Washington, D. C.: World Bank.

Bergevin, P. & McKinley, J. (1965). Participation training for adult education. St. Louis, Mo.: Bethany Press.

Cross, K. P. (1981). Adults as learners. San Francisco: Jossey-Bass.

Davies, I. K., Hudson, B. H., Dodd, B., & Hartley, J. (1973). The organization of training. London: McGraw-Hill.

Dejnozka, E. L. & Kapel, D. E. (1982). American educators' encyclopedia. Westport, Conn.: Greenwood Press.

Fasokun, T. O. (1981). Relationship between learning climate and academic achievement in adult extra-mural classes in Nigeria (ERIC document ED 217123). Ife, Nigeria: Dept. of Continuing Education, Faculty of Education, University of Ife.

Gardner, J. E. (1980). Training the new supervisor. New York: American Management Association, AMACOM.

Gibbs, G. (1981). Teaching students to learn. Milton Keynes, England: Open University Press.

Harris, B. M. (1980). Improving staff performance through in-service education. Boston: Allyn and Bacon.

Houle, C. (1972). Design of education. San Francisco: Jossey-Bass.

Howell, R. E. (1982). Development of rural leadership: problems, procedures, and insight (ERIC document ED 214748). Pullman, Wash.: Dept. of Rural Sociology, Washington State University.

Keeton, M. (Ed.) (1976). Experiential learning: Rational characteristics and assessment. San Francisco: Jossey-Bass.

Knowles, M. S. (1980). The modern practice of adult education. Chicago: Follett Publishing Company.

Kozoll, C. E. (1974). <u>Staff development in organizations: A cost evaluation manual for managers and trainers</u>. Reading, Mass.: Addison-Wesley.

Kulich, J. (Ed.) (1977). <u>The training of adult educators in East Europe</u>. Vancouver, B. C.: Centre for Continuing Education, University of British Columbia.

Mager, R. F. (1974). <u>Preparing instructional objectives</u>. Belmont, California: Fearon Publishers.

Mager, R. F. & Pipe, P. (1971). <u>Analyzing performance problems</u>. Palo Alto, Calif.: Fearon Publishers.

Malone, V. (1982). <u>Intructional strategies for teachers of adults</u>. Glenview, Ill.: Scott Foresman, Lifelong Learning Division.

Malone, V. (1979). <u>Comprehensive staff development plan</u>. Urbana, Ill.: Cooperative Extenson Service, University of Illinois.

Malone, V. M. (1973). <u>A plan of action to implement a job stability program for an extension adviser in an urban area</u>. Tallahassee: Florida State University.

Marks, S. E. and Davis, W. L. (1975). The experiential learning model and its application to large groups. In <u>The 1975 annual handbook for group facilitators</u>. LaJolla: California: University Associates Publishers, Inc.

Matsui, T. & Okada, A. (1983). Mechanism of feedback affecting task performance. <u>Organizational Behavior and Human Performance</u>, <u>31</u>, 114-121.

Okure, B. E. (1982). Issues in research and training relating to women. <u>Convergence</u>, <u>15</u>(4), 26-31.

Rhodes, D. & Hounsell, D. (Eds.) (1980). <u>Staff deveopment for the 1980's: International perspectives</u>. Papers presented at the 4th International Conference on Higher Education held at the University of Lancaster in 1978. Normal, Ill.: Illinois State University Foundation.

Rowat, R. (1979). <u>Trained manpower for agricultural and rural development</u> (FAO economic and social development paper no. 10). Rome: Food and Agriculture Organization.

Schein, E. (1978). <u>Career dynamics: Matching individual and organizational needs</u>. Reading, Mass.: Addison-Wesley.

Shipper, F. (1983). Quality circles using small group formation. <u>Training and Development Journal</u>, <u>37</u>(5), 80-84.

Stone, M. M., Sacco, A. F. S., Simison, P., Stewart, T. S., & Thompson, C. (1980). <u>In-service education: A research vocabulary</u>. Durham, England: School of Education, University of Durham.

Tyler, R. W. (1975). <u>Basic principles of curriculum and instruction</u>. Chicago: The University of Chicago Press.

Veri, C. & Malone, V. (in press). <u>Train the adult trainer</u>. Lydonville, Vt.: Lydon State University.

Vincent, P. (1981). <u>Cost effective training strategy at field level strengthens influence of extension agents in community</u> (Development communication report no. 34) (ERIC document ED 207519). Washington, D. C.: Clearinghouse of Development Communication, Agency for International Development.

Vroom, V. H. & Deci, E. L. (Eds.) (1972). <u>Management and motivation</u>. Harmondsworth, Middlesex, England: Penguin Books, Ltd.

Warren, M. (1979). <u>Training for results</u>. 2nd ed. Reading, Mass.: Addison-Wesley.

Chapter 16
Information Sources to Strengthen Agricultural Extension and Training

F. W. Lancaster and A. Sattar

The phenomena associated with the flow of information to, from, and within particular communities have received increasing interest and attention within the past 30 years. Studies performed in this period have consistently demonstrated the importance of certain key individuals who perform the role of "information providers" within various communities. Such individuals may be referred to generically as information gatekeepers (Allen, 1971). In effect, the gatekeeper expedites the flow of appropriate information from selected sources to selected audiences. Agricultural extension services are probably the most visible and clear-cut manifestation of the gate-keeper concept, because the key role that they play is one of selecting, disseminating, and interpreting information generated by one community (that of agricultural research and development) for the benefit of a second community (those who are actually able to apply the results of this research). Because extension is essentially an activity involving the dissemination of information, it follows that the effectiveness of extension services must depend heavily on the quality, reliability and efficiency of the information sources they themselves draw upon.

It is a well-established fact, of course, that information in all fields (and agriculture is certainly no exception) is proliferating rapidly. While more and more is generated, however, the time that any one individual has to absorb this information remains relatively constant. Therefore, services designed to make information more readily accessible, and at the same time to aid identification of the information most relevant and appropriate to particular needs, are becoming increasingly important throughout society as a whole. The objective of this chapter is to identify some major information sources and information services that can provide essential input to extension activities.

THE INFORMATION TRANSFER CYCLE IN AGRICULTURE

Information flows through many channels, with many different types of individuals and organizations involved. In Figure 16.1 an attempt is made to reduce a rather complex set of interactions to a simple diagram.

The diagram is divided into a global level and a national level. The national level depicts the role of the extension service in the information transfer cycle. Various types of agricultural information (mostly the results of research and experimentation) serve as input to the service. The output is the delivery of this information, in whatever form is most appropriate, to the farming/rural community. The output functions of the extension service, which involve changing the presentations and the delivery of information, are dealt with in Chapters 9 and 10 and are outside the scope of the present chapter. The purpose of this chapter is to look at information sources as input to extension activities.

Before information can be delivered to the farming community it must be judged to be appropriate, valid, and useful. Within each country, therefore, there must exist an organizational structure of research centres, government agencies, and institutions of higher education which are capable of performing this evaluation and validation function (i.e., keeping up with new developments in agriculture, and determining which are useful and appropriate to the local situation). This organizational structure is represented by the box labelled "national research system." The diagram shows that the national research system provides essential linkage between the national and global levels of information flow.

In general, an extension service will draw its information from the various components of the national research system. Much of this information will be obtained by word of mouth, although it will usually be supplemented by printed sources. It will be the components of the research system in a country, rather than the extension services themselves, that will be the major exploiters of the rich information sources depicted in the diagram at the global level. Nevertheless, extension personnel should be aware of the nature and variety of these resources, and it is to create this awareness that the present chapter was formulated.

Agricultural information is generated by different types of agencies, including agricultural colleges and universities, agribusiness and, of course, extension services themselves. A certain amount is generated by the farming community, at least to the extent that reports of farming practice appear in the published literature. The major source of agricultural information, however, is a multitude of agencies (mostly governmental or academic) involved in agricultural research and experimentation.

Figure 16.1   The information transfer cycle in agriculture

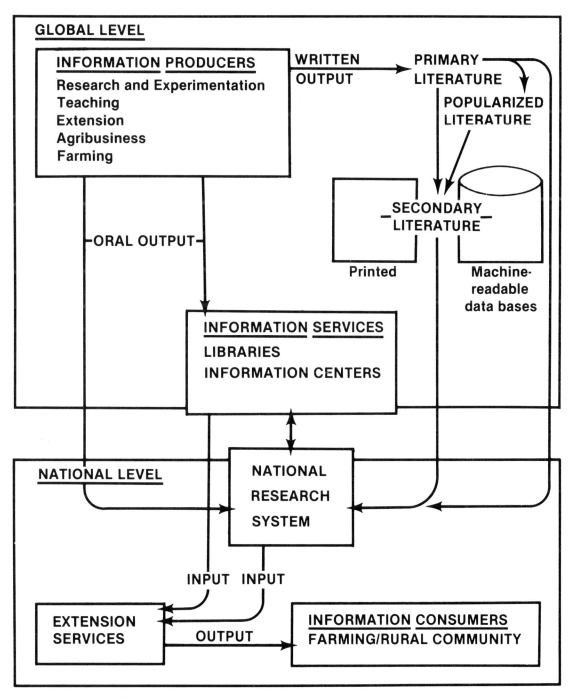

The information output from these agencies may be in oral or written form. The oral output may be formal (e.g., papers presented at meetings) or completely informal and transmitted by word of mouth.

INFORMATION SOURCES IN AGRICULTURE

The most visible sources of information, of course, are publications of various types, and it is convenient to make a distinction between primary publications and secondary publications. A primary publication is essentially one that presents original information. The most obvious example is the periodical (journal), but other primary forms include conference papers (including the published proceedings of conferences), reports (e.g., emanating from research groups or from extension services themselves), and publications in monographic form. A special form of primary literature is referred to as "popularized" in Figure 16.1. This consists of technical information rewritten for a different, less technical audience; it includes such things as "popular" farming magazines.

Secondary publications are those that provide guides to, and perhaps synopses of, the primary literature. Most important are the indexing and abstracting services, of which several exist in the field of agriculture. Such services increase in importance as the primary literature grows. There are now perhaps some 50,000 periodicals published world-wide in the broad fields of science and technology. While only several hundred of these are exclusively agricultural, the majority could occasionally carry papers of agricultural significance. The indexing and abstracting services are of critical importance because they are the only means available to keep informed on what is currently being published (in a wide variety of forms) on a particular topic, or to allow a comprehensive search on what has been written on this topic in the recent past.

In actual fact, agriculture presents certain features that make information retrieval in this field unusually difficult. Agriculture is perhaps the most inter-disciplinary of all the spheres of human activity drawing, as it does, from biology, medicine, chemistry, soil science, various branches of engineering, climatology, food technology, the environmental sciences, economics, management, and a whole host of other fields. This inter-disciplinary character brings with it some special problems in the provision of effective library and information services. In the first place, relevant literature can be widely scattered throughout a multitude of sources, making it difficult and costly to identify, to collect, to index, to abstract and to search. While virtually all the literature of interest to chemists is to be found in Chemical Abstracts, and almost all the literature of relevance to medicine in Index Medicus or Excerpta Medica, it is probably true to say that the secondary services of agriculture cover, by comparison, a much smaller proportion of the literature of interest to the agricultural community. This is not to suggest that the major agricultural sources are deficient (in point of fact, the core of agriculture is exceptionally well-served by three major services), but merely to point out that the information needs of this field are so diverse that no single service is likely to satisfy them all completely. Some needs, then, may involve the use of sources in other fields, such as chemistry, medicine, engineering, meteorology, food science, economics, biology and management. This great diversity of information needs poses further problems for librarians and other information specialists. It is difficult to become and to remain knowledgeable in all subject areas impinging on agriculture, and it is equally difficult to become and remain familiar with the myriad information sources that reflect this impingement.

A second characteristic relating to the literature of agriculture is that it seems more "fugitive" than that of any other field. In medicine, for example, periodicals, monographs, and conference proceedings constitute perhaps 99 percent of all the literature; other forms, including technical reports and patents, are relatively insignificant. This is not true of agriculture, where reports, pamphlets, patents, and other forms constitute a significant portion of the literature of interest. Agriculture literature is both fragmented by subject dispersion and scattered over a wide range of forms. Table 16.1 shows the great variety in agriculture-related publications, ephemeral and enduring, that has been well-illustrated by Lilley (1981), drawing upon a report by Craig (1979).

220

Table 16.1
Ephemeral and Enduring Agricultural Publications

(A) Ephemeral   (i)   local and national newspapers
                (ii)   farming press
                (iii) press releases
                (i v) market reports
                (v)   weather reports and forecasts
                (vi) disease intelligence reports

(B) Enduring, Non-Scientific

                (i)   statutory and official publications (e.g. parliamentary; official statistics; statutory and approved lists)
                (ii) publications from agricultural bodies, product manufacturers and the trade (e.g. promotion of products and services, technical specifications)
                (iii) popular articles, reports, reviews, books
                (i v) advisory publications (e.g. advisory leaflets, technical news-letters, farming notes)

(C) Enduring, Scientific
primary   (i)   technical notes, specification and patents
                (ii) theses and dissertations
                (iii) reports and research bulletins
                (i v) conference proceedings
                (v)   scientific research papers

(D) Enduring Scientific
secondary   (i)   textbooks
                (ii) reviews and monographs
                (iii) reference books
                (i v) bibliographies, indexes, abstracts

Note: From "Agricultural literature: Scope--characteristics of publication--information content" by G. P. Lilley, In: G. P. Lilley (Ed.) Information sources in agriculture and food science, 1981, London: Butterworths.

A third characteristic of agriculture, and its literature, is universality. There is virtually no country in the world that is not dependent, to a greater or lesser extent, on its own agricultural production. Agriculture and the literature of agriculture are international in the truest sense of the term. Improvements in agricultural methods, as reflected in the literature, could be of critical and immediate importance to a small developing country for which advances in other branches of science and technology may have no immediate interest or relevance. By the same token, agricultural research institutes in one developing country may produce reports that could be of potential interest and value to many other nations. The literature of agriculture is unique in its universality. Not only is it of concern to all countries, but virtually any of these, large or small, developed or developing, has the potential for making a significant contribution to this literature. In this respect, agriculture differs markedly from all other fields. Dominance of a small group of languages is much less true in the literature of agriculture than in virtually any other area of activity. Hence the need for truly international services in agriculture.

Fortunately, the computer has proved to be a tool of immense value in the provision of information services. Since early in the 1960s, computers have been applied to the publication of the major indexing and abstracting services. As a result, machine-readable data bases, used to generate the printed publications, have become available. Such data bases can be used as an effective and efficient means of locating publications dealing with specific topics, as well as to keep organizations and individuals up-to-date on what is newly published in areas of current concern. Other data bases, sometimes referred to as data banks, provide access to numerical or statistical information of a commercial or scientific nature, and some of these have relevance to the agricultural community.

Access to information sources has increased by leaps and bounds in the past decade as more and more data bases and data banks have become available through various on-line computer networks. These sources are now readily accessible throughout most of the developed countries, and they are beginning to become accessible within the developing nations as telecommunications and computer networks are established.

Within the last 20 years there has occurred a growing recognition of the importance of adequate library services to support the needs of agriculture, and many libraries devoted to agriculture have sprung up in research institutes, universities, government agencies, and the commercial sector. In several countries, national libraries devoted to agriculture are already in place.

A good modern library is more than a mere store of publications. Many offer information services of a high quality, answering factual questions and compiling bibliographies on requested topics (perhaps through on-line access to data bases), as well as aiding users in identifying and accessing the particular publications they need. The modern librarian should be much more than a custodian. He or she is a true information specialist. With the massive store of agricultural information to which users have potential access, the information specialist becomes essentially an information consultant, whose role is to help users define their information needs, and then to link them with relevant sources of information (which may be key personnel, primary or secondary scientific literature, factual data banks, management information systems, or ephemeral literature, such as reports, pamphlets, proceedings, trade literature, and extension materials).

Information centres specializing in various facets of agriculture, and incorporating appropriate library facilities, have sprung up in many parts of the world. Some of the major centres of this type, along with other information sources of potential relevance to extension services, are identified in the next sections of this chaper.

PUBLISHED SOURCES IN PRINTED AND MACHINE-READABLE FORM

A comprehensive listing of information sources in agriculture is well beyond the scope of this chapter. Such a bibliography already exists in Lilley (1981), and a useful guide to major reference books and data bases is provided in an article by Polach (1979). The most that can be done here is to mention some important sources that can support the information needs of a national research system and, ultimately, of national extension services.

Three major indexing/abstracting services deal with agriculture in its entirety; the Bibliography of Agriculture, a product of the National Agricultural Library (NAL) in the United States; Agrindex, produced by FAO; and a series of 26 journals of abstracts generated by the Commonwealth Agricultural Bureaux (CAB). This last source is probably the most comprehensive. Each journal is devoted to a different facet of agriculture, and three of them have special relevance to the needs of agricultural extension services: Rural Development Abstracts, Rural Extension Education and Training Abstracts, and World Agricultural Economics and Rural Sociology Abstracts, all three published quarterly.

The other two services should not be overlooked. Agrindex, because it is based on input received from almost 100 participating countries, is extremely useful in the identification of some of the more obscure reports generated in the developing countries. Moreover, some input is also received from NAL and CAB. The Bibliography of Agriculture is especially useful for its coverage of the publications issued by the many agricultural extension services in the United States.

All three of these publications have roughly equivalent data bases in machine-readable form: AGRICOLA (NAL), CAB Abstracts, and the AGRIS data base (FAO). In various parts of the world, one or other (or all three) of these are accessible through on-line networks.

In addition to these general sources, a number of specialized indexing/abstracting services deal with various facets of agriculture of potential relevance to extension activities. Among these are IRRICAB, covering irrigation, a product of the International Irrigation Information Centre; Abstracts on Tropical Agriculture (published monthly by the Royal Tropical Institute, the Netherlands, and accessible on-line as TROPAG); and Nutrition Abstracts and Reviews (section A, Human and Experimental, section B, Livestock Feeds and Feeding), which is actually one of the publications of CAB. A more complete listing of these specialized sources can be found in a book by Lendvay (1980) and a chapter by Johnston (1981). Several complete directories of machine-readable data bases exist, and any one of these can be used to identify sources applicable to various specific problems.

Two major regional services also exist. Agriasia is a quarterly publication of the Agricultural Information Bank for Asia, SEARCA, College, Laguna 3720, Philippines, and the Indice Agricola de América Latina y el Caribe is a quarterly publication of the Inter-american Agricultural Information System, CIDIA, Turrialba, Costa Rica. Both publications are extremely useful for their coverage of the agricultural literature generated in their respective regions.

There are probably several hundred periodicals of direct relevance to agriculture, as well as many others that carry occasional articles of agricultural interest. Perhaps the best way to keep abreast of what is being published in these journals is to subscribe to Current Contents: Agriculture, Biology and Environmental Sciences, which reproduces the contents pages of several hundred journals. It is published by the Institute for Scientific Information, 3501 Market Street, Philadelphia, Pa., 19104, U.S.A., which guarantees to be able to supply a copy of any article listed.

Periodicals having special relevance to extension activities do exist (e.g., the Extension Review, a publication of the Department of Agriculture in the United States, and the Indian Journal of Extension Education), but these tend to have a national focus rather than a truly international one.

SPECIALIZED INFORMATION CENTRES

Over the past several years a number of specialized information centres, each devoted to a particular facet of agriculture, have been established in various parts of the world. While not necessarily devoted to extension per se, these centres are important sources of authoritative information. Several generate publications of potential value to extension activities (e.g., Irrinews, produced by the International Irrigation Information Centre, contains brief accounts of new developments in irrigation technology), and most are willing to answer questions, and perhaps to supply relevant bibliographies or publications themselves, in their areas of specialization. In some cases, these centres will provide services free, or at greatly reduced costs, to the developing countries.

Because these centres are sometimes overlooked or forgotten, a listing of some of the more important is given below. Agricultural research centres having significant information or publication programmes, as well as information centres per se, are included in the list.

1. The Commonwelath Agricultural Bureaux, Farnham House, Farnham Royal, Slough, Berkshire, United Kingdom SL2 3 BN (This is a unique organization that co-ordinates the work of 14 information centres on the following topics: entomology, mycology, biological control, helminthology, agricultural economics, animal breeding and genetics, animal health, dairy science and technology, forestry, horticulture and planta-tion crops, nutrition, pastures and field crops, plant breeding and genetics, and soils).

2. Centro Internacional de Agricultura Tropical (CIAT), Apartado Aereo 6713, Cali, Colombia (Cassava, field beans, rice and tropical pastures).

3. Centro Internacional de Mejoramiento de Maiz y Trigo (CIMMYT), Londres 40, Mexico 6, D.F., Mexico (Maize, wheat).

4. Centro Internacional de la Papa (CIP), Apartado 5969, Lima, Peru (Potato).

5. International Center for Agricultural Research in Dry Areas (ICARDA), P.O. Box 5466, Aleppo, Syria (Cereals, food legumes, forage crops, and farming systems in general for arid regions).

6. International Crops Research Institute for the Semi-Arid Tropics (ICRISAT), Patancheru, A.P., 500016, India (Chick pea, pigeon pea, pearl millet, sorghum, ground-nut, farming systems in semi-arid areas).

7. International Livestock Center for Africa (ILCA), P.O. Box 5689, Addis Ababa, Ethiopia.

8. International Laboratory for Research on Animal Diseases (ILRAD), P.O. Box 30709, Nairobi, Kenya.

9. International Rice Research Institute (IRRI), P.O. Box 933, Manila, Philippines.

10. International Institute of Tropical Agriculture (IITA), P.M.B. 5320, Ibadan, Nigeria (Maize, rice, roots and tubers, food legumes, farming systems for tropical areas).

11. West Africa Rice Development Association (WARDA), P.O. Box 1019, Monrovia, Liberia.

12. International Irrigation Information Centre (IIIC), Bet Dagan, Israel.

13. Asian Vegetable Research and Development Center (AVRDC), Box 42, Shanhua, Tainan 741, Taiwan.

14. International Center for Insect Physiology and Ecology (ICIPE), 1176 Massachusetts Ave., N.W., Washington, DC 20036, USA, and P.O. Box 30772, Nairobi, Kenya.

15. International Fertilizer Development Centre (IFDC), Box 2040, Muscle Shoals, Alabama 35660, USA.

16. International Food Policy Research Institute (IFPRI), 1776 Massachusetts Ave., N.W., Washington, D.C. 20036, USA.

17. The Interamerican Centre for Agricultural Information and Documentation, Apartado 10281, San Jose, Costa Rica.

18. International Buffalo Information Center, Kasetsart University, Bangkok, Thailand.

19. African Food and Nutrition Research Organization (ORANA), Dakar, Senegal.

20. Comprehensive Pig Information Centre (CPIC), Lane End House, Shinfield, Reading, RG2 9BB, UK

21. Soybean Insect Research Information Center (SIRIC), International Soybean Program, College of Agriculture, University of Illinois, Urbana, Illinois 61801, USA.

22. Dairy Society International (DSI), 3008 McKinley St., N.W., Washington, DC 20015, USA.

23. Nigerian Institute of Oil Palm Research, PMB 1030, Benin City, Bendal State, Nigeria.

24. Rubber Research Institute of Malaysia, POB 150, Kuala Lumpur, Malaysia.

25. Rubber Research Institute of Sri Lanka, Telawala Road, Ratmalana, Mt. Lavinia, Sri Lanka.

26. Tropical Pesticides Research Institute, POB 3024, Arusha, Tanzania.

27. Centre for Overseas Pest Research, College House, Wrights Lane, London W8 5SJ, UK

28. Institute of Tropical Forestry, POB AQ, Rio Piedras, 00928, Puerto Rico.

29. International Documentation Centre on Abaca, University of the Philippines, Los Baños, Philippines.

30. Institut Francais du Cafe et du Cacao, 42 rue Scheffer, Paris 16e, France.

31. Coconut Research Institute, Bandirippuwa Estate, Lunuwila, Sri Lanka.

32. Centro de Documentacão de Cafe, Missão de Estudos Agronomicos de Ultramar, Apartado 3014, Lisboã, Portugal.

33. Servico de Documentacão Economica, Instituto Brasileiro do Cafe, av. Rodriguez Alves 129, 20.000 Rio de Janeiro, R.J., Brazil.

34. Institut Francais de Recherches Fruitiéres Outre-Mer, 6 rue de Général Clergerie, 75 Paris 16e, France.

35. Institut de Recherches pour les Huiles et Oléagineux (IRHO), 8 sq. Pétrarque, Paris 75016, France (Oil seeds).

36. Instituto do Acucar e do Alcool, Praca 15 de Novembro 42, 20.000 Rio de Janeiro, R.J., Brazil.

37. Sugarcane Breeding Institute, ICAR, Coimbatore 641007, India.

38. Tropical Products Institute, 56-62 Gray's Inn Road, London WC1X 8LU, UK

39. Documentation Center on Tropical Forestry, University of the Philippines, Los Baños, Philippines.

40. Centre Technique Forestier Tropical (CTFT), 45 bis Avenue de la Belle Gabrielle, 94130 Nogent sur Marne, France.

While not a specialized information centre per se, mention should be made of Volunteers in Technical Assistance (VITA). VITA is a private non-profitmaking development agency based in the United States. Throughout the developing countries, since 1960 VITA has used the mail to inform farmers and extension officers about technical improvements they may not know about. A farmer's questions are answered by sending a photocopy from one of VITA's publications, which are written by experts in their fields and who are experienced in work in developing countries. Subjects of particular interest to VITA are agriculture and animal husbandry, alternative energy systems, water and sanitation, food processing, and small-scale industries. They maintain a collection of 45,000-50,000 published and unpublished documents concerning small-scale technologies suitable for developing countries.

VITA also produces many appropriate technology manuals, including some in French, Spanish, and Arabic. The Village Technology Handbook is a collection of plans and designs for the support systems necessary for a small community's survival. Specific topics include windmills, water wheels, and rabbit raising. In addition a number of technical bulletins are available.

The headquarters of VITA is located at 3706 Rhode Island Ave., Mt. Rainier, MD 20712, USA; regional field offices also exist in various parts of the world.

INFORMATION NEEDS IN AGRICULTURAL EXTENSION

The information needs of extension workers, of course, are somewhat different from those of agricultural scientists. The extension worker's role is to aid in transforming the results of research into such tangible benefits as increased crop yields and improved living conditions for rural populations. The information needed by the extension worker, then, can be characterized as practical, current, proven, and locally relevant. In general, the extension worker will want a specific item of information, or a few highly relevant publications, rather than a lengthy bibliography on some subject. Many extension workers will prefer to receive information from an oral source, an agricultural scientist or other subject expert, rather than spending much of their time searching the published literature. It is obvious, therefore, that extension services must maintain close and effective relationships with sources of appropriate information and expertise, agricultural research centres, local research stations, and teachers/researchers associated with agricultural faculties.

Nevertheless, extension workers should familiarize themselves with the published literature, and not only those sources in which agricultural information has already been repackaged for consumption by the farming community. Extension agencies must work closely with the components of the national research system to decide what information is relevant and appropriate to local conditions, what should be repackaged for local application, and what form this repackaging should take. This requires familiarity with sources of published information as well as oral sources.

We began the chapter by pointing out that extension services can be looked upon as primarily gate-keepers of information. Our focus has been on sources of information as input to extension services rather than on methods by which extension services themselves disseminate information (i.e., their output).

Because this manual is addressed to extension services of all types and sizes, and is without geographic restrictions, it is difficult to do more than deal in generalities. One purpose of this chapter has been to emphasize the important role that library and information services can play in support of the activities of extension personnel.

A national extension service should be supported by a strong information centre (under the direction of a qualified librarian/information specialist,) which will identify and collect published literature appropriate and relevant to local needs, as well as sources of information on the locality itself, e.g., natural resources, human resources and socio-economic conditions. In addition to collecting basic reference sources in the field of agriculture (the book by Lilley (1981) is a useful guide to these), the centre should be prepared to build its own information resources (e.g., directories of specialists and research units, with addresses and telephone numbers).

Where an extension agency is unable to maintain its own library/information centre it will be particularly important that the extension workers familiarize themselves with the information resources available to them in their own countries and that they establish relationships with libraries or information centres (in universities, research institutions or government agencies) that can act as entry points to the vast resources of agricultural information that are available worldwide, both in printed and machine-readable form. Because library services can now be linked through networks established at national, regional, and international levels, there should be no major barriers preventing access to any publication or information needed in support of any extension activity.

In countries where no adequate library support is yet established, the possibility exists of forming a "scientific literature service", which can provide copies of the contents pages of periodicals selected by some user group and then supply photocopies of articles selected from these contents pages. Services of this kind have already been set up in East Africa, the Philippines, Ireland, and Tanzania, and possibilities exist for international funding to support such a service. Further information on this approach can be obtained form Mr. S. Cooney, The Agricultural Institute, 19 Sandymount Avenue, Dublin 4, Ireland.

REFERENCES CITED

Allen, T. J. (1971). The international technological gatekeeper. Technology Review, 55, 36-43.

Craig, G. M. (1980). Information systems in UK agriculture. Aslib Proceedings, 32, 201-206.

Johnston, S. M. (1981). Use of computer-based bibliographic services. In: G. P. Lilley (Ed.), Information sources in agriculture and food science (pp. 55-101). London: Butterworths.

Lendvay, O. (1980). Primer for agricultural libraries. 2nd ed. Wageningen: Centre for Agricultural Publishing and Documentation.

Lilley, G. P. (1981). Agricultural literature: Scope--characteristics of publication--information content. In: G. P. Lilley (Ed.), Information sources in agriculture and food science (pp. 3-25). London: Butterworths.

Lilley, G. P. (Ed.) (1981). Information sources in agriculture and food science. London: Butterworths.

Polach, F. (1979). Current survey of reference sources in agriculture. Reference Services Review, 7, 7-14.

Epilogue
Utilization of Technology: The Corner-Stone of Agricultural Development Policy and Programmes

V. A. Sigman and B. E. Swanson

Extension is again moving into a position of central responsibility in the agricultural development process. The reason is clear; most national governments, donors and development agencies have given high priority to agricultural development, and the corner-stone of this policy is technological change. International agricultural research centres (IARCs), working jointly with national research programmes, have made considerable progress in increasing crop yields and total food crop production, particularly in irrigated and high-potential areas of many developing countries. This progress has been achieved largely through improved genetic technology (i.e., improved, high-yielding varieties), combined with improved agronomic practices, including increased usage of fertilizer and other agricultural chemicals.

The effective use of improved agricultural technology by farmers is the immense challenge facing extension. This very difficult task results from the millions upon millions of small producers to be served, the relatively poor resources (both human and material) available to national extension organizations, and the potentially disastrous social consequences of pursuing inappropriate extension strategies or disseminating inappropriate technology.

Clearly agricultural extension has a central role to play in any system using and developing effective technology. But let the critics not forget what Mosher (1966) said nearly two decades ago, that other factors were essential for agricultural development. In particular, he indicated that (a) improved technology, (b) adequate markets, (c) available supplies and inputs, (d) access to adequate transport, and, of course, (e) sufficient incentives to motivate the farmer to innovate, were all essentials. This list of essentials is as applicable today as it was in the mid-sixties.

However, the reality of development is that the poorly trained and equipped extension worker is on the firing line in the battle line against poverty and inadequate food production. Most critics pay too little attention to the appropriateness of the technology, to the timely availability of inputs, or whether the farmer can sell his or her produce at a favourable price. The extension worker is viewed as the "change agent," and if meaningful change does not occur, the responsibility somehow rests on his or her shoulders. But when success occurs, all participants in the development process are eager to claim credit, with extension workers and farmers largely being the "unsung heroes and heroines".

The road ahead for agricultural extension will be difficult, complex, and uphill. The need to minimize the socially disruptive consequences of technological change must be of central concern to extension. The need to effectively serve small farmers, many of whom are women, as well as the young men and women who are just entering farming, all pose a special and complex assignment for extension personnel. It is very important that none of these important categories of farm people be disenfranchised from the development process. Therefore, extension must proceed carefully, sensitive to the goal of growth with equity as outlined by the World Conference on Agrarian Reform and Rural Development (WCARRD) in 1979.

THE NEED TO STRENGTHEN AGRICULTURAL EXTENSION

This entire manual has focused on agricultural extension and how to increase its effectiveness in the agricultural development process. But where should national governments begin in strengthening national extension systems? In 1980, 59 directors of national extension organizations in developing countries responded to a survey to rank their perception of the seriousness of selected problems that could effect agricultural extension. The results of this study (Sigman & Swanson, 1984) are summarized here to suggest some areas of concern that should be further investigated to determine where additional resources are needed to strengthen national extension institutions.

The nine problems selected from the literature for inclusion in the study were as follows:

1. Technological Problems. Appropriate technology is not available to extend to farmers.

2. Linkage Problems. A continuing two-way flow of information between extension services and national agricultural research institutions is lacking.

3. Technical Training Problems. Field-level extension personnel lack practical agricultural training about improved technology.

4. Extension Training Problems. Extension personnel lack training in extension methods and communication skills.

5. Mobility Problems. Field-level extension personnel lack adequate transportation to reach farmers efficiently.

6. Equipment Problems. Extension personnel lack essential teaching and communications equipment.

7. Lack of Teaching Aids. Extension personnel lack essential teaching aids, bulletins, demonstration materials, and so forth.

8. Organizational Problems. Extension personnel are assigned many other tasks besides extension work.

9. Miscellaneous Problems. Specific other problems identified by respondents.

The survey questionnaire sent to directors asked about problems facing extension. Respondents were asked to rate categorically the seriousness of the problems on a three-point scale as a serious problem, somewhat of a problem, or not a problem. Respondents were then requested to place (rank), in order of importance, the five most serious problems.

FINDINGS

The findings presented here are based on respondents' rank-ordering of problems. Table 17.1 shows the combined results of rank-ordering problems for all the participating countries combined:

Table 17.1

Serious Problems Ranked by Extension Directors (All Countries)

RANKING IN ORDER OF IMPORTANCE OF SERIOUSNESS	PROBLEMS	MEAN RANKING
1st	Mobility	3.86
2nd	Extension Training	4.05
3rd	Equipment	4.56
4th	Organizational	4.61
5th	Technical Training	4.91
6th	Teaching Aids	5.37
7th	Linkage	5.75
8th	Technological	5.95
9th	Other Problems	6.00

NOTE: Table based on responses from 50 extension directors.

Overall, mobility and extension training problems are ranked as the first- and second-most important problems in order of seriousness, respectively. Linkage, technological, and "miscellaneous" problems are ranked low on the scale.

Table 17.2 shows the rank-ordering of problems by geographic location.

Table 17.2

Importance Ranking in Order of Seriousness of Problems by Geographic Location

RANK	AFRICA	LATIN AMERICA, CARIBBEAN	ASIA	OCEANIA	RANK
1st	Extension Training	Mobility	Mobility	Technical Training	1st
2nd	Equipment	Extension Training	Organizational	Extension Training	2nd
3rd	Mobility	Equipment	Technical Training	Organizational	3rd
4th	Teaching Aids	Organizational	Extension Training	Other Problems	4th
5th	Organizational	Technical Training	Equipment	Mobility & Teaching Aids	5th
6th	Technological & Other Problems	Linkage	Teaching Aids		6th
7th		Teaching Aids	Linkage	Linkage	7th
8th	Linkage	Technological	Technological	Technological	8th
9th	Technical Training	Other Problems	Other Problems	Equipment	9th

Note: Table based on director's responses categorized by region as follows, Africa 16, Latin America and Caribbean 15, Asia 14, and Oceania 5.

As shown in Table 17.2, extension training and mobility problems are consistently ranked as two of the more important problems. Respondents consistently ranked linkage and technological problems as among the least important.

The rank-order of problems by per capita income level is presented in Table 17.3. Per capita income figures are expressed in United States dollars and are for 1978 (World Bank, 1980).

In Table 17.3, extension training problems are consistently ranked within the three most important problems. Mobility problems are also ranked high, especially among the lower income countries. Technology problems ranked consistently low among the lower middle, middle, and high income countries, while organizational problems increased in seriousness as the per capita income level increased.

Table 17.3

Importance Ranking in Order of Seriousness by Country Per Capita Income Level

RANK	LOW ($1-$300)	LOWER MIDDLE ($301-$699)	MIDDLE ($700-$2999)	HIGH ($3000-$16000)	RANK
1st	Extension Training	Mobility	Mobility	Organizational	1st
2nd	Mobility	Extension Training	Extension Training	Technical Training	2nd
3rd	Equipment	Equipment	Organizational	Extension Training	3rd
4th	Technology	Organizational	Technical Training	Linkage & Mobility	4th
5th	Teaching Aids	Teaching Aids	Equipment		5th
6th	Linkage & Organizational	Technical Training	Other Problems	Equipment & Teaching Aids	6th
7th		Linkage	Linkage		7th
8th	Other Problems	Technology	Technology	Technology	8th
9th	Technical Training	Other Problems	Teaching Aids	Other Problems	9th

Note: Table based on directors responses categorized by country income level as follows: Low 18, Lower Middle 16, Middle 16, and High 8.

CONCLUSIONS AND RECOMMENDATIONS

All findings and conclusions reported in the study are influenced by the assumptions and limitations of the study. In particular, the scales utilized measure perceptions of problems from subtly differing viewpoints; therefore, all findings should be interpreted as suggestive, rather than conclusive.

The study of extension directors' perceptions of selected problems tentatively concludes that the eight specific problems examined appear to be actual problems facing most national agricultural extension organizations in the developing world. The relative importance of these problems in differing economic and geographic regions of the world is less conclusive and requires further study.

It does appear that the lack of mobility and extension training are perceived to be very serious problems worldwide. These problems are closely followed in importance by the lack of equipment (particularly for teaching and communications) and organizational problems.

Based on the findings of the study, more attention must be given to developing and testing alternative approaches to extension that effectively decrease extension workers' reliance on transportation as the primary means of gaining proximity to individual farmers. Teaching and communications tools and aids must be available to extension workers so that they can use more group and mass media approaches.

The importance of extension training problems, as compared to technical training problems, indicates a widespread concern for the educational role of extension. Research on the suitability of a more balanced approach to training, integrating both "what" to extend (technical training) as well as "how" to extend it educationally (extension training), may be in order.

In summary, a full two-thirds of the developing world's population and the majority of the world's poorest people still gain their livelihoods through agriculture. Meeting basic human needs and accelerating rural development are high priorities. Agricultural extension organizations have the potential to address these priorities. However, to do so, some of the more serious problems that are constraining agricultural extensions' ability to facilitate effective agricultural development will need to be solved.

REFERENCES

Mosher, A. T. (1966). Getting agriculture moving. New York: Agricultural Development Council.

Sigman, V. A. & Swanson, B. E. (1984). Problems facing national agricultural extension organizations in developing countries (INTERPAKS research series no. 3). Urbana, Illinois: University of Illinois at Urbana-Champaign.

World Bank. (1980). 1980 World Bank Atlas. Washington, D.C.: World Bank.

Bibliography

SECTION 1 -- GENERAL EXTENSION

Adams, M. E. (1981). Agricultural extension in developing countries. Harlow. Essex: Longman Group.

Axinn, G. H. & Thorat, S. (1972). Modernising world agriculture: A comparative study of agricultural extension education systems. New York: Praeger.

Ban, A. W. van den (1979). Inleiding tot de voorlichtingskunde [Introduction to extension education]. Meppel, The Netherlands: Boom.

Boyce, J. K. & Evenson, R. E. (1975). National and international agricultural research and extension programs. New York: Agricultural Development Council, 1975.

Claar, J. B. & Watts, L. H. (Eds.) (1984). Proceedings of a conference on international extension held at Steamboat Springs, Colorado, July, 1983. Urbana, Illinois: International Program for Agricultural Knowledge Systems, University of Illinois at Urbana-Champaign and Fort Collins, Colorado: Office of International Programs, Colorado State University.

Crouch, B. R. & Chamala, S. (Eds.) (1981). Extension, education and rural development. Vol. 1, International experience in communication and innovation. Vol. 2, International experience in strategies for planned change. Chichester, England: John Wiley.

Ekpere, J. A. (June, 1981). Achieving maximum impact with what you've got: The challenge of rural agricultural extension in the 1980s. Keynote address at the 29th International Course on Rural Extension. Wageningen, Netherlands: International Agricultural Centre.

International Agricultural Centre (in press). Proceedings of the Seminar on Strategies for Agricultural Extension in the Third World, 18-20 January 1984. Wageningen, Netherlands: International Agricultural Centre.

Kelsey, L. D. & Hearne, C. C. (1963). Cooperative extension work. Ithaca, N.Y.: Comstock.

Maalouf, W. D. (May, 1983). International experiences in agricultural extension and its role in rural development. Paper presented at the Regional Seminar on Extension and Rural Development Strategies, Serdang, Selangor, Malaysia. Rome: Food and Agriculture Organization of the United Nations.

Maunder, A. H. (Ed.) (1973). Agricultural extension: A reference manual. Abridged edition. Rome: Food and Agriculture Organization of the United Nations.

Mosher, A. T. (1978). An Introduction to agricultural extension. Singapore: Singapore University Press.

Orivel, F. (1983). The impact of agricultural extension services: A review of the literature. In Basic education and agricultural extension: Costs, effects, and alternatives (World Bank staff Working papers no. 564) (pp. 1-57). Washington, D.C.: World Bank.

Sanders, H. C. (Ed.) (1966). The cooperative extension service. New York: Prentice-Hall.

Savile, A. H. (1965). Extension in rural communities. Oxford: Oxford University Press.

Stavis, B. (1979). Agricultural extension for small farmers (MSU rural development series, working paper no. 3). East Lansing, Mich.: Michigan State University, Department of Agricultural Economics.

232

SECTION 2 -- EXTENSION CONCEPTS AND PHILOSOPHY

Benor, D. & Baxter, M. (1984). Training and visit extension. Washington, D.C.: The World Bank.

Benor, D. & Harrison, J. Q. (1977). Agricultural extension: The training and visit system. Washington, D.C.: The World Bank.

Benor, D., Harrison, J. Q., & Baxter, M. (1984). Agricultural extension: The training and visit system. Washington, D.C.: World Bank.

Blanckenburg, P. von (1982). Basic concepts of agricultural extension in developing countries. Agricultural Administration, 10, 35-43.

Blanckenburg, P. von (1982). The training and visit system in agricultural extension--a review of first experiences. Quarterly Journal of International Agriculture, 21, 6-25.

Claar, J. B., Dahl, D. T., & Watts, L. H. (1983). The cooperative extension service: An adaptable model for developing countries (INTERPAKS series no. 1). Urbana, Illinois: University of Illinois.

Coward, E. W. Jr. (1973). Authority innovation-decisions in rural society: Expanding the concept. Paper presented at the Rural Sociological Meetings, College Park, Maryland.

Duft, K. D. (1971). The team approach and extension economics. American Journal of Agricultural Economics, 53, 47-52.

Hulme, D. (1983). Agricultural extension: Public service or private business? Agricultural Administration, 14, 65-79.

Röling, N. (1979). Basic extension strategies for small farmer development. Approach (Netherlands) 5, 3-11.

Röling, N. (1979). The logic of extension. Indian Journal of Extension Education, 15, 1-8.

Simpson, I. G. (1982). Are agrobureaucracies essential? The need for direct farmer research links. Agricultural Administration, 9, 211-220.

Timmer, W. J. (1982). The human side of agriculture. Theory and practice of agricultural extension (with special reference to developing rural communities in the tropics and sub-tropics). New York: Vantage Press.

SECTION 3 -- GENERAL RURAL DEVELOPMENT

Ankers, D. L. W. (1973). Rural development problems and strategies. International Labour Review, 108, 461-468.

Cohen, J. & Uphoff, N. (1979). Rural development participation: Concepts and measures for project design, implementation and evolution. Ithaca, NY: Rural Development Committee, Cornell University.

Food and Agriculture Organization (1980). Key principles for operational guidelines in the implementation of WCARRD programme of action. Rome: Food and Agriculture Organization.

Holdcroft, L. E. (1978). The rise and fall of community development in developing countries: A critical analysis and an annotated bibliography (MSU rural development paper no. 2). East Lansing, Mich.: Michigan State University, Department of Agricultural Economics.

Johnston, B. F. & Clark, W. C. (1982). Redesigning rural development. Baltimore: Johns Hopkins Press.

Lele, U. (1975). The design of rural development: Lessons from Africa. Baltimore, Md.: Johns Hopkins University Press.

Lowdermilk, M. K. & Laitos, W. R. (1981). Towards a participatory strategy for inte-grated rural development. Rural Sociology, 46, 688-702.

Moris, J. (1982). Managing induced rural development. Bloomington, Indiana: Inter-national Development Institute.

Morss, E. R. & Gow, D. D. (Eds.) (1983). Implementing rural development projects: Nine critical problems. Boulder, Colo.: Westview Press.

Morss, E. R. Hatch, J. K., Mikkelwait, D. R., & Sweet, C. F. (1976). Strategies for small farmer development: An empirical study of rural development projects. Boulder, Colo.: Westview Press.

Mosher, A. T. (1976). Thinking about rural development. New York: Agricultural Development Council.

Mosher, A. T. (1966). Getting agriculture moving. New York: Agricultural Development Council.

Röling, N. & de Zeeuw, H. (1983). Improving the quality of rural poverty alleviation. Wageningen, Netherlands: International Agricultural Centre.

Smith, W. E., Lethem, F. J., & Thoolen, B. A. (1980). The design of organizations for rural development projects--a progress report (World Bank staff working paper no. 375). Washington, D.C.: The World Bank.

Vries, J. de (1978). Agricultural extension and development--Ujamaa villages and the problems of institutional change. Community Development Journal, 13, 11-19.

Wilson, P. N. (1983). Integrated rural development: Additional lessons from Nicaragua. Agricultural Administration, 14, 137-150.

Zavala, H. L. M. (1980). Rural development, science and political decision-making: Diverging or converging tendencies. Impact of Science on Society, 30, 167-177.

SECTION 4 -- EXTENSION ORGANIZATION AND ADMINISTRATION

Ekpere, J. A. (1974). A comparative study of job performance under two approaches to agricultural extension organization (Research paper no. 61). Madison, Wisc.: Land Tenure Center, Univeristy of Wisconsin.

Extension Development Around the World (1972). Guidelines for building extension organ-izations and programs. Washington, D.C.: Federal Extension Service, U.S. Depart-ment of Agriculture in cooperation with USAID and U.S. Department of State.

Gordon, J. (1979). Institutional management and agricultural development (Occasional paper 3). London: Agricultural Administration Unit, Overseas Development Institute.

Hardjasoemantri, K., Fussel, D., & Quarmby, A. (1976). KKN: Indonesia's national study-service scheme. The Hague, Netherlands: Centrum voor de Studie van het Onderwijs in Verenderende Maatschappijen.

Howell, J. & Hunter, G. (Eds.) (1982). Providing services to small farmers. Agricultural Administration, 11, 253-318.

Howell, J. (1982). Managing agricultural extension: The T and V system in practice. Agricultural Administration, 11, 273-284.

Hunter, G. (1979). The organisation and administration of agricultural development in relation to rural development (Meeting papers, WCARRD). Rome: Food and Agri-culture Organization of the United Nations.

Kingshotte, A. (1979). The organization and management of agricultural extension and farmer-assistance in Botswana (Agricultural administration network discussion paper 1). London: Overseas Development Institute Agricultural Adjustment Unit.

Leagans, J. P. (1971). Extension education and modernization. In J. P. Leagans and C. P. Loomis (Eds.) Behavioral change in agriculture (pp. 101-147). Ithaca, N.Y.: Cornell University Press.

Leonard, D. K. (1977). Reaching the peasant farmer: Organization theory and practice in Kenya. Chicago: University of Chicago Press.

Lever, B. G. (1970). Agricultural extension in Botswana (Development study no. 7). Reading: University of Reading, Department of Agricultural Economics.

Mosher, A. T. (1981). Three ways to spur agricultural growth. New York: International Agricultural Development Service.

Mosher, A. T. (1971). To create a modern agriculture: Organization and planning. New York, Agricultural Development Council.

Overseas Development Institute (1977). Extension, planning and the poor (Occasional paper, no. 2). London: Agricultural Administration Unit, Overseas Development Institute.

Qamar, M. K. (1979). Why not an inter-disciplinary rural extension service? Journal of Administration Overseas, 18, 256-268.

Schulz, M. (1977). Organizing extension services for integrated rural development in West and East African countries--the commodity approach. Sociologica Ruralis, 17, 87-106.

Singh, R. (1978). A suggested model of approach for extension services in developing countries. Indian Journal of Adult Education, 39(2), 22-26.

Swanson, B. E. & Rassi, J. (1981). International directory of national extension systems. Urbana, Ill.: Bureau of Educational Research, College of Education, University of Illinois at Urbana/Champaign.

SECTION 5 -- EXTENSION AND WOMEN AND YOUTH

Ames, G. C. W. (1975). Involving rural youth in economic development. Agricultural Administration, 2, 63-68.

Ashby, J. A. (1981). New models for agricultural research and extension: The need to integrate women. In B. C. Lewis (Ed.), Invisible farmers: Women and the crisis in agriculture (pp. 144-195). Washington, D.C.: Office of Women in Development, Agency for International Development.

Beneria, L. (Ed.) (1982). Women and development: The sexual division of labor in rural societies. New York: Praeger.

Boserup, E. (1970). Women's role in economic development. New York: St. Martin's Press.

Buvinic, M., Lycette, M. A., & McGreevey, W. R. (Eds.) (1983). Women and poverty in the third world. Baltimore, Maryland: Johns Hopkins University Press.

Food and Agriculture Organization (1979). The legal status of rural women: Limitations on the economic participation of women in rural development (FAO economic and social development paper no. 9). Rome: Food and Agriculture Organizaton.

Food and Agriculture Oganization (1980). WCCARRD: A turning point for rural women. Rome: Food and Agriculture Organization.

Jiggins, J. (1982). Extension and rural women. Approach (Netherlands), 12, 3-13.

Lewis, B. C. (Ed.) (1981). Invisible farmers: Women and the crisis in agriculture. Washington, D.C.: Office of Women in Development, U.S. Agency for International Development.

Mickelwait, D. R., Riegelman, M. A., & Sweet, C. F. (1976). Women in rural development. Boulder, Colo.: Westview Press.

Rogers, B. (1980). The domestication of women: Discrimination in developing societies. London: Tavistock.

Smithells, J. E. (1972). <u>Agricultural extension work among rural women</u>. Reading, England: University of Reading, Agricultural Extension and Rural Development Centre.

Staudt, K. A. (1975-76). Women farmers and inequitites in agricultural services. <u>Rural Africana</u>, <u>29</u>, 81-93.

SECTION 6 -- EXTENSION PROGRAM DEVELOPMENT, EVALUATION AND RESEARCH

Ad Hoc Committee on Small Farms (1979). <u>Research, extension and higher education for small farms</u>. Washington, D.C.: United States Department of Agriculture, Joint Council on Food and Agricultural Science.

Ascroft, J., Röling, N., Kariuki, J., & Chege, F. (1973). <u>Extension and the forgotten farmer: First report of a field experiment</u> (Bulletin no. 37). Wageningen, Netherlands: Afdelingen voor sociale wetenschappen aan de Landbouwhogeschool.

Belshaw, D. (1979). The appraisal, monitoring, and evaluation of agricultural extension programmes. In <u>Institutions, management and agricultural development</u> (Occasional paper, no. 3) (pp. 39-43). London: Agricultural Administration Unit, Overseas Development Institute.

Bennett, C. F. (1977). <u>Analyzing impacts of extension programs</u>. Washington, D.C.: U.S. Department of Agriculture, Extension Service.

Blencowe, J. P., Engel, A. E., & Potter, J. S. (1974). A technique of involving farmers in planning extension programmes. <u>Agricultural Record</u> (Adelaide, Australia) <u>1</u>, 18-23.

Boyle, P. G. (1981). <u>Planning better programs</u>. New York: McGraw-Hill.

Byrn, D. (Ed.) (1967). <u>Evaluation in extension</u>. Topeka, Kansas: H. M. Ives and Sons.

Cernea, M. M. & Tepping, B. J. (1977). <u>A system of monitoring and evaluating agricultural extension projects</u> (World Bank staff working paper no. 272). Washington, D.C.: The World Bank.

Chesterfield, R. & Ruddle, K. (1976). A case of mistaken identity: Ill-chosen intermediaries in a Venezuelan agricultural extension programme, <u>Community Development Journal</u>, <u>11</u>, 53-59.

Collison, M. & Keane, S. (1978). <u>Demonstration of an interdisciplinary approach to planning adaptive agricultural research programmes. (Part of Serenje District, Central Province, Zambia)</u>. Nairobi: CIMMYT Eastern Africa Economics Programme.

Contado, T. E. (n.d.). <u>Pedagogical and didactical aspects of developing an extension program</u>. Rome: Food and Agriculture Organization. Unpubished manuscript.

Coombs, P. H. & Ahmed, M. (1974). <u>Attacking rural poverty: How non-formal education can help</u>. Baltimore, Md.: Johns Hopkins University Press.

Honadle, G. H. (1982). Supervising agricultural extension: Practices and procedures for improving field performance. <u>Agricultural Administration</u>, <u>9</u>, 29-45.

Jamison, D. T. & Lau, L. J. (1982). <u>Farmer education and farm efficiency</u>. Baltimore, Md.: Johns Hopkins University Press.

Kulp, E. M. (1977). <u>Designing and managing basic agricultural programs</u>. Bloomington, Ind.: PASITAM Publications.

Lindt, J. H., Jr. & Armour, R. (1980). Agricultural research and extension: Experience with T & V and other research and extension systems. In <u>Proceedings of the Agricultural Sector Symposia, January 7-11, 1980</u> (pp. 346-470). Washington, D.C.: The World Bank.

236

Lowdermilk, M. K., Dong Wan Shin, & Russell, J. F. A. (1982). Approaches to agricultural extension in different production systems (3 papers). In Promoting increased food production in the 1980's. Proceedings of the second annual Agricultural Sector Symposia, January 5-9, 1981 (pp. 76-137, 138-150, 151-168). Washington: D.C.: The World Bank.

Nagel, U. J. (1979). Knowledge flows in agriculture: Linking research, extension and the farmer. Zeitschrift für Auslandische Landwirtschaft, 18, 135-150.

Rice, E. B. (1971). Extension in the Andes--an evaluation of official US assistance to agricultural extension services in Central and South America (AID evaluation paper no. 3). Washington, D.C.: Agency for International Development.

Russell, J. (1981). Adapting extension work to poorer agricultural areas. Finance and Development, 18(2), 30-33.

Sigman, V. A. & Swanson, B. E. (1983). Problems facing national agricultural extension organizations in developing countries. (INTERPAKS series, no. 3). Urbana, Illinois: University of Illinois at Urbana-Champaign, 1983.

Small farmers development manual (1978-1979). Vol. I, Field action for small farmers, small fishermen and peasants. Vol. II, The field workshop: A methodology for planning, training and evaluation of programmes for small farmers/fishermen and landless agricultural labourers (RAFE 36). Bangkok, Thailand: Regional Office for Asia and the Far East, Food and Agriculture Organization of the United Nations.

Staples, R. C. & Kuhr, R. J. (Eds.) (1980). Linking research to crop production. New York: Plenum Press.

SECTION 7 -- EXTENSION TEACHING METHODS

Ekpere, J. A. & Patel, A. U. (1976). Demonstration farms in agricultural extension teaching: A case study. Journal of the Association for the Advancement of Agricultural Sciences in Africa, 3, 1-4.

Haeggblom, S. (1982). Extension methods used in indoor meetings. (International Course on Rural Extension). Wageningen, The Netherlands: International Agricultural Centre.

Moore, H. (1980). Promoting behavioral change by structural discussion in small groups using a change model: Teaching farmers agricultural extension in the developing world. Essex: Longman.

Perraton, H. (1982). Agricultural extension and mass media. Media in Education and Development, 15, 159-162.

Sica, I. (1981). Communication in extension work, Training for Agriculture and Rural Development (FAO economic and social development series, no. 24). 1981, 103-113.

Tully, J. (1971). Teaching extension principles by a comparative approach (A/D/C, The spread of innovation, no. 2). New York: Agricultural Development Council.

United Nations Economic and Social Commission for Asia and the Pacific (1981). Information systems. Agricultural Information Development Bulletin, 3(3), 2-26.

SECTION 8 -- TRAINING EXTENSION PERSONNEL

Allo, A. V. & Schwass, R. H. (1982). The farm advisor. A discussion of agricultural extension for developing countries (Book series, no. 23). Food and Fertilizer Technology Center, Asian and Pacific Council.

Byrnes, F. C. (1974). Agricultural production training in developing countries. New York: The Rockefeller Foundation.

Colle, R. D., Esman, M. J., Taylor, E., & Berman, P. (1979). Paraprofessionals in rural development (Concept paper, Rural Development Committee, Center for International Studies). Ithaca, NY: Cornell University.

Goodell, G. E. (1983). Improving administrators' feedback concerning extension, training and research relevance at the local level: New approaches and findings from Southeast Asia. Agricultural Administration, 13, 39-56.

Hartzog, D. H. (September, 1983). World Bank involvement in agricultural education and training, 1977-83: Issues and constraints. Paper presented at the Inter-Agency Workshop on Agricultural Education and Training, Education Department, World Bank, Washington, D.C.

Howell, R. E. (1982). Development of rural leadership: Problems, procedures, and insights (ERIC document ED 214748). Pullman, Wash.: Department of Rural Sociology, Washington State University.

Jiggins, J. (1977). Motivation and performance of extension field staff. In Extension, planning and the poor (Occasional paper no. 2) (pp. 1-19). London: Overseas Development Institute, Agricultural Adjustment Unit.

Moss, G. & Treadwell, D. (1977). Teaching extension skills. A tutor's aid for training agricultural extension workers. Wellington, New Zealand: Ministry of Foreign Affairs.

Oakley, P. & Garforth, C. (1984). Guide to extension training. 2nd ed. Rome: Food and Agriculture Organization of the United Nations.

Rhodes, K. (1980). The development of rural leadership. Agricultural Administration, 7, 147-154.

Rowat, R. (1980). Trained manpower for agricultural and rural development (FAO economic and social development paper no. 10). Rome: Food and Agriculture Organization of the United Nations.

Rowat, R. (1981). Future targets for training programs. Ceres, 14(6), 15-20.

Sam, P. D. (1976). The group training unit: Key to village development. International Development Review--Focus, 2, 17-20.

Schwass, R. H. & Allo, A. V. (1982). Professional requirements of the extension worker (Extension bulletin, no. 173) (pp. 1-10). Taipei: Food and Fertilizer Technology Center.

Swanson, B. E. (1977). Impact of the international system on national research capacity: The IRRI and CIMMYT training programs. In T. M. Arndt et al. (Eds.), Resource allocation and productivity in national and international agricultural research (pp. 336-363). Minneapolis, Minn.: University of Minnesota Press.

Swanson, B. E. (1976). Coordinating research, training and extension--a comparison of two projects. Training for Agriculture and Rural Development, 1976, 1-13.

Uwakah, C. T. (1980). The training needs of agricultural extension staff in Eastern Nigeria. Agricultural Administration. 7, 79-86.

Vincent, P. (1981). Cost effective training strategy at field level strengthens influence of extension agents in community. Diffusion, 4(4), 16-21.

Westermarck, H. (1980). Selection and training of agricultural advisory service personnel in the 1980's and the years to come. Helsinki, Finland: Center for Extension Education and In-service Training, University of Helsinki.

SECTION 9 -- EXTENSION AND TECHNOLOGY DIFFUSION

Brady, N. C. (1981). Significance of developing and transferring technology to farmers with limited resources. In N. R. Usherwood (Ed.), Transferring technology for small-scale farming (pp. 1-21). Madison, Wisc.: American Society of Agronomy.

Busch, L. & Lacy, W. B. (1983). Information flows in research and extension: An alternative perspective. Rural Sociologist, 3, 92-97.

Chamala, S., Ban, A. W. van den, & Röling, N. (1980). A new look at adopted categories and an alternative proposal for target grouping of farming community. Indian Journal of Extension Education, 16(1/2), 1-18.

Das, R. (1981). Appropriate technology. New York: Vantage Press.

Eklund, P. (1983). Technology development and adoption rates. Systems approach for agricultural research and extension. Food Policy, 8, 141-153.

Foster, G. M. (1973). Traditional societies and technological change. 2nd ed. New York: Harper and Row.

Havens, A. E. & Flinn, W. L. (1975). Green revolution technology and community development: The limits of action programs. Economic Development and Cultural Change, 23, 469-481.

Institute of Development Studies. Agricultural extension and farmers' training: Results of SRDP experimentation in agricultural extension (Occasional paper no. 12). Nairobi, Kenya: Institute of Development Studies.

Leagans, J. P. (1979). Adoption of modern agricultural technology by small farm operators: An interdisciplinary model for researchers and strategy builders. Ithaca, N.Y.: Program in International Agriculture, Cornell University.

Lionberger, H. F. (1974). Organizational issues in farm informational systems for modernizing agriculture. Journal of Developing Areas, 8, 395-408.

Moran, M. J. (1978). Transfer of post-harvest technologies to small farmers. Desarrollo Rural en las Americas, 10, 143-152.

Pachico, D. & Ashby, J. (1983). Stages in technology diffusion among small farmers: Biological and management screening of a new rice variety in Nepal. Agricultural Administration, 13, 23-38.

Rogers, E. (1983). Diffusion of innovations. 3rd ed. New York: Free Press.

Rogers, E. M. & Shoemaker, F. F. (1971). Communication of innovations. New York: Collier Macmillan.

Rogers, E. M. (1976). Where we stand in the understanding of the diffusion of innovations. In W. Schramm & D. Lerner (Eds.), Communication and change (pp. 204-222). Honolulu: University of Hawaii Press.

Swanson, B. E. (1975). Organizing agricultural technology transfer: The effects of alternative arrangements. Bloomington, Ind.: Programme of Advanced Studies in Institution Building and Technical Assistance Methodology.

Woods, J. L. (1980). Agricultural extension and communication--why, what and for whom (Research bulletin, no. RB 372). Bangkok, Thailand: Development Training and Communication Planning, United Nations Development Programme.

SECTION 10 -- SOCIAL AND CULTURAL ASPECTS OF EXTENSION

Dixon, G. (1982). Agricultural and livestock extension. Vol. 1, Rural sociology. Canberra: Australian Universities International Development Programme.

Gabriel, T. (1981). Agricultural extension as social change. Abstract. In Development Studies Association annual conference, Oxford, September, 1981. A volume of abstracts. Oxford: Queen Elizabeth House.

Jiggins, J. & Hunter, G. (1976). Institutional and cultural problems of criteria for rural and agricultural development projects. London: Overseas Development Institute.

Jiggins, J. & Röling, N. (1982). The role of extension in people's participation in rural development. Rome: Human Resources, Institutions, and Agrarian Reform Division, Food and Agriculture Organization.

Jones, G. E., & Rolls, M. J. (Eds.) (1982). Progress in rural extension and community development. Vol. I: Extension and relative advantage in rural development. Chichester, UK: John Wiley.

Molho, S. (Ed.) (1970). Agricultural extension: A sociological appraisal. Jerusalem: Keter Publishing House.

Saguiguit, G. F. (1981). The social laboratory: Some experiences and perspectives, Training for Agriculture and Rural Development, 1981.

Schwarzweller, H. (Ed.) (in press). Research on rural sociology and rural development. Vol. 1. Greenwich, Conn.: JAI Press.

Uphoff, N. (1981). Farmer participation in the development process. In Small farmers in a changing world: Prospects for the eighties (pp. II/2/1-16). Farming Systems Research Symposium. Manhattan, Kansas: International Agricultural Program, Kansas State University.

APPENDIX 1
Questionnaire on Maize Production and General Farming Methods
in the Black Sea Regions, Turkey

Farmer's Name: _____

Village: _____

Bucak: _____

Ilce: _____

I. We would like to ask you some questions about the maize that you produced this year (1983).

    1.   How many maize fields did you plant this year? _____

       (If more than one, we would like to know about the largest maize field that you planted this year.)

    2.   Size: _____ da.

    3.   Altitude: _____ meters (approx.)

    4.   Slope: _____ Flat (0-5°) _____ moderate slope (6°-25°) _____ Steep slope (26°+)

    5.   Ownership: _____ own _____ rented _____ share-cropped.

    6.   What crop(s) did you plant in this field in:
       1982 _____
       1981 _____
       1980 _____

    7.   When did you harvest the crop that was in this field in 1982?

       _____ Month _____ early _____ mid _____ late.

    8.   After that harvest, did you do any kind of land preparation during the fall or winter of 1982?

Operation	Date*	Method**	Labour***

       * 1 = First part   ** T = Tractor   *** F = Family
         2 = Middle   A = Animal   E = Exchange
         3 = End of month H = Hoe   H = Hired
                            F = Fork   (Can check more than one)

    9.   When you began preparation of this field in spring 1983, what did the weeds in the field look like? _____

   10. What were all the activities that you performed on this field in spring 1983 in relation to planting?

Operation	Date*	Method**	Labour***
PLANTING		****	

       **** D = drill, L = line, B = broadcast

11.　About how many days passed between the first tillage and planting?

　　　_____

12.　If the answer to (11) is more than a few days, what is the cause of the delay?

　　　_____
　　　_____

13.　If you used animals for plowing were they: _____ oxen _____ horses

　　　_____ mules _____ other (specify) _____ _____ don't use

15.　If you used a tractor, was it: _____ own _____ borrowed _____ rented
　　　_____ don't use

16.　What type of maize was planted in this field? (Show farmer samples.)

　　　　　Area in da.　Type*　Name (optional)　kg. seed　Source of seed**

Most important _____
Secondary _____

　　　* 1 = White flint　　　3 = Yellow flint　　** F = Own farm
　　　　2 = White dent　　　4 = Yellow dent　　　N = Neighbour
　　　　　　　　　　　　　　　　　　　　　　　　　E = Extension

17.　What types of beans did you plant in this field?

　　　Name　　Colour　Habit+　Use++　Planting Method+++　kg. Seed

　　　_____
　　　_____
　　　_____

　　　+ = D = determinate　　　++ = G = green　　+++ = D = drill
　　　　　C = climbing　　　　　　　D = dry　　　　　　L = line
　　　　　　　　　　　　　　　　　　　　　　　　　　　　B = broadcast

18.　What other crops were planted with the maize (throughout the field, not just on
　　　the borders).

　　　　　_____
　　　　　_____
　　　　　_____

19.　What weeding and thinning operations did you perform on this field?

Operation° # Days after planting　Description°°　Type Labour°°° Person-days

_____
_____
_____
_____

° W = weed (only)　　°° W = weeding only　　°°° F = family
　T = thin (only)　　　　C = cultivation　　　　　E = exchange
　B = both　　　　　　　　H = hilling up　　　　　H = hired

20.　What are the most common weeds on this field?

　　　Order　　　　Name　　　　　Broad-leaf or Grass?
　　　1. _____
　　　2. _____
　　　3. _____

21.　Fertilization (including manure)

　　　　　　　　　　What other operations were
Fertilization　you performing at this time?　Product(s)　Quantity (kg)　Method*
　　　1. _____
　　　2. _____
　　　3. _____
　　　　　* B = broadcast
　　　　　　L = line
　　　　　　X = around individual plants

242

22. When did you harvest the maize in this field?

    _____ month _____ 1o. _____ 2o._____ 3o.

23. What type of labour did you use for harvest? _____ family
    _____ exchange _____ hired

24. What was the yield of maize from this field? (See #16)

    Type of maize   Kgs.
    _____
    _____

25. What was the yield of <u>dry beans</u> from this field? (See #18)

    Type of bean   Kgs.
    _____
    _____

II. <u>We would now like to ask you some questions about your maize production in general, and about your farm.</u>

26. In the last ten years, how many times was your maize yield seriously reduced by:

    - Dry weather
    - Flooding
    - Hail
    - Lodging caused by wind
    - Insects
    - Frost
    - Other (specify) _____

27. If insects are a problem in your maize crop, which are most serious?

Insect	Stage of plant growth at attack	Control method
Cutworm		
Mole cricket		
Borer		
Wire worm		

28. For your maize field, what is the best time for planting, to make sure the crop develops properly? _____

29. Are you always able to plant at that time? _____ Yes _____ No

30. <u>If no</u>. Why not? _____
    _____

31. Do you feel you are able to do your weeding and thinning on time?
    _____ Yes _____ No

32. <u>If no</u>. Why are you late? _____

33. Most farmers seem to plant a lot of seed, then thin the plants afterward. What would be the problem with planting less seed to begin with?

    _____ Poor germination
    _____ Insect attack
    _____ Thinned plants are used for animals
    _____ Other (Specify) _____

    (If farmer has more than one answer, Use "1", "2" to indicate order of importance).

34. Do you ever plant wheat or oats? _____ Yes _____ No

35. If yes, have you ever planted maize as a second crop after harvesting the wheat or oats? _____ Yes _____ No

36. If yes, what was your experience? _____
_____

37. If no, why have you not tried this? _____
_____

38. What problems do you have in bean production? (In each case, specify the variety that has this problem.)

_____   Poor germination
_____   Diseases (describe disease) _____
_____
_____   Insects (which insects) _____
_____
_____   Other _____
_____

39. Have you ever planted the maize variety K 3/74? (the variety from extension).

_____ Yes _____ No

40. If yes. What was your experience?

	Acceptable	Unacceptable
Germination	_____	_____
Yield	_____	_____
Taste of grain	_____	_____
Fodder quality	_____	_____
Other comments	_____	

41. What type of maize do you prefer for food? (Show samples)

42. What type of maize do you prefer for sale? (Show samples)

43. What proportion of the maize you grew last year was:

Yellow _____% White _____%

44. Among the following three uses, what is the proportion of your maize production that you use for:

_____% Food _____% Feed on the farm _____% Sale

45. Does your family eat more:
_____ Maize
_____ Wheat

46. Do you use all of the maize stalks for fodder?

_____ Yes
_____ No - leave some in the field
_____ No - sell some
_____ No - Other _____

47. What are your animals eating during summer? _____

48. What other types of fodder do you provide your animals in winter?

Type	Source°
1. _____	
2. _____	

° F = Farm
P = Purchased

244

49. What are the principal crops you grew last year?

Crop	Area (da)
Maize	_____
Hazelnut	_____
Rice	_____
Pasture	_____
Tobacco	_____
_____	_____
_____	_____

50. How many animals do you have?

	# Adult	# Young
Cattle	_____	_____
Buffalo	_____	_____
Horses	_____	_____
Donkeys	_____	_____
Sheep	_____	_____
Goats	_____	_____

51. What are the busiest times of year for you and your family on the farm?

	Month	Activity
1.	_____	_____
2.	_____	_____

52. Does the cash income your family earns come more from:

_____ Farming (crops and animals)   _____ Off-farm work

53. List the three crops, animals, or animal products that provide most income:

Order

1. _____
2. _____
3. _____

54. Have you ever used:

	Crop	Own	Borrowed	Rented	From extension
Knapsack sprayer	___	___	_____	_____	_____
Tractor-driven pump	___	___	_____	_____	_____

55. In 1984 we would like to plant some experiments with farmers, in order to test new techniques and new varieties. We are looking for farmers who would be willing to lend us a small piece of land (½-1 da.). All of the production from the experiment would stay with the farmer. Would you be interested in loaning us a piece of land?

_____ Yes
_____ No

Example of Field Trial/Demonstration Book

(1) PLANTING DATA

Location_____ Farmer_____ Trial_____

Planting date_____

Type of trial_____ Design_____

Size of trial_____

Length of rows_____ Distance between rows._____

No. seeds per hole: Maize._____ Beans._____ _____
(other)

Method of planting beans._____

Crop	Variety	Source of seed
Maize	_____	_____
	_____	_____
	_____	_____
Beans	_____	_____
	_____	_____
	_____	_____

Fertilization of plot (including that of the farmer)_____

_____

_____

Weed control_____

_____

Insect control_____

_____

Other operations at planting_____

_____

Soil moisture_____

_____

(2) FIELD PLAN

Location_____ Farmer_____ Trial_____

_____

KEY

247

(3) CHARACTERISTICS OF THE PLOT

Location _____ Farmer _____ Trial _____

Cropping History

Year	Crop(s)	Fertilization

Preparation of the plot

Activity	Method
Plow 1	
2	
Harrow 1	
2	
Furrow	

Slope _____

Irrigation?   Yes/No   Frequency of irrigation _____

Soil type _____

Soil analysis: N _____ P _____ K _____ Zn _____ Mn _____

Altitude _____

(4) MANAGEMENT OF THE TRIAL

Location _____ Farmer _____ Trial _____

ACTIVITY	DATE	METHOD
Replanting		
Weed control 1		
2		
3		
4		
5		
Fertilization 1		
2		
Other 1		
2		
3		
4		
Irrigation 1		
2		
3		
4		
5		
6		

Other observations _____

_____

_____

(5) OBSERVATIONS ON THE FARMER'S CROP

Location_____ Farmer_____ Crop_____ Trial_____

Location of the field under observation _____

Slope_____ Soil type _____

Cropping history

Year	Crop(s)	Fertilization

Planting Density

Crop	Variety	No. of seeds per hole:	Distance between	
			Plants	Rows

Management

Activity		Date	Method
Weed control	1		
	2		
	3		
	4		
Fertilization	1		
	2		
Other	1		
	2		

Irrigation during year_____
_____

What are the principal insect problems in the field? _____
_____

What are the principal disease problems in the crop?_____
_____

What are the most common weeds? _____
_____

Source of seed used in the field:_____
_____

Other observations:

(6) CHARACTERISTICS OF THE FARMER

Number of hectares worked this year:

Own_____ Rented_____ Sharecropped_____

Principal crops: 1)_____ 3)_____

2)_____ 4)_____

In which crops does farmer hire labor?_____

_____

In which crops does farmer use fertilizer?_____

_____

Use of other agro-chemicals _____

_____

Which crops are sold? 1)_____

2)_____

3)_____

4)_____

Off-farm employment or activities:_____

_____

_____

Problems in storage of crops:_____

_____

_____

Other observations:

(7) AGRONOMIC DATA

Location_____ Farmer_____ Trial_____

Crop_____

No.	% Germ.	Days to Flowering	Diseases				Damage				Other observations	
			Date	Classification	Date	Classification	Date	%	Date	%	Date	Date

8) HARVEST DATA

Location _____ Farmer _____ Trial _____

Each treatment harvested is _____ rows of _____ meters.

Area of each treatment harvested: _____ square meters.

No.	Number of plants	Number of ears			Weight			% humidity	% shelling	dry weight	Beans			
		Good	Rotten	Total	Good	Rotten	Total				Number of plants	Weight	% humidity	Dry weight

(9) OBSERVATIONS OF THE TRIAL
(To be filled in on each visit to the trial)

Location_____ Farmer _____ Trial_____

Date_____ Technician_____

Was farmer present? Yes / No

Work on the plot since last visit: _____
_____

Development of the crop_____
_____

Insects _____
_____

Diseases _____
_____

Weeds_____
_____

Weather conditions _____
_____

Other observations:

APPENDIX 3
SUGGESTED OUTLINE OF A TEACHING PLAN
WHEN USING THE PROBLEM SOLVING METHOD OF TEACHING

I.   UNIT OR ENTERPRISE: Maize

II.   PROBLEM AREA: Planting and growing maize

III.   SITUATION

    A.   Indicate the type of students to whom the problem area will be taught.

    B.   Types of local information and data needed to appraise the situation adequately.

    C.   District or provincial information, and data pertinent to the problem area.

    D.   Research data with which local, district, and provincial information may be compared.

IV.   TEACHER OBJECTIVES

    A.   Statements of what the teacher hopes to accomplish.

    B.   Student objectives should be stated as behavioural or performance objectives, not as agricultural production objectives.

    C.   Objectives may be stated in terms of developing specific types of knowledge, skills and attitudes.

V.   SUGGESTED INTEREST APPROACHES: (possible questions to ask)

    A.   How many of you grow maize on your home farm?

    B.   How many hectares of maize do you grow?

    C.   What have been your average yields for maize over the past few years?

    D.   Do you have any idea about how much it costs you to produce a quintel of maize? How many quintels of maize per hectare does it take you to recover costs or break even?

    E.   What kind of problems or difficulties, if any, have you experienced in growing maize?

VI.   ANTICIPATED GROUP (student and teacher) OBJECTIVES

    A.   Question: Why is it important that we do a good job of maize production?

         Anticipated Rationale:
         1.   Maize is our major food crop.

         2.   There is a high demand for maize, therefore it has been our highest profit crop.

         3.   Most of our farmers grow maize.

    B.   Question: Why is it important that farmers in our province do a good job of maize production?

         Anticipated Objectives:
         1.   To achieve an average yield of 3-5 tons/hectare.

         2.   To keep production costs as low as possible, to ensure a profit.

         3.   To maintain soil fertility.

255

4. Soil and water conservation.

5. To produce high-quality grain for home family consumption and to command a good market price.

VII. ANTICIPATED PROBLEMS AND CONCERNS

A. Question: What do we need to know and be able to do in order to do a good job of maize production?

Anticipated Problems:
1. What are the soil fertility requirements for maize?

2. What tillage practices should I use for maize?

3. What variety of maize should I plant?

4. How can I control weeds, insects, and diseases of maize?

5. Should I use supplemental irrigation on maize?

6. What plant population should I use, and when is the best time to plant maize?

B. Special events or activities, such as field trips, demonstrations, the use of resource persons, special projects, and so forth.

VIII. REFERENCES AND TEACHING AIDS

IX. APPROVED PRACTICES TO BE DEVELOPED: Recommended technology or package of practices to be used on student plots or to be tested on farmers fields.

X. SUPPLEMENTARY INFORMATION

A. Mimeographed materials prepared for the problem area.

B. Plans for demonstrations, field trips, and so forth.

C. Teaching aids prepared for the problem area.

APPENDIX 4
THE PROBLEM-SOLVING APPROACH TO TEACHING:
AN OUTLINE OF STEPS, AND SOME SUGGESTIONS FOR TEACHERS

A. Write name of problem on chalkboard

1. Problem areas should stem from students' experience and their on-the-job situations.

2. Use verbal "-ing" form in stating problem areas (Example: planting and growing maize).

3. Problem area should be broad enough to challenge students, but not so broad that the class gets lost in attempting to solve it. A problem area which can be completed in 2-7 days (assuming 2 hour class periods) is recommended.

B. Conduct an interest approach

1. For example, in a technical problem area, the teacher may wish to determine the present practices used by class members on their home farms by opening with the following leading questions:

   a. How many of you have done this job? (Show of hands)

   b. How did you do it? (Call on several individuals)

   c. How does your father, mother or neighbour do it? (Call on those who have not done it themselves.)

   d. What results are you, your parents or your neighbour getting?

2. Promote interest by raising questions, creating suspense, opening up new avenues of thought and pointing out differences in present practices.

3. Avoid giving answers at this time.

C. Conduct discussion to arrive at group objectives (teacher and student)

1. Use leading questions such as, "Why is it important that we do a good job (of planting and growing maize)"? or, "Why do we need to study (planting and growing maize)"?

2. As students volunteer objectives, the teacher raises other questions such as, "What results should you get in carrying out this job"?

3. Teacher lists group objectives on chalkboard, and students list them in their notebooks.

D. Problems and Concerns: Let students identify things they need to know and things they need to be able to do

1. Use leading question, "What do we need to know (knowledge and information) and/or what do we need to be able to do in order to do a good job of (planting and growing maize)"?

2. As students volunteer problems, questions (things we need to know), and jobs (things we need to be able to do), the teacher should list them on chalkboard. Students should also list them in their notebooks.

3. Problems, questions, and jobs should ideally come from the students, but the teacher must contribute if the students fail to come up with a good list.

E. Let students help re-order and re-number the list of problems, questions and jobs

1. This step is necessary because students do not always identify problems and questions in the order they need to be considered.

2. The teacher should plan to conduct classroom activities first, so he or she will have time to plan field trips and/or laboratory exercises for later in the unit, where skill training and other "learning to do" activities can be more effectively conducted.

F. Select a problem or question

1. Erase other material and rewrite the first question or problem to be considered on chalkboard.

G. Conduct a trial solution

1. Lead a discussion about the problem, to see if the class can solve it (the teacher must not provide the answer).

2. If an adequate solution is reached by the class, then the teacher writes the solution on chalkboard, and students write it in their notebooks.

3. Proceed to the next problem or question.

H. Conduct supervised study of unknown problems

1. If class discussion about a particular problem or question does not result in a complete and satisfactory solution, supervised study is conducted.

2. Before supervised study is begun, teacher needs to make sure that the problem is stated in such a way that students can find a solution. In some cases, sub-questions may have to be identified.

3. The source of information for supervised study may be a reference book or bulletin, films, charts, resource person, field trip, filmstrip, or the teacher.

4. Students usually record their tentative conclusions on scratch paper.

I. Arrive at a solution to the problem

1. Teacher asks students to indicate what tentative solutions or conclusions they found during supervised study.

2. Teacher records conclusions on chalkboard.

3. If there is disagreement, then through discussion and consideration of the source of information, a conclusion can generally be reached, and then the students record the appropriate conclusion(s) in their notebook.

J. Repeat Steps F - I for each of the other problems and questions

K. Proceed to list things that students need to be able to do (jobs or skills identified under Step D)

1. Teacher demonstrates correct procedures (skills) for performing the job.

2. Teacher provides opportunities for students to practise at school or on-the-job (field trip in the project area).

3. Teacher gives individual instruction while students practice.

L. Let the class summarize the problem area by listing the approved practices for the problem area

1. Teacher records on chalkboard, and students record list in their notebooks.

M. Students develop written plans for the application of the approved practices

For example, students are asked to indicate which approved practices they will actually use in their individual plots, and when they plan to apply each practice.

N. Teacher and students evaluate student progress

1. Teacher observes progress of students on individual plots, in the laboratory or shop, and in the classroom.

2. Teacher gives pencil and paper tests.

3. Students record in their farm management field book which of the actual approved practices they have tried and either adopted or rejected, and why they reached this conclusion.

Appendix 5
STAFF DEVELOPMENT PROFILE*

Name _____   Supervisor _____
Location _____   Date of Employment _____
Job Title _____   Position Description _____
Education: Number of years of job-related experience:

Degree	Field	Institution	Experience - Where	Nature
_____	_____	_____	_____	_____
_____	_____	_____	_____	_____
_____	_____	_____	_____	_____

Objective: The objective of the local extension service is, through educational programmes in agriculture, home economics, and related subjects, to help people, men and women, solve their problems in a way that is both socially desirable and personally satisfying and to grow in their knowledge and competence as individuals.

I. IN TERMS OF YOUR POSITION DESCRIPTION, STATE BRIEFLY THE OBJECTIVES YOU HAVE SET FOR YOURSELF:

    a. Professional objectives: _____

    _____

    b. Personal objectives: _____

    _____

II. STATE YOUR INTERESTS AND STRENGTHS REGARDING FUTURE PROFESSIONAL AND PERSONAL GROWTH:

    _____
    _____
    _____

* This is a brief evaluation form for professional staff development. It should be started within the first week of employment and completed before the end of the first year.

III. PRESENT NEED ASSESSMENT, BASED ON ENTRY BEHAVIOUR:
From your educational background and professional experience assess and check (√) your level of competency in understanding the objectives.

	Educational Background	Previous Professional Experience
Staff Development Objectives:	Remarks	Remarks
1. Understand the extension service		
2. Understand my position description		
3. Develop and use the programme development process		
4. Understand the benefits associated with my position		
5. Understand the privileges associated with my position		
6. Develop my understanding and skill in using the teaching and learning processes		
7. Develop my understanding and skill in interpersonal relationship		

I V.   PERSONAL PLAN FOR PROFESSIONAL IMPROVEMENT:
This section is to be developed annually during your tenure on the job.
Consultation with your supervisor should help you to identify educational experiences which will best meet your personal and programme
needs for professional growth.

Professional Growth	Learning Experiences*	When	Progress
19__			
19__			
19__			
19__			
19__			
19__			

* Anticipated: formal/informal courses